SOCIETY FOR EXPERIMENTAL BIOLOGY
SEMINAR SERIES · 21

CHLOROPLAST BIOGENESIS

CHLOROPLAST BIOGENESIS

Edited by

R. J. ELLIS

Professor of Biological Sciences, University of Warwick

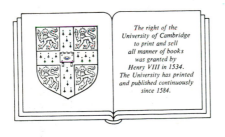

The right of the
University of Cambridge
to print and sell
all manner of books
was granted by
Henry VIII in 1534.
The University has printed
and published continuously
since 1584.

CAMBRIDGE UNIVERSITY PRESS

Cambridge

London New York New Rochelle

Melbourne Sydney

Published by the Press Syndicate of the University of Cambridge
The Pitt Building, Trumpington Street, Cambridge CB2 1RP
32 East 57th Street, New York, NY 10022, USA
296 Beaconsfield Parade, Middle Park, Melbourne 3206, Australia

© Cambridge University Press 1984

First published 1984

Printed in Great Britain by Cambridge University Press

Library of Congress catalogue card number: 84-7734

British Library Cataloguing in Publication Data

Chloroplast biogenesis. – (Society for
  Experimental biology seminar series; 21)

  1. Chloroplasts
  I. Ellis, R. J.   II. Series
  581.87′33   QK882

ISBN 0 521 24816 7

CONTENTS

CONTRIBUTORS

Bennett, J.
Department of Biological Sciences, University of Warwick, Coventry CV4 7AL, UK. (Present address: Biology Department (Bldg 433), Brookhaven National Laboratory, Upton, New York 11973, USA.)

Benz, J.
Botanisches Institut, Universität München, Menzinger Strasse 67, 8000 München 19, Federal Republic of Germany.

Block, M. A.
Physiologie Cellulaire Végétale, ERA au Centre National de la Recherche Scientifique No. 847, DRF/BV-CENG & USMG-85X, 38041 Grenoble-Cedex, France.

Bohnert, H. J.
EMBL, Postfach 10.2209, D-6900 Heidelberg, Federal Republic of Germany. (Present address: Department of Biochemistry, BioSciences West, University of Arizona, Tucson, Arizona 85721, USA.)

Broglie, R.
Laboratory of Plant Molecular Biology, The Rockefeller University, 1230 York Avenue, New York, New York 10021, USA.

Chua, N.-H.
Laboratory of Plant Molecular Biology, The Rockefeller University, 1230 York Avenue, New York, New York 10021, USA.

Coruzzi, G.
Laboratory of Plant Molecular Biology, The Rockefeller University, 1230 York Avenue, New York, New York 10021, USA.

Crouse, E. J.
Institut de Biologie Moléculaire et Cellulaire de Centre National de la Recherche Scientifique, Université Louis Pasteur, 15 rue René Descartes, F-67084 Strasbourg-Cedex, France.

Cuming, A. C.
Department of Biological Sciences, University of Warwick, Coventry CV4 7AL, UK.

Dorne, A. J.
Physiologie Cellulaire Végétale, ERA au Centre National de la Recherche Scientifique No. 847, DRF/BV-CENG & USMG-85X, 38041 Grenoble-Cedex, France.

Douce, R.
Physiologie Cellulaire Végétale, ERA au Centre National de la Recherche Scientifique No. 847, DRF/BV-CENG & USMG-85X, 38041 Grenoble-Cedex, France.

Edelman, M.
Department of Plant Genetics, Weizmann Institute of Science, Rehovot 76100, Israel.

Ellis, R. J.
Department of Biological Sciences, University of Warwick, Coventry CV4 7AL, UK.

Gallagher, T. F.
Department of Biological Sciences, University of Warwick, Coventry CV4 7AL, UK.

Gray, J. C.
Botany School, University of Cambridge, Downing Street, Cambridge CB2 3EA, UK.

Griffiths, W. T.
Department of Biochemistry, The Medical School, University of Bristol, Bristol BS8 1TD, UK.

Groot, G. P. S.
Biochemical Laboratory, Free University, de Boelelaan 1083, 1081 HV Amsterdam, The Netherlands.

Hartley, M. R.
Department of Biological Sciences, University of Warwick, Coventry CV4 7AL, UK.

Jenkins, G. I.
Department of Biological Sciences, University of Warwick, Coventry CV4 7AL, UK.

Joyard, J.
Physiologie Cellulaire Végétale, ERA au Centre National de la Recherche Scientifique No. 847, DRF/BV-CENG & USMG-85X, 38041 Grenoble-Cedex, France.

Lichtenthaler, H. K.
Botanisches Institut (Pflanzenphysiologie) der Universität, Kaiserstrasse 12, D-7500 Karlsruhe 1, Federal Republic of Germany.

Marder, J. B.
Department of Plant Genetics, Weizmann Institute of Science, Rehovot 76100, Israel.

Mattoo, A. K.
Department of Plant Genetics, Weizmann Institute of Science, Rehovot 76100, Israel. (Present address: Plant Hormone Laboratory (Bldg 002), Agricultural Research Center, USDA, Beltsville, Maryland 20705, USA.)

Meier, D.
Botanisches Institut (Pflanzenphysiologie) der Universität, Kaiserstrasse 12 D-7500, Karlsruhe 1, Federal Republic of Germany.

Oliver, R. P.
Department of Biochemistry, The Medical School, University of Bristol, Bristol BS8 1TD, UK.

Phillips, A. L.
Botany School, University of Cambridge, Downing Street, Cambridge CB2 3EA, UK.

Rüdiger, W.
Botanisches Institut, Universität München, Menzinger Strasse 67, 8000 München 19, Federal Republic of Germany.

Schmitt, J. M.
Botanisches Institut I, Universität Würzburg, Mittlerer Dallenbergweg 64, D-8700 Würzburg, Federal Republic of Germany.

Silverthorne, J.
Department of Biology and Molecular Biology Institute, University of California, Los Angeles, CA 90024, USA.

Smith, A. G.
Botany School, University of Cambridge, Downing Street, Cambridge CB2 3EA, UK.

Smith, S. M.
Department of Biological Sciences, University of Warwick, Coventry CV4 7AL, UK. (Present address: Department of Botany, University of Edinburgh, The King's Buildings, Edinburgh EH9 3JH, UK.)

Stiekema, W. J.
Department of Biology and Molecular Biology Institute, University of California, Los Angeles, CA 90024, USA.

Tilney-Bassett, R. A. E.
Department of Genetics, University College of Swansea, Singleton Park, Swansea SA2 8PP, UK.

Tobin, E. M.
Department of Biology and Molecular Biology Institute, University of California, Los Angeles, CA 90024, USA.

Williams, R. S.
Department of Biological Sciences, University of Warwick, Coventry CV4 7AL, UK.

Wimpee, C. F.
Department of Biology and Molecular Biology Institute, University of California, Los Angeles, CA 90024, USA.

PREFACE

This book discusses recent research on the biochemistry of chloroplast development. The chapters were written by experts who spoke at a symposium entitled Chloroplast Biogenesis held at Leiden in 1982 under the auspices of the Society for Experimental Biology, with financial assistance from the Dutch Biochemical Society. In most cases the articles review the more important findings published up to the summer of 1983.

Six years separate this volume from the publication of the second edition of *The Plastids* by John Kirk and Richard Tilney-Bassett. This splendid book more than any other is responsible for the current upsurge of interest in the biology of plastids. I agree with its authors that for too long chloroplasts, the most studied plastids, have been equated with photosynthesis, to the detriment of the study of their other aspects. Indeed I have heard a noted researcher of photosynthesis exclaim that his interest in chloroplasts ceased one millisecond after the light was switched off! The narrow-mindedness of such views has become all too apparent during the last decade when the most innovative research on chloroplasts has been concerned not with their role in photosynthesis but with their function as an extranuclear genetic system. It is now clear that the possession of genetic systems outside the nucleus is a fundamental feature of eukaryotic cells. One of the most fascinating areas of modern plant biochemistry is concerned with unravelling the molecular interplay between the nuclear genome and its counterparts in mitochondria and plastids. Moreover, this interplay must be understood if the genetic manipulation of chloroplast metabolism for agricultural purposes is to be achieved. This volume discusses some of the more interesting discoveries about chloroplast development that have emerged since the publication of the second edition of *The Plastids*.

To my co-authors I express my heartfelt thanks for their help in creating this book, and for their patience in helping me overcome unexpected problems in its production. We are all indebted to Mrs Sandi Irvine for her painstaking correction of errors and infelicities.

March 1984   John Ellis
Editor for the Society for Experimental Biology

ABBREVIATIONS

ACP   acyl-carrier protein
ALA   5-aminolaevulinic acid
bacteriochlide bacteriochlorophyllide
cDNA   copy DNA
chl   chlorophyll
chlide   chlorophyllide
CFI   coupling factor, F1-particle of plastid ATP synthase
CP   chlorophyll-protein
CPI   P700-chlorophyll *a* apoprotein
cp   chloroplast
cp-DNA   chloroplast DNA
cp-RNA   chloroplast RNA
D   Dalton: unit of molecular weight equal to one-twelfth the mass of a ^{12}C atom
DGDG   digalactosyldiacylglycerol
DCMU ⎫
Diuron ⎭   3-(3,4-dichlorophenyl-1,1-dimethyl)urea
EDTA   ethylene diamine tetra-acetic acid disodium salt
ER   endoplasmic reticulum
FNR   ferredoxin–$NADP^+$ reductase
FPP   farnesyldiphosphate
GG   geranylgeraniol
GGPP   geranylgeranyldiphosphate
HL   high light
IgG   immunoglobulin G
IPP   isopentenyldiphosphate
kbp   kilobase-pairs
kD   kiloDaltons
LDS   lithium dodecylsulphate
LHC   light-harvesting chlorophyll *a/b*-binding complex
LHCP   light-harvesting chlorophyll *a/b*-binding protein (apoprotein of LHC)
LL   low light
LSU   large subunit of ribulose bisphosphate carboxylase/oxygenase
MGDG   monogalactosyldiacylglycerol
M_r relative mass
MD   megaDalton (10^6 Dalton)
mRNA   messenger RNA

PC   phosphatidylcholine
PCR   photosynthetic carbon reduction
PEP   phosphoenolpyruvate
PG   phosphatidylglycerol
PGA   phosphoglyceric acid
PPP   phytyldiphosphate
P_r, P_{fr} red and far-red absorbing forms of phytochrome
PSI   photosystem I
PSII   photosystem II
pchlide $\Big\}$ protochlorophyllide
protochlide
p16S (ps23S) rRNA precursor molecule of 16S (23S) rRNA
rbcL   gene for LSU
rDNA   ribosomal DNA
rRNA   ribosomal RNA
RuBP   ribulose-1,5-bisphosphate
SDS   sodium dodecylsulphate
SDS–PAGE sodium dodecylsulphate–polyacrylamide gel electrophoresis
SSU   small subunit of ribulose bisphosphate carboxylase/oxygenase
16SrDNA gene for 16S RNA
tRNA   transfer RNA
trnL-UAA gene for $tRNA^{Leu}_{UAA}$

R. J. ELLIS

Introduction: principles of chloroplast biogenesis

The importance of chloroplasts

Chloroplasts are the best-studied examples of a related group of intracellular organelles termed plastids. It is the possession of plastids more than any other feature that distinguishes plant cells from animal cells. Chloroplasts are important because they contain the entire enzymic machinery for the process of photosynthesis which provides all the organic carbon and gaseous oxygen required by most living organisms. The extensive literature on the mechanism of photosynthesis is witness to the interest of biochemists and physiologists in this process. An International Congress has been held, every three years from 1968, solely devoted to research on photosynthesis. No other metabolic pathway has received such singular attention. It has also become clear in recent years that several other metabolic pathways essential for plant growth are located in chloroplasts and related plastids, particularly those pathways concerned with the synthesis of amino acids and lipids. Thus the *metabolic* importance of chloroplasts cannot be overemphasised.

However, an entirely distinct thread of interest in chloroplasts can be discerned. This thread stretches back to Strasburger (1882), who observed that, in some algae, chloroplasts divide and are passed to the daughter cells in cell division. This observation suggested that chloroplasts do not arise *de novo* in each generation, but come from pre-existing plastids. This idea was strengthened by the work of Baur (1909) and Correns (1909) at the start of this century. These geneticists found that mutations affecting chloroplasts in variegated plants were sometimes inherited in a fashion different from that expected from Mendelian principles. Often the defect was inherited via the maternal line only. This maternal mode of inheritance is found in about two-thirds of all the genera of flowering plants studied, and is explicable in terms of the absence of plastids from the pollen tube. The concept thus arose that chloroplasts themselves contain genetic material controlling at least part of their development, and that this genetic material is handed on from parent to offspring via the physical transmission of plastids during reproduction. The

1

discovery in 1962 that chloroplasts contain both DNA (Ris & Plaut, 1962) and ribosomes (Lyttleton, 1962) opened the modern era in which the biogenesis of chloroplasts is as active an area of research as is the mechanism of photosynthesis.

The explosive advances in biochemical techniques that began in the 1950s have brought about an intertwining of the two threads of interest in chloroplasts. Thus for the modern researcher it is as important to enquire about the synthesis and assembly of a chloroplast component as it is to probe its role in metabolism. Indeed it could be argued that it is *more* important, since this knowledge is vital if attempts genetically to manipulate chloroplasts for agricultural purposes are to be successful.

Structure and composition of chloroplasts

A brief account of the structure and composition of chloroplasts is given here to orient the unfamiliar reader. Reviews on this topic should be consulted for more detailed information (Kirk & Tilney-Bassett, 1978; Reid & Leech, 1980; Schnepf, 1980; Thomson & Whatley, 1980). In higher plants chloroplasts are green bodies with approximately the shape of biconvex lenses; the long diameter is usually 4–10 μm. In the algae a wide variety of shapes and sizes is found. Each chloroplast is delimited from the cytoplasm by an *envelope* consisting of two membranes. The envelope surrounds a hydrophilic matrix or *stroma* which in intact cells appears to be in motion. The stroma contains DNA, ribosomes, plastoglobuli, amino acids, nucleotides, organic acids, starch grains, saccharides and other intermediates, inorganic ions and at least 200 proteins. The protein complement of the stroma is unusual in that one protein is present in much larger amounts than any other; this so-called Fraction I protein can account for up to 50% of the total soluble protein in a leaf extract. Fraction I protein has been shown to be identical to ribulose-1,5-bisphosphate carboxylase/oxygenase, the enzyme which catalyses the first steps in the pathways of both photosynthesis and photorespiration (Lorimer, 1981; Miziorko & Lorimer, 1983). The abundance of this stromal enzyme has made it a favourite molecule for researchers interested in chloroplast biogenesis (Ellis, 1979, 1981a).

Embedded in the stroma is a continuous membrane system which contains DNA, DNA polymerase, RNA polymerase, ribosomes, chlorophylls, carotenoids, quinones, phospholipids, glycolipids, sulpholipids, and the proteins concerned with photon capture, electron transport and photophosphorylation. These membranes resemble flattened sacs in that the opposite sides are close together and enclose a narrow space. Each flattened sac is termed a *thylakoid*. Thylakoids often occur in regular closely-packed stacks or *grana*.

Grana are not always found in chloroplasts, and the extent of their formation can be altered by varying the growth conditions. Thylakoid membranes connect individual grana, forming an intergranal fretwork (see Fig. 1).

Both the stromal and thylakoid-bound ribosomes resemble those of prokaryotes in their sedimentation coefficient (about 70S), the sizes and sequences of their constituent RNA molecules (about 16S, 23S and 5S), their use of *N*-formylmethionine as the initiating amino acid in protein synthesis, and their sensitivity to several bacterial antibiotics. Chloroplast ribosomes can represent up to 60% of the total ribosomes in a leaf extract, and thus, unlike mitochondrial ribosomes, must be regarded as a major component of the protein-synthesising systems of photosynthetic eukaryotes.

Chloroplast DNA occurs as a covalently-closed circular molecule of largely unique base sequence. Each molecule has a potential coding capacity for about 125 proteins, each of molecular weight 50 000. Chloroplasts are polyploid, since each chloroplast contains multiple copies of this DNA molecule. These circles lack histones but resemble bacterial genomes in possessing other bound proteins.

It should be clear from this brief description that chloroplasts are almost as complex as cells in the number and complexity of their components. A full understanding of chloroplast biogenesis will require the elucidation of the origin and assembly of all these myriad components.

Fig. 1. Cut-away diagram of a young chloroplast of a higher plant showing the main structural features. (Reprinted with permission from Reid & Leech, 1980.)

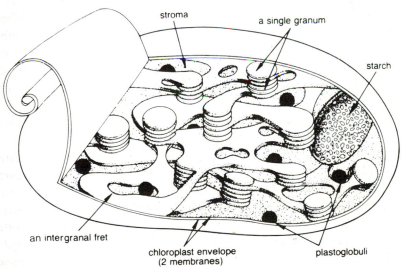

Why be interested in chloroplast biogenesis?

There are two answers to this question. It can be argued that the major conceptual challenge in biology at the present time is to unravel the *molecular basis of differentiation*. The leaf is a differentiated tissue partly because of the presence of chloroplasts. Moreover chloroplasts are easy to isolate and contain large amounts of particular proteins. The synthesis and accumulation of these proteins is under developmental control and is responsive to changes in the external environment, especially to the quality and intensity of light. Thus the regulation of the biogenesis of chloroplast components is an aspect of plant differentiation that is an attractive area for research.

The second reason for being interested in chloroplast biogenesis stems from the fact that these organelles constitute an *extranuclear genetic system*. Chloroplasts, like mitochondria, contain an entire genetic system additional to the one located in the nucleus and cytoplasm. The term 'genetic system' is used here to embrace the four linked components necessary to express genetic information, viz. DNA, DNA polymerase, RNA polymerase, and a protein-synthesising apparatus. It appears that the possession of such extranuclear genetic systems is a fundamental feature of the eukaryotic mode of cell organisation. There are no well-authenticated cases of eukaryotic cells which lack extranuclear genetic systems under natural conditions, although they can be generated in the laboratory. The ubiquity of such systems raises questions both as to their evolutionary origin and their significance for the success of eukaryotic cells. In the case of chloroplasts the evidence that these genetic systems originated as endosymbionts from free-living prokaryotic cells has become stronger in the last few years (Margulis, 1981; Ellis, 1983*a*).

It is clearly established that, despite the relatively large size of their genetic system, chloroplasts are not autonomous entities in genetic terms. Instead the biogenesis of chloroplasts results from a complex interaction between the nuclear genetic system and that based in the plastid. Thus there is little prospect that chloroplasts will ever be cultured *in vitro*. The cells of photosynthetic eukaryotes are the most complex in genetic terms that have been discovered, since their development involves the interplay of three distinct genetic systems located in the nucleus, plastid and mitochondrion, respectively. The fascination of this subject lies in trying to unravel the molecular basis of this interplay.

The origin of this book

In 1982 a special Symposium of the Society for Experimental Biology was held at Leiden in The Netherlands under the title 'Chloroplast Biogenesis'.

The invited speakers at this Symposium have since written articles describing the current state of research in their field. It is these articles that are presented in this book. In most cases the literature survey in each chapter covers the more important publications up to the summer of 1983.

The design of the Leiden Symposium was based around five basic 'principles' of chloroplast biogenesis which the editor has suggested (Ellis, 1983*b*). These principles summarise much of our current knowledge about the nature of the interaction between the nuclear and chloroplast genetic systems during the light-induced biogenesis of chloroplasts. It must be emphasised that these suggested principles are an attempt to impose some order on a mass of emerging information, and must be regarded as provisional summary statements of current understanding rather than as basic laws of nature. To provide a context for the articles in this book, these principles will now be stated and briefly discussed. A more extensive discussion has been published (Ellis, 1983*b*).

Five principles of chloroplast biogenesis

A diagrammatic representation of these principles is presented in Fig. 2. The *first* principle states that the majority of chloroplast polypeptides are encoded in nuclear genes and are synthesised on cytoplasmic ribosomes. There are two aspects to this principle; the first concerns the site of *encoding* of chloroplast polypeptides, and the second the site of *synthesis* of these polypeptides. It is theoretically possible that a given polypeptide could be encoded in the nuclear genome but that its mRNA is translated by chloroplast ribosomes. However, the current evidence suggests that those chloroplast proteins encoded in the nucleus are synthesised by cytoplasmic ribosomes, while those encoded in the chloroplast genome are synthesised by chloroplast ribosomes. In other words there is no evidence that mRNA traverses the chloroplast envelope in either direction.

The evidence for the first principle comes from a variety of different approaches, including the analysis of mutants by classical genetic tests, the identification of proteins synthesised *in vitro* by subcellular fractions, the application of selective inhibitors of protein synthesis to intact cells, and the study of heat-treated plants which lack chloroplast ribosomes. The evidence from all these approaches is consistent and complementary in its support for this principle. The current status of the genetic evidence is presented by Tilney-Bassett, while Broglie *et al.* discuss a specific example of the differential expression of nuclear genes in the biogenesis of C_4 photosynthesis.

The significance of this principle is that it shows that the nucleus dominates chloroplast biogenesis. This conclusion suggests that those people attempting to manipulate chloroplasts in the hope of benefiting agriculture should put more effort into studying the nuclear genes for chloroplast proteins; current research is strongly biassed towards a study of the chloroplast genome because it is technically easier to dissect. Added weight to this argument is provided by the fact that all the transformation systems currently being developed for plant cells insert DNA into the nucleus or cytoplasm, not into the chloroplast.

One consequence of the first principle is that a vast traffic of nuclear-encoded polypeptides must flow across the chloroplast envelope during the biogenesis of chloroplasts. This traffic represents the only way so far discovered by which the two genomes interact at the macromolecular level. The *second* principle states that those chloroplast polypeptides synthesised by cytoplasmic ribosomes are transported into the chloroplast by a post-translational mechanism based on the chloroplast envelope. Protein transport across membranes is a fundamental activity of all cells. Proteins which are to be secreted from the cell or which are inserted into membranes are usually synthesised by membrane-bound ribosomes in a co-translational manner, i.e. transport is

Fig. 2. Representation of the principles governing the biogenesis of chloroplasts. The numbers in circles refer to the five principles stated in the text. The thickness of the arrows indicates the relative contributions of the nuclear and chloroplast genomes to the synthesis of chloroplast proteins. The wavy dotted lines indicate the stimulatory effect of light on the transcription of nuclear and plastid genes. (Reprinted with permission from Ellis, 1983*b*.)

PRINCIPLES OF CHLOROPLAST PROTEIN SYNTHESIS

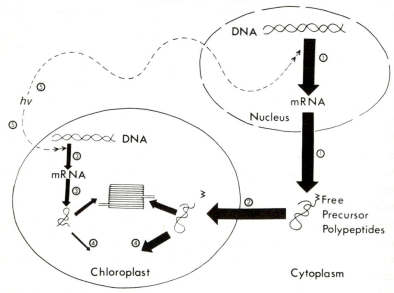

⸲ The cleaved N-terminal extension of precursor polypeptides made in the cytoplasm.

concomitant with the elongation of the polypeptide chain by the ribosome. In contrast, proteins which are destined to enter chloroplasts or mitochondria are synthesised by free cytoplasmic ribosomes, and enter these organelles only after complete synthesis and release from these ribosomes (Ellis, 1981*b*).

In many cases chloroplast polypeptides are synthesised in the cytoplasm as higher molecular weight precursors with an aminoterminal extension: this extension is removed after entry into the chloroplast by a specific stromal metalloendoprotease (Ellis & Robinson, 1984). This extension has been termed the 'transit peptide', but the use of this term is not recommended since it can be taken to imply that the role of this peptide is to bind the protein directly to the chloroplast envelope, in much the same way that the signal peptide mediates binding to the endoplasmic reticulum during co-translational protein transport. However, if this were the case, the chloroplast envelope would be studded with cytoplasmic ribosomes during chloroplast biogenesis, and this is not observed. Instead, the role of the extension appears to be to confer on the rest of the molecule the appropriate conformation to allow recognition by the transport machinery. If this view is correct, it predicts that there will be no similarity between the extensions of different chloroplast precursor proteins, except possibly around the cleavage site recognised by the stromal endoprotease. One consequence is that it will not prove possible to induce novel proteins to enter chloroplasts simply by adding a known 'transit peptide', and this limitation should be borne in mind by those hoping to alter chloroplast metabolism by inserting new proteins.

The *third* principle states that the chloroplast genetic system is essential for chloroplast biogenesis and contributes of the order of 100 polypeptides. The identification and structure of the genes located inside chloroplasts is currently a very active area of research, and aspects of this are discussed by Groot, Crouse *et al.* and Gray *et al.* We can look forward with some confidence to the complete identification of the structural genes located in the chloroplast genome within the next decade. However the essential mystery of the chloroplast genetic system still remains – why does it exist at all? It seems a high price for the cell to specify all those extra polypeptides involved in the replication and expression of the chloroplast genome merely to ensure that a rather small number of polypeptides is synthesised inside the chloroplast rather than outside it. The recent discovery of nucleotide sequences common to more than one genetic compartment, the so-called 'promiscuous DNA', lends some support to the view that all the genes for chloroplast proteins were originally present inside the primordial endosymbiont, and that what we see today is a snapshot of an evolutionary process which is transferring these genes to the nucleus (Ellis, 1983*a*).

The *fourth* principle states that chloroplast-encoded polypeptide and RNA molecules function entirely within the organelle. It is the most controversial of the five principles, since the evidence is sparse. It could be argued that polypeptides may be exported from chloroplasts either to function in the cytoplasm as enzymes in metabolic pathways interacting with chloroplast metabolism, or to act as regulatory signals informing the nucleocytoplasmic system of the state of affairs within the chloroplast. However, no rigorous evidence to support any of these possibilities has yet appeared.

The *fifth* principle concerns the role of light in chloroplast biogenesis. It states that the synthesis, transport and assembly of the majority of chloroplast polypeptides can occur in both the light and the dark, but that illumination has a stimulatory effect at several different levels on the accumulation of these polypeptides. When seeds of most flowering plants are germinated in the dark, the seedlings possess no chlorophyll, and are termed *etiolated*. In more primitive plants such as gymnosperms, ferns, mosses and algae, the ability to make chlorophyll in the dark is usually, but not invariably, present. In flowering plants grown from seed for several days in darkness, plastids occur in the form of *etioplasts*, which contain characteristic structures called prolamellar bodies. Etiolated seedlings also differ in morphology from light-grown seedlings; dicotyledons develop long stems with leaves which fail to expand, while monocotyledons produce long leaves which are narrower than normal. Exposure of such etiolated seedlings to light results in the rapid conversion of etioplasts to photosynthetically active green chloroplasts. Many studies have been carried out on the molecular changes that occur during this 'greening' process, and as might be expected there is an increase in the accumulation of all those polypeptides characteristic of chloroplasts. However, it is also clear that etioplasts contain detectable amounts of most of the polypeptides required for photosynthesis.

The essence of the fifth principle is that the dramatic effect of light on chloroplast polypeptide accumulation is largely a stimulation of processes going on at lower rates in the dark, and not the initiation of entirely novel processes. The only well-established process which requires light in an absolute sense is the reduction of protochlorophyllide to chlorophyllide. Etioplast membranes contain an enzyme which reduces protochlorophyllide from NADPH only when the enzyme–substrate complex is illuminated; Griffiths and Oliver discuss what is known about this enzyme. The remaining chapters of this book discuss recent findings about how light stimulates the accumulation of chloroplast polypeptides and pigments, the emphasis being on the effect of light on the transcription of nuclear and plastid genes.

References

Baur, E. (1909). Das Wesen und die Erblich-keitsverhaltnisse der 'Varietates albomarginatae hort' von *Pelargonium zonale. Z. VererbLehre*, **1**, 330–51.

Correns, C. (1909). Vererbungsversuche mit blass (gelb) grunen und bunt blattrigen sippen bei *Mirabilis jalapa, Urtica pilulifera*, und *Lunaria annua. Z. VererbLehre*, **1**, 291–329.

Ellis, R. J. (1979). The most abundant protein in the world. *Trends Biochem. Sci.*, **4**, 241–4.

Ellis, R. J. (1981*a*). Chloroplast proteins: synthesis, transport and assembly. *A. Rev. Pl. Physiol.* **32**, 11–37.

Ellis, R. J. (1981*b*). Protein transport across membranes: an introduction. In *Molecular Mobility and Migration*, ed. P. B. Garland & R. J. P. Williams, Biochem. Soc. Symp. No. **46**, pp. 223–34. Biochemical Society, London.

Ellis, R. J. (1983*a*). Mobile genes of chloroplasts and the promiscuity of DNA. *Nature*, **304**, 308–9.

Ellis, R. J. (1983*b*). Chloroplast protein synthesis: principles and problems. In *Subcellular Biochemistry*, vol. 9, ed. D. B. Roodyn, pp. 237–61. Plenum Publishing Corporation, New York.

Ellis, R. J. & Robinson, C. (1984). Post-translational transport and processing of cytoplasmically-synthesized precursors of organellar proteins. In *The Enzymology of Post-Translational Modifications of Proteins*, vol. II, eds. R. B. Freedman & H. Hawkins. Academic Press, London. (In press.)

Kirk, J. T. O. & Tilney-Bassett, R. A. E. (1978). *The Plastids: their Chemistry, Structure, Growth and Inheritance*, 2nd edn. Elsevier/North-Holland Biomedical Press, Amsterdam.

Lorimer, G. (1981). The carboxylation and oxygenation of ribulose-1,5-bisphosphate. *A. Rev. Pl. Physiol.*, **32**, 349–83.

Lyttleton, J. W. (1962). Isolation of ribosomes from spinach chloroplasts. *Expl Cell Res.*, **26**, 312–17.

Margulis, L. (1981). *Symbiosis in Cell Evolution: Life and its Environment on the Early Earth*. Freeman, San Francisco.

Miziorko, H. M. & Lorimer, G. (1983). Ribulose-1,5-bisphosphate carboxylase-oxygenase. *A. Rev. Biochem.*, **52**, 507–35.

Reid, R. A. & Leech, R. M. (1980). *Biochemistry and Structure of Cell Organelles*. Blackie and Son Ltd, London.

Ris, H. & Plaut, W. (1962). Ultrastructure of DNA-containing areas in the chloroplast of *Chlamydomonas. J. Cell Biol.*, **13**, 383–91.

Schnepf, E. (1980). Types of plastids: their development and interconversions. In *Chloroplasts, Results and Problems in Cell Differentiation*, vol. 10, ed. J. Reinert, pp. 1–27. Springer-Verlag, Berlin.

Strasburger, E. (1882). Ueber den Theilungsvorgang der Zellkerne und das Verhaltniss der Kerntheilung zur Zelltheilung. *Arch. mikrosk. Anat. EntwMech.*, **21**, 476–590.

Thomson, W. W. & Whatley, J. M. (1980). Development of non-green plastids. *A. Rev. Pl. Physiol.*, **31**, 375–94.

PART I

The nuclear contribution

R. A. E. TILNEY-BASSETT

The genetic evidence for nuclear control of chloroplast biogenesis in higher plants

The common expectations of nuclear versus extranuclear control

The genetic quest for the informational seat of chloroplast biogenesis is based on the analysis of wild-type forms associated with differences between related species or cultivars, and of spontaneous or induced mutants. The genetic analysis seeks to specify the informational site to within the nuclear or plastid genome according to the pattern of inheritance – nuclear or extranuclear. The nuclear pattern occurs whenever a mutant, or morph, is inherited according to the principles of Mendelian segregation (Fig. 1). The extranuclear pattern, characteristic of plastid inheritance, occurs when the F_1 and F_2 generations do not conform to the Mendelian rule. There are now two alternative

Fig. 1. The expectations of nuclear control. The homozygous dominant, dark-green wild-type, crossed with the homozygous recessive, pale-green mutant, gives a uniform F_1 dark-green, irrespective of the direction of the cross. The F_2 and back-cross segregate in characteristic Mendelian 3:1 and 1:1 ratios.

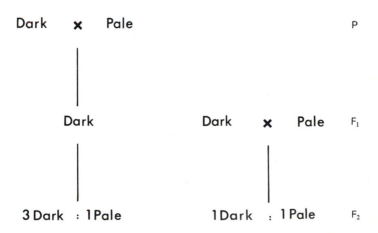

NUCLEAR

Dark ✗ Pale			P
Dark	Dark ✗ Pale		F_1
3 Dark : 1 Pale	1 Dark : 1 Pale		F_2

expectations. The more frequent result is the maternal pattern of inheritance (Fig. 2); the less frequent result is the biparental pattern of inheritance (Fig. 3). About two-thirds of the approximately 60 higher plant genera classified so far have a maternal, and one-third a biparental, plastid inheritance (Kirk & Tilney-Bassett, 1978; Hagemann, 1979; Tilney-Bassett & Abdel-Wahab, 1979; Sears, 1980).

Fig. 2. The expectations of maternal control. The F_1 phenotype, after reciprocal crosses, is solely dependent upon that of the maternal parent. There is no F_2 segregation.

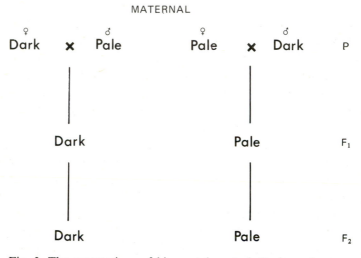

Fig. 3. The expectations of biparental control. Reciprocal crosses give a mixture of F_1 maternal, biparental and paternal progeny in dissimilar proportions and with considerable variance. F_2 segregation is limited to biparental (variegated) progeny and does not accord with Mendelian ratios.

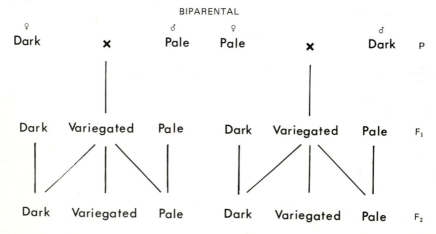

Sometimes the expected segregation ratios are modified. For example, when a nuclear mutant is partially dominant, owing to a semi-dominant or co-dominant allelic interaction, the heterozygote displays a phenotype distinct from either homozygote and instead of a 3:1, a 1:2:1 ratio is obtained. Another common modification is when the homozygous recessive is semi-lethal, in which case the ratio greatly exceeds 3:1. Such modifications, or indeed more involved ones caused by the joint segregation of two or more independent or linked genes, are easily recognized and quite distinct from the extranuclear inheritance patterns.

Each new nuclear mutant ought to be assessed as to whether it is the expression of a mutation in an entirely new gene, or of a new allele of a previously known gene. This is achieved by crossing the new homozygous recessive mutant with other phenotypically similar homozygous recessive stocks (Fig. 4). Should the homozygous recessives be non-viable, or infertile, tests of allelism are made by crossing their corresponding heterozygotes. Allelic mutants segregate in a ratio of 3 wild-type to 1 mutant phenotype, whereas non-allelic mutants produce a wholly wild-type progeny.

Finally, like every newborn baby, each new mutant needs a name. All too often insufficient attention is given to identity, and important research is lavished upon inadequately described and subsequently unidentifiable mutants. Less confusing, but no less irritating, are the plethora of names and numbers having no relation to the usual concepts of genetic nomenclature. I do not want to elaborate on these. Rather, I should like to recommend to the reader the excellent system adopted by von Wettstein and his associates in the naming of their large collection of barley (*Hordeum vulgare*) mutants.

Fig. 4. Allelism test. Left, two homozygous recessive, allelic mutants *vir-x*[1] and *vir-x*[9] produce a heteroallelic, homozygous recessive, mutant progeny. Right, two homozygous, non-allelic mutants *vir-x* and *vir-y* produce a double heterozygous F_1, wild-type progeny.

ALLELISM TEST

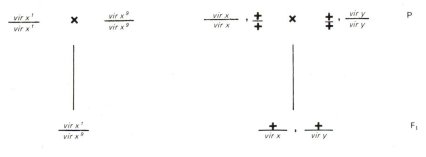

Allelic : mutant   Non-allelic : wild-type

Each mutant is first classified, according to its external phenotype, into one of a few categories – *albina*, *xantha*, *viridis*, *tigrina* – and numbered consecutively upon isolation (Wettstein *et al.*, 1971). The mutant is tested for allelism by diallel crosses between existing mutant stocks of the same category. When a new mutant is allelic with an existing mutant, it is given the same letter, representing the locus of the gene and, when it is allelic with none, the next available letter is used. For example, *tig-b*[23] and *tig-o*[34] are the 23rd and 34th isolates of the *tigrina* series of mutants, which are caused by mutations at the *b* and *o* gene loci respectively.

Nuclear control of plastid inheritance and replication in *Pelargonium*

The survey of plastid inheritance in higher plants has provided no evidence for a split into maternal or biparental patterns along major taxonomic and evolutionary divides. Rather, it appears as if this division has occurred on many occasions at the level of order, family or genus. This flexibility suggests that the switch from biparental to maternal, or vice versa, is easily achieved and is therefore probably controlled by few genes. The study of biparental inheritance in the zonal pelargonium (*Pelargonium × Hortorum*) suggests that the major genetic control is nuclear. Crosses between pelargonium cultivars, in which the plastids carry either the wild-type allele (green phenotype (G)) or a mutant allele (white phenotype (W)) in their germ cells, produce progeny with a mixture of maternal zygotes (MZ), biparental zygotes (BPZ), and paternal zygotes (PZ), as defined by the presence or absence of green or white plastids in the young embryos into which the zygotes develop. There are two distinct segregation patterns: the type I pattern breeds true and is therefore homozygous ($Pr_1 Pr_1$), while the type II pattern, which does not breed true, is heterozygous ($Pr_1 Pr_2$). In general, reciprocal backcrosses between the two types confirm these genotypes (Tilney-Bassett & Abdel-Wahab, 1982). The nuclear gene which controls these alternative patterns of plastid segregation is called *Pr* as it presumably has an effect on the replication of plastid DNA (Tilney-Bassett, 1973). After G × W crosses, type I female parents have a high frequency of embryos with only maternal alleles, and are intermediate for biparental, and low for paternal, embryos (MZ > BPZ > PZ). By contrast, the type II female parents have an equally high frequency of maternal and paternal embryos and a generally low biparental frequency (MZ > BPZ < PZ). These correspond to L-shaped and U-shaped gene frequency distributions (Fig. 5) (Tilney-Bassett & Birky, 1981).

Hence it appears as if the $Pr_1 Pr_1$ homozygotes preferentially select the maternal plastid for replication, whereas with the $Pr_1 Pr_2$ genotype selection in favour of the paternal plastid allele is equally successful. Modifying genes account for differences between cultivars within the alternative patterns (Tilney-Bassett, 1976) and among the progeny of type I plants can push the inheritance of some siblings into 100% maternal transmission (Fig. 6). The variance between siblings, both in respect of the progeny of type I and type II plants, is typical of what we might expect from the polygenic segregation of nuclear genes and can have no relationship to the wild-type plastid, which remains constant (Fig. 7). Thus, in these G × W crosses, the constant mutant plastid is transmitted by the male pollen tube into the environment of the constant wild-type plastid within the female egg cell, where it multiplies either moderately successfully, or not at all, depending largely on the female nuclear genome. It is not difficult to imagine that, with a little selection pressure, the population could easily become fixed in the maternal mode. The mutant plastid does have some effect, and some mutants compete more successfully with the wild-type plastid in the zygote than others (Abdel-Wahab & Tilney-Bassett, 1981; Vaughn, 1981), but it is the female nuclear genotype that makes the big switch in the overall inheritance pattern.

Once effective male plastid replication is blocked in the zygote, it is not illogical to suppose that further selection may sort out genotypes that debilitate the plastid in the pollen tube (Vaughn, DeBonte & Wilson, 1980; Vaughn, Kimpel & Wilson, 1981), or physically exclude plastids from the generative cell of the pollen grain (Hagemann, 1976, 1979; Kirk & Tilney-Bassett, 1978; Sears, 1980).

Some mutant plastids have no plastid ribosomes and so cannot synthesize

Fig. 5. Left, L-shaped, and right, U-shaped gene frequency distributions after G × W type I and type II crosses respectively, in which the only modes are of 100% (maternal embryos) and 0% (paternal embryos) with no mode corresponding to the population mean and no sign of a Gaussian distribution.

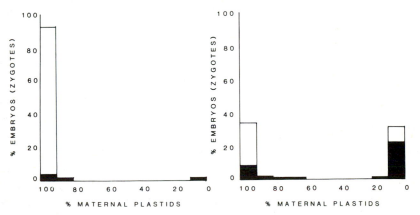

plastid proteins. Yet they replicate, which indicates that their DNA-polymerase is encoded in the nucleus and translated on the cytoplasmic ribosomes (Hagemann & Börner, 1978). It is unlikely that *Pr* alleles affect the polymerase directly as, at least eventually, the plastids do replicate. Unfortunately, the genetic control of organelle replication is not well understood, but if plasmid biology should prove to be a good model, it may be quite complicated. The justification for considering such a model is that the relationship between a plasmid and its prokaryotic bacterial host is somewhat analogous to the relationship between an organelle and its eukaryotic plant host cell. The evidence from plasmid biology suggests that plasmid replication involves at least five separate events, each controlled by bacterial or plasmid genes. These genes control the enzymology of DNA replication, the number of plasmid copies within the host cell, the partitioning of the plasmid into the daughter cells during host cell division, the process of conjugation of the bacterial host,

Fig. 6. Variation between individual cultivars and hybrids, after G × W and isogenic W × G reciprocal crosses, indicative of the effects of modifying genes upon the basic type I pattern (upper group) or type II pattern (lower group). The symbols represent the following cultivars: Type I: MBC = 'Miss Burdette-Coutts', LG = 'Lass O'Gowrie', DV = 'Dolly Varden', FH = 'Frank Headley'. Type II: JCM = 'J. C. Mapping', FS = 'Flower of Spring', FOS = 'Foster's Seedling', HS = 'Hills of Snow'. In the six hybrids, G or W is placed adjacent to the cultivar that was the source of green or white plastids respectively.

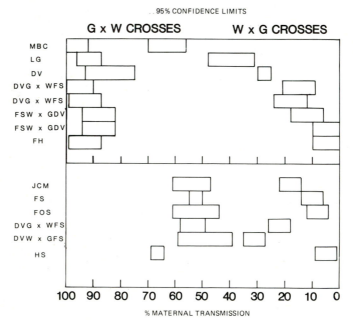

and the ability of two kinds of plasmid to be compatible, or incompatible, in the same host cell, which therefore determines whether they both persist or not. Several of these kinds of events are possible candidates for the action of the *Pr* gene.

The probable nuclear control of plastid inheritance gives rise to the paradox that the maternal inheritance of chloroplasts, which is so important for the localization of genes within the chloroplast, is not itself under chloroplast control. Another paradox (Ellis, 1981) is that some spontaneous plastome mutants, which have non-Mendelian inheritance, may have originated as nuclear mutations that caused the plastids to lose their ribosomes (Walbot & Coe, 1979). Once lost, the plastids are unable to regain their ribosomes even if returned to a normal nuclear background, and therefore behave as mutant plastids for ever after even though their DNA is unaltered. This seems a very plausible explanation for a group of gene-induced plastome mutants that produce variegated plants with only one kind of mutant plastid, rather than many kinds (Kirk & Tilney-Bassett, 1978).

Fig. 7. Variation between families with similar maternal nucleus and plastids after G × W and W × G crosses, in which the female parent is type I (G_1 and W_1) or type II (G_2 or W_2). The considerable variance within families – the number of families in each of the five crosses is stated at the right-hand end – follows an approximately skewed or normal distribution indicative of the effects of modifying genes upon the basic pattern of plastid inheritance.

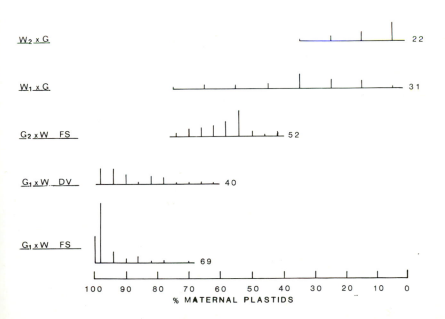

VARIANCE BETWEEN FAMILIES WITH SIMILAR MATERNAL NUCLEUS & PLASTIDS

Evidence from variant proteins

In the genetic code, minor changes leading to alterations in the amino acid composition of proteins frequently give rise to protein variation among related species. After suitable extraction and purification procedures, the protein variants can be distinguished by a range of techniques – chromatography, gel electrophoresis, and iso-electrofocussing – and revealed by appropriate stains. The consequent banding patterns serve as a marker both of the proteins present and, in comparisons between species, of any differences between them. Whenever there is variation the differences in position, or the presence or absence, of alternative variants are usable as markers of their inheritance; in particular to determine whether the protein is encoded by nuclear or chloroplast DNA (Fig. 8). This approach has been extensively used by Wildman and his associates (Gillham, 1978) to locate the coding site for a number of chloroplast proteins of tobacco (*Nicotiana*) species in which the plastids are maternally inherited (Kirk & Tilney-Bassett, 1978; Sears, 1980).

Fig. 8. The two species Z (4 band) and W (5 band) differ in respect of the precise position of band N, and the presence or absence of band C. With respect to the N bands, which represent different variants of the same protein, both reciprocal crosses produce an identical F_1 in which both N bands are present, which indicates that the N protein is encoded by nuclear DNA. With respect to the C band, this is observed in the F_1 after the cross W♀ × Z♂ (6 band), when the maternal parent has the C band, but not in the F_1 after the reciprocal cross Z♀ × W♂ (5 band), when the maternal parent has no C band, which indicates that the C protein is encoded by extranuclear DNA.

ELECTROPHORETIC ANALYSIS OF HYBRIDS

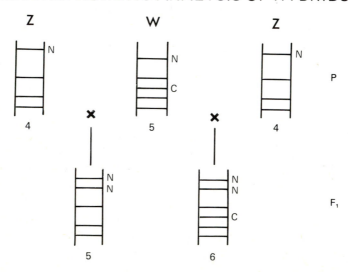

Fraction I protein, ribulose-1,5-bisphosphate (RuBP) carboxylase, is a very abundant oligomeric protein (550 kD) responsible for fixing atmospheric carbon dioxide: it consists of large and small subunits. Chan & Wildman (1972) found that in crosses between *Nicotiana gossei* and *N. tabacum* the large subunit peptide of the F_1 hybrids followed that of the maternal parent, indicating that it was coded by chloroplast DNA. By contrast, Kawashima & Wildman (1972) found that a peptide of the small subunit from the *N. tabacum* enzyme, which was absent from *N. glutinosa* or *N. glauca*, was present among the F_1 hybrids irrespective of the direction of the cross, indicating that the small subunit was coded by nuclear DNA, later confirmed by others (Gray *et al.*, 1974; Kung *et al.*, 1975 *b*). Furthermore, cloned cDNA sequences made from mRNA for the small subunit hybridized to nuclear DNA (Cashmore, 1979), and to nuclear RNA, but not to chloroplast RNA (Smith & Ellis, 1981). The large subunit is resolvable into three polypeptides each of 55 kD, and the small subunit into one to four polypeptides of 12.5 kD, depending on the species (Kung, Sakano & Wildman, 1974; Sakano, Kung & Wildman, 1974). Indeed, there appears to be a close correlation between diploid species of tobacco, with $2n = 24$ chromosomes, and never more than two, and usually one, small subunit polypeptide, and tetraploid species, with $2n = 48$ chromosomes, and two to four subunit polypeptides (Kung *et al.*, 1975 *b*). After examining diploid, tetraploid and hexaploid wheat species, Dean & Leech (1982) found that the quantity of RuBP carboxylase per mesophyll cell increased in step with each increase in nuclear ploidy, so that the ratios of enzyme to nuclear DNA per mesophyll cell were almost identical in the three species. A similar finding was reported from a diploid, tetraploid, octaploid series in alfalfa (*Medicago sativa*) with isogenic complements (Meyers *et al.*, 1982). The ability to differentiate between the large and small subunits of Fraction I protein, in addition to the polymorphism of their peptides, has enabled several workers to use this protein as a unique genetic marker (Kung, 1976) in the analysis of other hybrids, and in somatic hybrids (Kung *et al.*, 1975 *a*; Chen, Wildman & Smith, 1977; Belliard *et al.*, 1978; Melchers, Sacristan & Holder, 1978; Kumar, Wilson & Cocking, 1981).

A similar approach to the biochemical and genetic analysis of other proteins has shown that at least one subunit of the oligomeric protein coupling factor I (325 kD), required for photosynthetic phosphorylation, is encoded by nuclear DNA, while at least two of the five subunits are encoded by chloroplast DNA (Kwanyuen & Wildman, 1978). The genetic information for the primary structure of the electron transfer protein cytochrome *f* is also located in the chloroplast genome (Gray, 1980). Other proteins transcribed from nuclear DNA include at least one from the light-harvesting chlorophyll *a/b* protein complex (Kung, Thornber & Wildman, 1972), and two from the chloroplast ribosomes (Bourque & Wildman, 1973).

Anderson & Levin (1970) found that the chloroplast enzyme fructose-1,6-bisphosphate aldolase produced isozymes in pea (*Pisum sativum*) with an isoelectric point 0.15 pH units higher in cv. 'Laxton's Progress' than in cv. 'Little Marvel'. The analysis of the reciprocal F_1s between these cultivars, and of the segregating progeny derived from backcrosses between the F_1 and one parent, indicated that the structural gene for the chloroplast aldolase resides in the nuclear DNA. The same conclusion was reached for the *Aldo*-1 gene in pea by Weeden & Gottlieb (1980); unfortunately the heterozygotes between the fast and slow moving homozygotes on starch gels were too blurred to decide whether the enzyme was monomeric or dimeric. A second chloroplast enzyme variation detected in peas and in *Clarkia williamsonii*, and shown to be under nuclear control, was the dimeric enzyme phosphoglucose isomerase, *Pgi*-1. Another dimeric enzyme from pea was aspartate amino transferase, *Aat*-1, and two monomeric enzymes from pea chloroplasts were phosphoglucomutase, *Pgm*-2, and shikimate dehydrogenase, *Skdh*-1; all three enzymes are encoded by codominant alleles of nuclear genes. For some of these enzymes, there were genetically independent isozymes, *Pgi*-2, *Pgm*-1 and *Aat*-3; these are controlled by nuclear genes, but are localized in the cytoplasm and absent from the chloroplasts.

Corroborative evidence for the nuclear control of any of these proteins from mutants is weak. A number of barley mutants with defects in photosynthetic carbon dioxide fixation, which might affect the Fraction I protein, were examined physiologically by Carlsen (1977). Five mutants (*vir-c*[12], *vir-1*[27], *vir-zb*[63], *vir-zd*[69] and *xan-m*[3]) were completely blocked in light-dependent CO_2 fixation, but they did fix in darkness, or in light, small amounts of labelled $^{14}CO_2$ into malate, aspartate, glutamate and citrate, as did the wild-type. Another group (*vir-m*[29], *vir-t*[45], *vir-u*[46], *xan-c*[47], *xan-d*[49] and *xan-t*[50]) yielded reduced photosynthetic rates on a gram fresh weight basis, but differed little from wild-type on a chlorophyll basis. The photosynthetic products of these mutants were qualitatively the same as the wild-type, and the distribution patterns of the carbon tracer gave no suggestion of any defects in the reaction following the fixation of CO_2. Carlsen suggested that the lower photosynthetic rates were the result of a reduced light absorption and were not caused by any genetic lesion affecting the biochemistry of photosynthesis. Owing to the activity of its RuBP carboxylase *in vivo* being much lower than that in the wild-type, a recessive nuclear mutant of *Arabidopsis thaliana* has been called 'regulation of carboxylase activation' (*rca*). The mutant was green and generally healthy in appearance when grown under high carbon dioxide levels, but became chlorotic after several days of illumination in standard atmospheric conditions in which photosynthesis was severely impaired (Somerville, Portis & Ogren, 1982). The precise position of the five polypeptides of coupling factor I among the many proteins of wild-type chloroplast thylakoids, after electrophoretic separation, was determined by the use of crossed immuno-electrophoresis obtained by injecting the purified enzymes into rabbits to produce antibodies (Hoyer-Hansen, Moller & Pan, 1979). Three barley mutants (*vir-e*[64], *xan-b*[12] and *xan-d*[49]) were then checked for the presence of the coupling factor peptides, which produced a very weak reaction with the antibody, proving that the protein was highly deficient.

Evidence from mutants

The advantage of studying natural variation is that we usually start with a recognizable protein of known function, for which it is generally simple to determine the mode of inheritance. By contrast, with mutants, although it is equally simple to determine the mode of inheritance, it has often proved difficult to locate the damaged polypeptide and to assess its true function, owing to the many pleiotropic effects. In practice, therefore, mutants are generally classified into broad phenotypic categories, e.g. cotyledon or leaf colour, high fluorescence and temperature-sensitivity. After considerable analysis, they are more narrowly classified by various schemes which seek to relate the properties of the mutant to a probable disturbance within a limited functional activity of the plastid. Accordingly, it is argued that the mutation has probably occurred in a gene that codes for a polypeptide associated with the control of chlorophyll, carotenoid, lipid or starch synthesis, for example; or which is associated with the replication of chloroplasts, the electron transport pathway, or the fixation of carbon dioxide; or which is concerned with chloroplast biogenesis through the assembly of protein complexes associated with photosystems I and II, or the assembly of the chlorophyll *a/b* light-harvesting pigment complex. In a review of limited length, it is not practical to examine all these categories of plastid mutants, especially if there is little to add to the last review (Kirk & Tilney-Bassett, 1978). I shall therefore concentrate on examples from higher plants of current interest that vividly illustrate the progress and problems in unravelling the role of the very numerous nuclear genes that control chloroplast biogenesis.

The control of chlorophyll synthesis

It is reasonable to expect that mutations in the chlorophyll bio-synthetic pathway should be located with some precision, as the main biochemical steps are well known (Fig. 9). A number of mutants with lowered chlorophyll content are the white and yellow, respectively *albina* and *xantha*, recessive lethals in barley. Of 61 *xantha* mutants tested at 20 loci and two *albina* mutants (Wettstein *et al.*, 1971; Henningsen, Nielsen & Smillie, 1974), the analysis of a few has been followed in detail. To facilitate identification of the biosynthetic block the mutant seedlings were supplied in the dark with the chlorophyll precursor δ-aminolaevulinic acid (ALA) to promote the accumulation, and hence simplify the characterization, of the intermediate preceding the block. The accumulated porphyrins were identified by thin-layer chromatography and spectroscopy (Gough, 1972). Among the wild-type seedlings, about 90% of the accumulated porphyrin was protochlorophyllide and 10% protoporphyrin; *xantha* mutants accumulated as much porphyrin as the wild-type, *albina* mutants rather less, but the proportions and constituents changed.

The leaky mutant *xan-u²¹* accumulated 35% of its porphyrins as uropor-

Fig. 9. Nuclear mutants assumed to control steps in the pathway of chlorophyll synthesis.

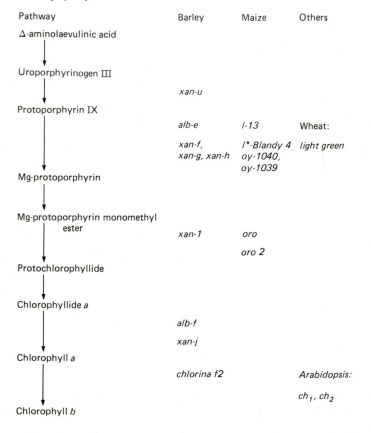

Pathway	Barley	Maize	Others
Δ-aminolaevulinic acid			
↓			
Uroporphyrinogen III			
↓	*xan-u*		
Protoporphyrin IX			
	alb-e	*l-13*	Wheat:
	xan-f,	*l*-Blandy 4*	*light green*
	xan-g, xan-h	*oy-1040,*	
		oy-1039	
Mg-protoporphyrin			
↓			
Mg-protoporphyrin monomethyl ester	*xan-1*	*oro*	
		oro 2	
Protochlorophyllide			
↓			
Chlorophyllide *a*			
	alb-f		
↓	*xan-j*		
Chlorophyll *a*			
	chlorina f2		*Arabidopsis:*
↓			ch_1, ch_2
Chlorophyll *b*			

N.B. The four *tigrina* genes are excluded as they are assumed to have a regulatory role. The temperature-sensitive and virescent genes are excluded as they have not been shown to have direct control over steps in the pathway of chlorophyll synthesis.

phyrin and protoporphyrin IX (Gough, 1972). It had an amorphous and abnormally structured prolamellar body (Wettstein *et al.*, 1971) and, owing to a block in β-carotene synthesis, accumulated aliphatic polyenes (Henningsen *et al.*, 1974; Nielsen & Gough, 1974; Wettstein *et al.*, 1974). Gough (1972) concluded that there was a major lesion in the pathway between uroporphyrin and protoporphyrin (Fig. 9).

The mutants *alb-e* and *xan-f* (6 alleles: f^{10}, f^{26}, f^{27}, f^{41}, f^{59}, f^{60}), *xan-g* (2 alleles: g^{37}, g^{45}), and *xan-h* (2-alleles: h^{36}, h^{56}) accumulated 80–97% protoporphyrin and 20–3% protochlorophyllide, and were assumed to be blocked between these two compounds (Gough, 1972; Henningsen *et al.*, 1974; Wettstein *et al.*, 1974). Non-leaky alleles of the *xantha* mutants grown in the dark formed no prolamellar bodies, just widely spaced parallel discs which were unperforated and traversed a major portion of the plastid, whereas leaky alleles of *xan-f* and *xan-g* (and *xan-l*[35] below) made crystalline prolamellar bodies in the dark and spheroidal grana in the light (Wettstein *et al.*, 1971). Electrophoretic analysis of *xan-f*, *xan-g*, *xan-h*, and *xan-l*, yielded banding patterns lacking the proteins that normally appeared in response to light and band J was missing (N. C. Nielsen, 1975). A spontaneous mutant, which was partially dominant in the light green heterozygote and yellow in the homozygous state, was described in hexaploid wheat (*Triticum vulgare*) (Pettigrew, Driscoll & Rienits, 1969; Newell & Rienits, 1975). It also accumulated protoporphyrin on incubation with ALA in the dark, and was thought to affect the conversion of protoporphyrin to magnesium protoporphyrin. Three maize (*Zea mais*) mutants, the dark *luteus* and recessive *l*-Blandy*-4 and *l*-13, and the *oil yellow* (with several recessive alleles), also appeared to block the conversion of protoporphyrin IX to magnesium protoporphyrin after feeding with ALA in the dark (Mascia & Robertson, 1980). The mutants *l*-Blandy*-4 and *oy*-1040 accumulated almost exclusively protoporphyrin IX and were unable to develop prolamellar bodies (Mascia & Robertson, 1978). Two leaky mutants which responded to feeding with ALA far less than did the wild-type were not identical. The allele *oy*-1039 developed a little chlorophyll when light-grown and approximately 40% of the normal level of protochlorophyllide when dark-grown, whereas *l*-13 accumulated much chlorophyll when light-grown and a nearly normal level of protochlorophyllide when grown in the dark (Mascia, 1978).

The barley grana-deficient mutant *xan-l*[35], with a raised chlorophyll *a/b* ratio (Nielsen *et al.*, 1979), accumulated 48% of its total porphyrins as magnesium protoporphyrin and was assumed to produce a metabolic block before protochlorophyllide a (Gough, 1972; Henningsen *et al.*, 1974; Wettstein *et al.*, 1974; Kannangara, Gough & Wettstein, 1978). Photosystems I and II were active but the mutants lacked the light-harvesting protein complex

(Simpson, Moller & Hoyer-Hansen, 1978). Two recessive *orobanche* loci in maize, *oro* (6 alleles) and *oro*-2, were yellow to tan necrotic in the light, or cross-banded in the normal day–night cycle (Mascia & Robertson, 1980). When fed with ALA, they accumulated a mixture of protoporphyrin IX, protoporphyrin monomethyl ester, and their magnesium derivatives, and it was suggested that there was a block after the ester which they were unable to convert to protochlorophyllide (Mascia, 1978). Within the etioplasts, the overall symmetry of the prolamellar body was disturbed (Mascia & Robertson, 1978). A dominant, and genetically independent, modifier gene, *orom*, partly suppressed the *oro* phenotype, which then became more normal-looking under a light–dark regime and could develop a considerable amount of chlorophyll.

The mutants *alb-f*[17] and *xan-j*[59] in barley could make protochlorophyllide and photoreduce it to chlorophyllide, but appeared unable to phytolate the chlorophyllide to form chlorophyll *a* (Henningsen *et al.*, 1974; Henningsen & Thorne, 1974; O. F. Nielsen, 1975), and there was a block in the dispersal of the prolamellar body into the primary thylakoids on continuous illumination, as well as a failure to show the 'Shibata shift' of the absorption peak of newly formed chlorophyllide from 682 to 672 nm (Wettstein *et al.*, 1971). Mutants *xan-a* and *xan-m* were also unable to carry out prolamellar dispersal, but without the absence of the 'Shibata shift'.

Mutants making chlorophyll *a* but not chlorophyll *b* were isolated in *Arabidopsis thaliana* (Röbbelen, 1957). In one mutation the distinction was made between the alleles ch_1 and ch_2 which made, respectively, either no chlorophyll *b* or a reduced amount of it (Hirono & Rédei, 1963). Another well known mutant that is completely devoid of chlorophyll *b* is the *chlorina* *f*2 mutant in barley, which will be discussed in a later section.

An interesting group of chlorophyll deficiencies are the *tigrina* and *zonata* mutants derived from 15 barley loci. In the *tigrina* mutants, the seedlings develop pale green transverse necrotic bands, and in the *zonata* mutants transverse non-necrotic bands, akin to zebra-striping, when grown under normal light–dark cycles. The leaves have a uniform pale colour when grown in continuous light. Some of the mutants were formerly referred to as infra-red (Wettstein *et al.*, 1971). Most of them are lethal or semi-lethal recessives, while a few are partially dominant, exhibiting semi-dominance or co-dominance (Nielsen, 1974*a*). Mutations at the four loci *tig-b* (2 alleles: b^{19}, b^{23}), *tig-d*[12], *tig-n*[32] and *tig-o*[34] caused the accumulation of 1.3–15-fold as much protochlorophyllide in the dark as occurs in the wild-type. The high protochlorophyllide levels resulted from the 1.3–6-fold higher rates of ALA formation than in the wild-type (Gough & Kannangara, 1979). Despite the large differences in protochlorophyllide content, the amount which was photoconvertible was the same in all genotypes, indicating a constant number of conversion sites – corresponding to the pigment–protein complexes called protochlorophyll holochrome – bound to the membranes of the prolamellar body (Nielsen, 1973; Nielsen, 1974b). The plastid ultrastructure, either as etioplasts or chloroplasts, was somewhat modified (Nielsen, 1974a); there was a high frequency of particles on the PFs freeze-fracture face (see p. 31) of *tig-o*[34] when grown in continuous light (Simpson *et al.*, 1977). The mutants had, on a molar basis, approximately the same total content of carotenoids as the wild-type, but they differed in detail: *tig-b*[23] synthesized β-carotene and accumulated ζ-carotene, *tig-d*[12] was indistinguishable from wild-type, *tig-n*[32] accumulated lycopenic pigments, and *tig-o*[34] accumulated lycopene (Henningsen *et al.*, 1974; Nielsen & Gough, 1974; Wettstein *et al.*, 1974). It was suggested that the products of these genes had a regulatory role. The genes *tig-b*, *tig-n*, and *tig-o* regulated the coordinated functions of the chlorophyll and carotene pathways in plastid development, while the function of *tig-d* was restricted to the chlorophyll pathway. This hypothesis received support from the analysis of hybrids containing two of these genes with mutants of other genes known to cause the accumulation of intermediates other than protochlorophyllide (Wettstein, 1974; Kahn, Avivi-Bleiser & Wettstein, 1976). The double mutant containing both *tig-d*[12] and *xan-l*[35] increased the level of magnesium protoporphyrins several fold, and the double mutant containing both *tig-o*[34] and *xan-f*[10] greatly increased the level of protoporphyrin. Hence, whether combined with wild-type genes or with mutant genes blocking the pathways at specific points, the effect of the *tig* genes was to stimulate the biosynthesis of chlorophyll precursors; in fact these *tig* genes behave as constitutive mutants in regulatory genes for protochlorophyllide synthesis and, when grown in darkness, they mimic the effects of addition of ALA. More recent analysis of *tig-o*[34] has revealed its sensitivity to both temperature, and light intensity and wavelength. When grown in the light at a restrictive temperature of 20 °C, the barley seedlings contained wild-type levels of 80S cytoplasmic ribosomes, but were deficient in 70S chloroplast ribosomes and the associated chloroplast membrane proteins translated thereon (Hoyer-Hansen & Casadoro, 1982). In white light at 20 °C, the chloroplasts were first formed and subsequently destroyed during plant maturation, and red light above 600 nm was more destructive than the less intense red light above 650 nm (Casadoro *et al.*, 1983). These observations were thought to be related to defective β-carotene biosynthesis at the lower temperature, which gave rise to faulty membrane structures and led to the photoinstability of the plastids resulting in the destruction of the plastid ribosomes. By contrast, at the higher temperature, an increased β-carotene level in the mutant plastids protected the 70S ribosomes from photodamage and so the seedlings resumed a normal appearance. Nevertheless, as the homozygous seedlings were unable to fix CO_2, the primary defect in the mutant remained (Casadoro *et al.*, 1983).

There are quite a few mutants in which the development of chlorophyll pigments is sensitive to temperature, or is preceded by a lag period. Their consideration in this section is convenient on phenotypic grounds, without implying that the primary effect of their mutations is directly connected with the control of chlorophyll synthesis. In normal maize, the chlorophyll content increases progressively from about zero at 12 °C to high levels at 20 °C, whereas in the recessive mutant *M 11* there was no chlorophyll accumulation below 17 °C (Millerd & McWilliam, 1968). The etioplasts contained proto-chlorophyll and appeared to have a normal ribosomal content, but an abnormal prolamellar body (Millerd, Goodchild & Spencer, 1969). At 10 °C a recessive mutant of lucerne (*Medicago sativa*) produced scarcely any chlorophyll, carotenoid or RuBP carboxylase, and phosphoribulokinase was much reduced. At 18 °C the level of the two enzymes was largely restored and the pigments increased to 10% of the normal level, and at 27 °C the plants had a normal level of pigments and enzymes (Huffaker *et al.*, 1970).

Six virescent mutants in barley (*vir-y^{ts2}*, *vir-zf^{ts4}*, *vir-zg^{ts9}*, *vir-zh^{ts46}*, *vir-zi^{ts49}* and *vir-zj^{ts57}*) that green normally at permissive temperatures below 23 °C, had their chlorophyll development inhibited at the restrictive temperature of 32 °C (Smillie *et al.*, 1978). The mutants were distinguished by responses to five differences in temperature control; between the permissive and restricted temperatures they displayed differences in respect of chlorophyll production, chloroplast ultrastructure and photochemical activities. It was also found that etioplasts formed at the permissive temperature could develop into fully active chloroplasts at the restrictive temperature, which suggests that the genes code for products formed in the dark, or else for products not needed in the assembly of the photosynthetic membranes.

Other virescent mutants show a clear-cut lag period between germination, when the seedlings have little pigment, and some days later, when there is a rapid increase in chloroplast pigments to the wild-type levels, as in the *pale yellow*-1 mutant of corn (Kay & Phinney, 1956) and the *T-811* peanut (*Arachis hypogaea*) mutant (Benedict & Ketring, 1972). In the latter, if 6-day-old seedlings were illuminated with white light for 2 h and then put back in the dark for 3 days, the plastids developed into normal etioplasts with prolamellar bodies and, on exposure to light, chlorophyll and chloroplast enzymes were formed without a lag period (Benedict, Ketring & Tomas, 1974). The slow accumulation of chlorophyll in a virescent mutant of *Phaseolus vulgaris* was associated with the slow development of the plastid fine structure, especially the stacking of the grana; there was also the possibility that chlorophyll was breaking down under strong light (Dale & Heyes, 1970; Heyes & Dale, 1971). The authors suggested that the *v 18* mutant in maize had a block in the dark enzymatic reactions of photosynthesis (Chollet & Paolillo, 1972).

A virescent mutant of barley, called *yellow-viable*-2 (yv_2), and located on chromosome 1 (Walker *et al.*, 1963), was expressed when homozygous-recessive (Stephansen & Zalik, 1971). The mutant has been extensively studied because the time-lag in its development makes it a useful alternative to the examination of the greening of etioplasts, or to mutants of a fixed phenotype, for the analysis of the processes involved in the development of chloroplast structure and function. Young virescent seedlings were deficient in chlorophyll and carotenoids and had small, irregular-shaped chloroplasts that contained large vesicles but no normal lamellae or grana (Maclachlan & Zalik, 1963). Whether the seedlings died or survived and developed into fully green plants was dependent upon the quality and intensity of light and temperature (Miller & Zalik, 1965). Under low light intensity (58 lx, 20 °C), 4-day-old virescent seedlings were lacking in some lamellar proteins (Jhamb & Zalik, 1973); they contained predominantly single thylakoids, and their photoreductive activity was low. By 8 days the chlorophyll content had increased, the grana were well developed, and the photoreductive activity had reached normal levels, but under higher light intensity (130 lx, 23 °C), they were still not fully normal (Horak & Zalik, 1975). Closely correlated with the increase in chlorophyll content from 4 to 8 days, there was an increase in lipids (Thomson & Zalik, 1981*a*). The activity of acetyl coenzyme A carboxylase, an enzyme of the chloroplast stroma, which reached its peak by 4 days, was actually greater in the mutant than the wild-type (Thomson & Zalik, 1981*b*). Fully greened mutant seedlings contained both the light-harvesting chlorophyll–protein complex, and the reaction centres of photosystem I (PSI) and photosystem II (PSII) and their associated photochemical activities appeared functional (Kyle & Zalik, 1982*a*). An early suggestion for the primary effect of the mutation was that the synthesis of the chlorophyll holochrome protein was partially inhibited (Sane & Zalik, 1970). More recently, fluorescence studies have led to the suggestion that a certain population of the light-harvesting chlorophyll–protein complex was unconnected to the PSII reaction centres (Kyle & Zalik, 1982*b*).

Mutations affecting the chlorophyll–protein complexes

I propose to use the system of nomenclature suggested by Machold, Simpson & Moller (1979), and outlined in Table 1, in order to classify a wide range of chloroplast mutants within which the chlorophyll proteins and their complexes have been disturbed. At the same time, I shall endeavour to reveal the relationship between altered complexes and the pleiotropic effects of the

Table 1. *Nomenclature for chlorophyll-proteins and their complexes after their separation by sodium dodecyl sulphate polyacrylamide gel electrophoresis, and comparison with alternative systems of nomenclature (Adapted from Machold et al., 1979, in which full references are given)*

Chlorophyll proteins	Chlorophyll protein complexes	Uncharacterized bands	Alternative nomenclatures							Approximate molecular weights (kD)
			A	B	C	D	E	F	G	
Chl$_a$-P1	Chl$_a$-P1**		CPIa					Ia		110
	Chl$_a$-P1*		CPI	CPI	I	CPI	CPI	Ib	CPI	107
		Chl-P						Ic		92
Chl$_{a/b}$-P2	Chl$_{a/b}$-P2***		LHCP^1		II'	IIb		IIa		71
		Chl-P								62
	Chl$_{a/b}$-P2**		LHCP^2			IIa				50
		Chl-P						IIb_1		49
Chl$_a$-P2			CPa	CPIII	IV	A	CPIII	IIb_2		41
Chl$_a$-P3				CPIV						32
Chl$_{a/b}$-P1	Chl$_{a/b}$-P2*		LHCP^3	CPII	II	II	CPII	CPII	CPII	29

N.B. The apoproteins are indicated by the abbreviation Chl$_a$-AP1, Chl$_a$-AP2 etc.

mutation upon the essential features of ultrastructure and photosynthetic and biosynthetic properties of the plant.

Chlorophyll a–protein complexes. In the hope of predicting the nature of photosynthetic defects, Simpson & Wettstein (1980) classified 42 *viridis*, *chlorina* and *xantha* barley mutants by their fluorescence emission spectra. Unfortunately, useful though this analysis was, they did not find it possible to discern a usable relationship between chloroplast ultrastructure and specific photosynthetic deficiencies; hence there appears to be no short cut to individual mutant analysis. One of the mutants, *vir-n*[34], produced homozygous recessive, lethal pale green seedlings. They contained 76% of the normal chlorophyll content and had an almost normal chlorophyll a/b ratio (Hiller, Moller & Hoyer-Hansen, 1980), with no ultrastructural changes (Simpson & Wettstein, 1980). The mutant contained 40% of the wild-type level of Chl_a-P1 (Hoyer-Hansen *et al.*, 1982) and was deficient in three polypeptides – thought to be iron–sulphur proteins – of 13.8 kD, 15.6 kD and 16.5 kD. Analysis of the photosynthetic capacity of the mutants revealed a nearly normal PSII activity and a 90% blocked PSI (Hiller *et al.*, 1980; Moller, Smillie & Hoyer-Hansen, 1980; Moller, Hoyer-Hansen & Henry, 1982). The mutant was not deficient in cytochrome f or ferredoxin–$NADP^+$ oxido-reductase, but the P700 reaction centre of PSI, which is closely associated with Chl_a-P1, was 75% impaired (Hoyer-Hansen *et al.*, 1982) (Fig. 10). Freeze-fracturing of thylakoid membranes produces four fracture faces. PFs and PFu are the outer fracture faces (bordering the exterior of the thylakoid) of the membrane in stacked and unstacked regions of the thylakoids respectively. EFs and EFu are the inner fracture faces (bordering the interior of the thylakoids) in stacked and unstacked regions respectively. Analysis of the four fracture faces of freeze-fractured thylakoids from a PSI barley mutant led Simpson (1982) to conclude that the loss of P700 Chl_a-P1 caused a reduction in the average size of the large, heavily shadowed PFu particles without significant change in particle density. It was suggested that the absence of a functional P700 reaction centre may prevent the incorporation of the iron–sulphur proteins, or alternatively, the absence of these three proteins may prevent the formation of the reaction centre. The lethal seedlings of two similar mutants, *vir-h*[15] and *vir-zb*[63], contained 45% and 64% of the normal chlorophyll content, and had a slightly raised chlorophyll a/b ratio (Hiller *et al.*, 1980), and *vir-zb*[63] had a decreased β-carotene and a very low xanthophyll content (Henry, Mikkelsen & Moller, 1983). Their ultrastructure was similar to wild-type, except that *vir-h*[15] had rather few stroma lamellae (Simpson & Wettstein, 1980); *vir-zb*[63] had EF particles in grana thylakoids (Simpson *et al.*, 1978). There was a complete absence of Chl_a-P1, P700 was highly deficient (Moller, Nugent & Evans, 1981), and the three iron–sulphur peptides were much reduced. Electron transport through PSI was weak in *vir-h[15]* and completely blocked in *vir-zb[63]*, and in both cytochrome *f* was reduced (Hiller *et al.*, 1980). Hence all three genes are closely implicated in the formation of the P700 reaction centre of PSI (Fig. 10), but their precise role and the nature of the differences between them remains to be elucidated.

A group of recessive lethal mutants, selected for high chlorophyll fluorescence (Miles & Daniel, 1974), have been identified in maize. Five *hcf* mutants caused the loss of Chl_aP1 with a concomitant decrease in PSI electron transport. Three mutants had a reasonably active PSII, while in the other two it was either diminished or not detectable (Miles, 1980; Miles, Markwell & Thornber, 1979). One of the maize mutants, *hcf E 1481*, with a severely reduced PSI and near-normal PSII, had a measurable reduction in the diameter of particles on the unstacked protoplasmic face of freeze-fracture chloroplasts. Chl_a-P1 was missing, a 68 kD peptide severely reduced or eliminated, and an 18 kD peptide considerably weakened (Miller, 1980). It was thought that the reduced particle size might be caused by a crucial component missing from the PSI reaction centre, thereby causing a severe retardation in PSI electron transport. It was also suggested that the majority of PSI centres are in unstacked regions, and PSII and light-harvesting

Fig. 10. Nuclear mutants which appear partially or completely to block steps in the formation of Chl_a–protein complexes and associated polypeptides leading to the assembly of the photosystem I and II reaction centres. See text for details.

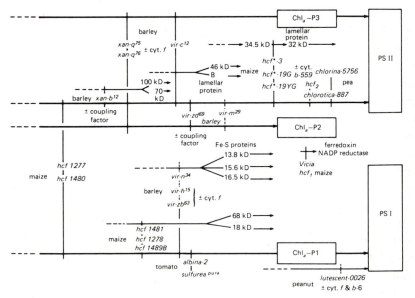

chlorophyll *a/b*–protein complexes in stacked regions of the membrane system. The absence of Chl_a-P1 was also reported for the nuclear mutants *albina-2* and *sulfurea*[pura] of the tomato (*Lycopersicon esculentum*) (Börner *et al.*, 1975).

The lutescent mutant, *0026*, in peanut had a 42% reduction in chlorophyll content and a slightly raised chlorophyll *a/b* ratio (Alberte, Hesketh & Kirby, 1976). These alterations in the chlorophyll protein complexes were accompanied by a reduction in the number of PSI reaction centre components and associated electron transport assemblages (P700 and cytochromes *f* and *b*-6) per unit leaf area, and a photosynthetic unit size 1.5 times larger than that of wild-type. These findings indicated that mutant leaves had many fewer, but larger, photosynthetic units than green leaves. As a consequence, it was suggested that the observed low photosynthetic rates, on both leaf area and chlorophyll bases, were probably caused by the much reduced capacity of the mutant chloroplasts to produce sufficient quantities of ATP and NADPH to support a high level of carbon dioxide fixation. Within a nuclear mutant of the broad bean (*Vicia faba*) PSI functioned to the extent that cyclic photo-phosphorylation was catalysed, but the chloroplasts did not photoreduce NADP, even when a source of ferredoxin was added, and so the mutation might possibly affect the ferredoxin–NADP+ reductase (Heber & Gottschalk, 1963). The leafy maize mutant hcf_1 is similar (Miles & Daniel, 1974).

Two recessive and rather similar mutants of barley were lethal *vir-zd*[69], and sub-lethal *vir-m*[29], with chlorophyll contents of 25% and 61%, respectively, and *vir-zd*[69] had a low β-carotene content and decreased levels of xanthophylls (Henry *et al.*, 1983). Chloroplasts of the lethal mutant, prior to seedling death, had a low chlorophyll *a/b* ratio, and were 'grana-rich' containing large grana with several thylakoids and narrow intra-thylakoidal spaces, and severely reduced stroma lamellae (Nielsen *et al.*, 1979), whereas those of the sublethal mutant had a fairly normal thylakoid system, apart from one or more giant grana with extremely large thylakoid discs (Simpson & Wettstein, 1980). *Vir-zd*[69] had normal PF faces, but few particles on the EFs face when freeze-fractured (Simpson *et al.*, 1977; Simpson *et al.*, 1978). PSI was not seriously disturbed, but PSII activity was lacking in *vir-zd*[69] (Nielsen, Henningsen & Smillie, 1974) and light-dependent carbon dioxide fixation was reduced (Carlsen, 1977). In both mutants, alterations to thylakoid structure and depression in photosynthetic activity were correlated with the lack of the Chl_a-P2 band (Fig. 10), and the absence or reduction of its associated apoprotein (Machold & Hoyer-Hansen, 1976); the coupling factor enzyme was also reduced in *vir-zd*[69] (Simpson *et al.*, 1978). It seems as if the *vir-zd*[69] mutation is more severe than the *vir-m*[29] mutation, but that both genes are associated with the incorporation of Chl_a-P2 into the thylakoid membranes.

Two barley mutants, $vir-c^{12}$ and the allelic pair $xan-q^{75}$ and $xan-q^{76}$, appeared to affect Chl_a-P3, if this is indeed the reaction centre of PSII (Machold *et al.*, 1979). Both mutants contained 25–30% of the normal amount of chlorophyll and were lethal. The thylakoids of $vir-c^{12}$ were organized into a few grana of extremely long diameter and consisting of several discs, plus stroma lamellae oriented parallel to the grana; there was a reduced number of EFs particles when freeze-fractured (Simpson *et al.*, 1978). The *xantha* mutants had large grana comprising many discs with relatively few stroma lamellae (Simpson & Wettstein, 1980). The $vir-c^{12}$ mutant was unable to fix carbon dioxide (Carlsen, 1977), had a poorly developed electron transport system – particularly PSII – and was deficient in lamellar protein B and a minor polypeptide (46 kD) believed to be components of the reaction centre (Machold & Hoyer-Hansen, 1976; Simpson *et al.*, 1977; Smillie *et al.*, 1977*a*); the coupling factor was also reduced (Simpson *et al.*, 1978). In the *xantha* mutants, PSI was slightly reduced and Chl_a-P1 slightly deficient, but PSII was completely absent and there was a possible deficit in the Chl_a-P2 and/or Chl_a-P3 bands, and there were only trace amounts of cytochrome f.

A number of other *viridis* and *xantha* mutants have been studied to a lesser extent. One of these, $xan-b^{12}$, has been used for a study of the development of photochemical activity in chloroplast membranes owing to its particular sensitivity to temperature and light intensity (Smillie *et al.*, 1977*b*). It responded to feeding with ALA by producing massive amounts of protochlorophyllide a (Wettstein *et al.*, 1971), and under suitable conditions could also accumulate substantial amounts of chlorophyll (Kannangara *et al.*, 1978). It had a reduced PSII activity (Simpson *et al.*, 1978), and high molecular weight bands of 100 kD and 70 kD were somewhat reduced in abundance (Nielsen *et al.*, 1979). It was a 'grana-rich' mutant with a mixture of normal and giant grana, but very few unpaired stroma lamellae (Henningsen *et al.*, 1974; N. C. Nielsen, 1975; Nielsen *et al.*, 1979), which is consistent with the view that coupling factor, deficient in this mutant, is localized in the stroma lamellae and the end granal discs.

Two pea mutants, *chlorotica-887* and *chlorina-5756*, have a normal PSI activity when grown at both 18 °C and 30 °C, whereas PSII is detectable only at the lower temperature. At 18 °C, *chlorotica-887* has about the same level of PSII activity as the wild-type on a chlorophyll basis, whereas *chlorina-5756* is considerably depressed. Although no absence of polypeptides was detected, it was suggested that these mutants were affected in a component required for the assembly of functional PSII complexes (Stummann *et al.*, 1980).

The recessive lethal mutants, hcf_2 and hcf_3, of maize have normal chlorophyll pigmentation, but show differences in the degree of the disturbance to the grana stacking with respect to wild-type. In hcf_2 variable fluorescence was greatly reduced and was missing in hcf_3. Both mutants were unable to fix carbon dioxide (Miles & Daniel, 1974). The high and low potential forms of cytochrome *b*-559 were missing or greatly reduced in *hcf**-3 and in a yellow-green maize mutant, *hcf*-19G*, while in the yellow-green mutant *hcf*-19YG* there was a reduction in the high potential form and a normal or slightly increased amount of the low potential form. There was a complete loss of the light-inducible C-550 signal from *hcf**-3 and *hcf*-19G*, and a reduction in *hcf*-19YG* (Leto & Miles, 1980). These changes, which were associated with a partial or complete block in PSII activity, were accompanied by a partial or total loss of a 32 kD lamellar polypeptide (Leto & Miles, 1980; Leto & Arntzen, 1981) derived from a 34.5 kD precursor after post-translational modification (Fig. 10). The polypeptide, which was thought to be an integral part of the PSII reaction centre complex, may have a structural role in the assembly of morphologically distinguishable and functionally active PSII complexes during plastid development (Leto, Keresztes & Arntzen, 1982). Since the structural gene for the 32 kD peptide resides in chloroplast DNA (Bedbrook *et al.*, 1978), its synthesis, or turnover, may be specifically regulated by a nuclear gene for which *hcf*-3* seems a probable candidate (Leto *et al.*, 1982). There are some *hcf* mutants of maize that appear to block both photosystems (Fig. 10), and others in which the seedlings have lost Chl_a-P1 together with $Chl_{a/b}$-protein complexes (Fig. 11), or in which one or other of these is lost on its own (Miles *et al.*, 1979). The further investigation of these mutants is awaited with interest.

Chlorophyll a/b-*protein complexes.* The most investigated of all chlorophyll-deficient mutants is the pale green *chlorina f2* in barley. The viable recessive mutant grew slower than wild-type, had approximately normal amounts of chlorophyll *a*, and no chlorophyll *b* (Highkin, 1950); there was also a high level of β-carotene and a decreased level of xanthophylls (Henry *et al.*, 1983). It photosynthesized (Highkin & Frenkel, 1962) and the electron transfer reactions of PSI and PSII were fully active (Boardman & Highkin, 1966; Boardman & Thorne, 1968). The average number of thylakoids per granum was somewhat reduced and there were few large grana (Goodchild, Highkin & Boardman, 1966). *Chlorina f2* has been extensively used as an aid to understanding the relationship between the molecular structure of the chloroplast, the organization of chlorophyll pigments within it, and the role of these pigments in photosynthesis (Boardman *et al.*, 1974; Anderson, 1975; Demeter, Sagromsky & Faludi-Daniel, 1976; Steinback *et al.*, 1978; Searle *et al.*, 1979). The detailed analysis of fluorescence curves has led Brown & Schoch (1982) to conclude that, as well as lacking chlorophyll *b*, the mutant is without the population of chlorophyll *a* molecules that are normally bound to the light-harvesting $Chl_{a/b}$–protein complex. Comparisons between the fracture faces of thylakoids from wild-type and mutant plastids have been especially useful in locating $Chl_{a/b}$–protein complexes, missing from the mutant, within the thylakoid membrane (Henriques & Park, 1976; Miller, Miller & McIntyre, 1976; Simpson *et al.*, 1977). Simpson (1979) concluded that in the wild-type granal thylakoids, the $Chl_{a/b}$–proteins were located in particles which were closely associated with the EFs face, but which cleaved with the PFs face. The electrophoretic separation of complexes showed that all Chl_a–protein bands were present, but the $Chl_{a/b}$–protein bands were absent (Genge, Pilger & Hiller, 1974; Thornber & Highkin, 1974; Machold *et al.*, 1979). Nevertheless, this did not amount to a complete loss of all the peptides associated with the $Chl_{a/b}$–protein complexes (Anderson & Levine, 1974; Henriques & Park, 1975; Machold *et al.*, 1977; Anderson, Waldron & Thorne, 1978; Burke, Steinback & Arntzen, 1979; Waldron & Anderson, 1979). It is thought that in the light, the nucleus transcribes massive amounts of mRNA coding for a 29.5 kD peptide precursor of the apoprotein of the $Chl_{a/b}$–protein. This readily separates into two major peptides of 25 kD and 23 kD and possibly a minor one of 28 kD. The *chlorina* mutant appears to

Fig. 11. Nuclear mutants which appear to partially or completely block steps in the formation of $Chl_{a/b}$–protein complexes and associated polypeptides leading to the assembly of the light-harvesting chlorophyll *a/b*–protein centres. See text for details.

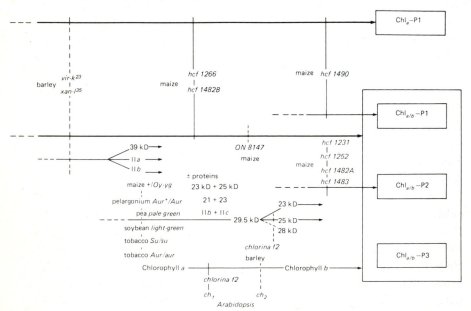

transcribe the mRNA that codes for the 29.5 kD precursor, but the further processing to the 25 kD peptide, but not the 23 kD peptide, is inhibited (Apel & Kloppstech, 1978 *a*, 1978 *b*). Hiller & Goodchild (1981) suggested that the *chlorina* mutation (Fig. 11) affected the chlorophyll *b* biosynthetic pathway and that it was the lack of chlorophyll *b* that prevented the correct processing of the peptide precursor.

The scarcely viable recessive barley mutant *vir-k*[23] had a 27% chlorophyll content, a high chlorophyll *a/b* ratio, and an enhanced rate of carbon dioxide fixation on a chlorophyll basis (Carlsen, 1977). The 'grana-deficient' chloroplasts contained long, parallel stroma lamellae, with limited regions of membrane pairing (Nielsen *et al.*, 1979; Simpson & Wettstein, 1980); the frequency of particles on the PF face was abnormally high, and the particles on the EF face smaller than normal (Simpson *et al.*, 1977; Simpson *et al.*, 1978), which may have been caused by a reduction in Chl_a-P1. Both photosystems were active. Electrophoretic analysis revealed a lack of three bands, of which two peptides, II*a* and II*b* (Fig. 11), were probably derived from $Chl_{a/b}$–protein complexes (Machold & Hoyer-Hansen, 1976); the mutant was strongly reduced in a band of approximately 39 kD (Nielsen *et al.*, 1979). The 'grana-deficient' *xan-l*[35] was very similar.

The lethal olive necrotic mutant of maize, *ON 8147*, had a 30% chlorophyll and 70% carotenoid content, and a chlorophyll *a/b* ratio 3-fold higher than wild-type. There was a greater proportion of stroma to grana lamellae, with which was associated an activity of PSI relatively higher than that of PSII; coupled with this difference was a preponderance of PSI pigments, a 2-fold smaller photosynthetic unit, and several differences in biochemical and biophysical properties (Bazzaz, Govindjee & Paolillo, 1974; Papageorgious, 1975).

Several nuclear mutants of cotton (*Gossypium hirsutum*), maize and soybean (*Glycine max*), which were virescent at low temperatures, had diminished levels of $Chl_{a/b}$–proteins (Benedict & Kohel, 1968; Alberte *et al.*, 1974). In one cotton mutant, the absence of any significant reduction of photosynthetic enzymes resulted in high rates of carbon dioxide fixation on a chlorophyll basis (Benedict & Kohel, 1969, 1970; Benedict, McCree & Kohel, 1972).

The *yellow-green* tobacco (Burk & Menser, 1964) – resembling sulphur deficiency (*Su/su*) – the *aurea* (*Aur/aur*) tobacco (Okabe, Schmid & Straub, 1977), the *oil-yellow*, *yellow-green* (+/*Oy-yg*) maize (Hayden & Hopkins, 1977), the *aurea* (*Aur*[+]/*Aur*) pelargonium cv. 'Cloth of Gold' (Herrmann & Hagemann, 1971), the *pale-green* pea (Highkin, Boardman & Goodchild, 1969), and the *light-green* soybean (Weber & Weiss, 1959) all arose from partially dominant mutations. In each case the heterozygotes segregated into a monohybrid 1:2:1 ratio. The heterozygotes contained from 20 to 60% as much chlorophyll as wild-type (Wolf, 1963), and had a raised chlorophyll a/b ratio, owing to a deficiency of chlorophyll b. Usually the homozygous lethal seedlings (Wolf, 1965), which contained little or no pigment, received scant attention. The sulphur tobacco had predominantly single thylakoids with only small grana stacks (Schmid, Price & Gaffron, 1966; Homan & Schmid, 1967; Schmid & Gaffron, 1967), the maize mutant had predominantly single thylakoids, and the pea and soybean mutants also had fewer grana, with mostly lower stacks, than in the wild-type. The yellow leaves of the homozygous soybean mutant had very few grana and only low stacks (Sun, 1963; Keck *et al.*, 1970). Two characteristic particles revealed by freeze-etching were about 15% smaller in the light green than in the dark green soybean leaves. There was no significant alteration to the Chl_a–protein complexes of the *sulphur* mutant, but the $Chl_{a/b}$–protein complexes were much reduced (Vernotte, Briantais & Remy, 1976). The maize mutant was similar, and in addition to the loss of $Chl_{a/b}$–proteins, the loss of associated 23 kD and 25 kD peptides was recorded (Hopkins, German & Hayden, 1980); in pelargonium, protein bands 21 and 23 were particularly weak in young yellow leaves before they greened up with age (Herrmann *et al.*, 1976), and in pea, peptides IIb and IIc were markedly reduced (Anderson & Levine, 1974).

At low light intensities in the *sulphur* tobacco, although the chlorophyll content rose and more grana developed, the energy supply supported only a slow growth rate. At high light intensities, the increased energy supply more than compensated for less chlorophyll and fewer, smaller grana, and so growth was much more vigorous. The mutation had apparently produced a sun plant in which the inherited loss of light-absorbing pigments was made up by an increased capacity for using light (Schmid, 1967a). On a leaf area basis, the rate of photosynthesis at high light intensities was similar to the wild-type, but since the chlorophyll content was approximately one-fifth, the rate was about 5-fold higher on a chlorophyll basis (Schmid, 1967a, 1967b; Homan, 1968; Okabe *et al.*, 1977). This enhanced efficiency was attributed to a reduced photosynthetic unit size in which the number of reaction centre chlorophyll molecules was increased in comparison to the light-harvesting chlorophyll present (Schmid, 1971; Schmid & Gaffron, 1971; Okabe & Schmid, 1978). Another important variable was photorespiration, which increased from a very low level in young leaves to a high level in old, fully expanded leaves (Zelitch & Day, 1968). This effect helped to explain why, under high light intensity, the mutant plants grew as fast as the wild-type during the first weeks after germination, but later slowed down (Salin & Homan, 1971). In conjunction with *sulphur*, *aurea* appeared either to repress photorespiration or to increase the number of functioning photosynthetic units (Okabe *et al.*, 1977; Okabe & Schmid, 1978).

The *oil-yellow, yellow-green* mutation in maize did not adversely affect photosynthetic activity (Hayden & Hopkins, 1977; Hopkins, Hayden & Neuffer, 1980), but it too was sensitive to light, and adjusted to increasing light intensity by a decrease in the photosynthetic unit size. The Chl_a–proteins remained relatively unchanged, and the capacity for PSI and PSII electron transport in mutant seedlings was 2–3-fold higher than in the wild-type (Hopkins, German & Hayden, 1980). At high light intensity all three $Chl_{a/b}$–protein complexes were lost, but at lower intensities only one of these was missing (Miles *et al.*, 1979). The pea mutant was at least twice as photosynthetically active as the green control on a chlorophyll basis, and it too probably had a smaller photosynthetic unit (Highkin *et al.*, 1969). Within the soybean, it was suggested that the *light-green* chloroplasts had a significantly faster turnover time for plastoquinone which, together with a large plastoquinone pool, accounted for the faster rates of electron transport and phosphorylation (Keck, Dilley & Ke, 1970).

Conclusion

The large number of nuclear mutants discussed in these pages, plus the many others as yet too little studied to merit inclusion, fully support Ellis's (1983) first principle that the majority of chloroplast polypeptides are encoded in nuclear genes. It is also clear that while these mutants have yielded considerable information about their pleiotropic effects on the structure, development and physiology of the plastids, they have not yet led to the recognition of most of the proteins primarily affected by the genetic lesions. Nor, in many cases, are we yet able to recognize, more than very approximately, the primary pathways of gene control. Nevertheless, there is good reason to expect an acceleration in the rate of progress, especially if more attention is given to the completion of the genetic analysis and to the use of appropriate symbolism, and also to a greater standardization in the nomenclature of chlorophyll–protein complexes. Analysis of these mutants has stressed their physiological behaviour, which, although of great interest in itself, has not generally led to a very full understanding of chloroplast biogenesis. This may be expected to progress with a rather more widespread attention to the form and frequency of particles on freeze-fracture planes, and in greater concentration upon the analysis of protein bands following chloroplast fractionation and electrophoretic separation. The improvements in the techniques of protein analysis, particularly the use of two-dimensional electrophoresis (Ellis, 1981) to separate many more individual proteins from composite bands – a technique that has not yet been applied to any mutants – is a particularly attractive development likely to play an important role in the analysis of nuclear mutations in chloroplast proteins.

I should very much like to express my thanks to Professor R. J. Ellis for inviting me to go to Leiden and to participate in the symposium on 'Chloroplast Biogenesis', and to the Society for Experimental Biology and the Netherlands Biochemical Society for their generous financial assistance and cordial hospitality. I should also like to thank Dr B. B. Sears for her invaluable scrutiny of my manuscript.

References

Abdel-Wahab, O. A. L. & Tilney-Bassett, R. A. E. (1981). The role of plastid competition in the control of plastid inheritance in the zonal *Pelargonium. Plasmid,* **6**, 7–16.

Alberte, R. S., Hesketh, J. D., Hofstra, G., Thornber, J. P., Naylor, A. W., Bernard, R. L., Brim, C., Endrizzi, J. & Kohel, R. J. (1974). Composition and activity of the photosynthetic apparatus in temperature-sensitive mutants of higher plants. *Proc. natn. Acad. Sci. USA,* **71**, 2414–28.

Alberte, R. S., Hesketh, J. D. & Kirby, J. S. (1976). Comparisons of photosynthetic activity and lamellar characteristics of virescent and normal green peanut leaves. *Z. Pfl. Physiol.,* **77**, 152–9.

Anderson, J. M. (1975). The molecular organization of chloroplast thylakoids. *Biochim. Biophys. Acta,* **416**, 191–235.

Anderson, J. M. & Levine, R. P. (1974). Membrane polypeptides of some higher plants. *Biochim. biophys. Acta,* **333**, 378–87.

Anderson, J. M., Waldron, J. C. & Thorne, S. W. (1978). Chlorophyll–protein complexes of spinach and barley thylakoids. Spectral characterization of six complexes resolved by an improved electrophoretic procedure. *FEBS Lett.,* **92**, 227–33.

Anderson, L. E. & Levin, D. A. (1970). Chloroplast aldolase is controlled by a nuclear gene. *Pl. Physiol.,* **46**, 819–20.

Apel, K. & Kloppstech, K. (1978*a*). The plastid membranes of barley (*Hordeum vulgare*). Light-induced appearance of mRNA coding for the apoprotein of the light-harvesting chlorophyll *a/b* protein. *Eur. J. Biochem.,* **85**, 581–8.

Apel, K. & Kloppstech, K. (1978*b*). Light-induced appearance of mRNA coding for the apoprotein of the light-harvesting chlorophyll *a/b* protein. In *Chloroplast Development*, ed. G. Akoyunoglou & J. H. Argyroundi-Akoyunoglou, pp. 653–6. Elsevier/North-Holland, Amsterdam.

Bazzaz, M. B., Govindjee & Paolillo, D. J. Jr (1974). Biochemical spectral and structural study of olive necrotic 8147 mutant of *Zea mays* L. *Z. Pfl. Physiol.,* **72**, 181–92.

Bedbrook, J. R., Link, G., Coen, D. M., Bogorad, L. & Rich, A. (1978). Maize plastid gene expressed during photoregulated development. *Proc. natn. Acad. Sci. USA,* **75**, 3060–4.

Belliard, G., Pelletier, G., Vedel, F. & Quetier, F. (1978). Morphological characteristics and chloroplast DNA distribution in different cytoplasmic parasexual hybrids of *Nicotiana tabacum. Molec. gen. Genet.,* **165**, 231–7.

Benedict, C. R. & Ketring, D. L. (1972). Nuclear gene affecting greening in virescent peanut leaves. *Pl. Physiol.*, **49**, 972–6.

Benedict, C. R., Ketring, D. L. & Tomas, R. N. (1974). Elimination of the lag period in chloroplast development in a chlorophyll mutant of peanuts. *Pl. Physiol.*, **53**, 233–40.

Benedict, C. R. & Kohel, R. J. (1968). Characteristics of a virescent cotton mutant. *Pl. Physiol.*, **43**, 1611–16.

Benedict, C. R. & Kohel, R. J. (1969). The synthesis of ribulose-1,5-diphosphate carboxylase and chlorophyll in virescent cotton leaves. *Pl. Physiol.*, **44**, 621–2.

Benedict, C. R. & Kohel, R. J. (1970). Photosynthetic rate of a virescent cotton mutant lacking chloroplast grana. *Pl. Physiol.*, **45**, 519–21.

Benedict, C. R., McCree, K. J. & Kohel, R. J. (1972). High photosynthetic rate of a chlorophyll mutant of cotton. *Pl. Physiol.*, **40**, 968–71.

Boardman, N. K., Björkman, O., Anderson, J. M., Goodchild, D. J. & Thorne, S. W. (1974). Photosynthetic adaptation of higher plants to light intensity: relationship between chloroplast structure, composition of the photosystems and photosynthetic rates. In *Proc. 3rd int. Congr. Photosynthesis*, ed. M. Avron, pp. 1809–27. Elsevier/North-Holland, Amsterdam.

Boardman, N. K. & Highkin, H. R. (1966). Studies on a barley mutant lacking chlorophyll *b*. I. Photochemical activity of isolated chloroplasts. *Biochim. biophys. Acta*, **126**, 189–99.

Boardman, N. K. & Thorne, S. W. (1968). Studies on a barley mutant lacking chlorophyll *b*. II. Fluorescence properties of isolated chloroplasts. *Biochim. biophys. Acta*, **153**, 448–58.

Börner, Th., Schumann, B., Krahnert, S., Pechauf, M., Herrmann, F. H., Knoth, R. & Hagemann, R. (1975). Struktur und Funktion der genetischen Information in den Plastiden. XIII. Lamellarproteine bleicher Plastiden von Plastom- und Genmutanten von *Hordeum* und *Lycopersicon*. *Biochem. Physiol. Pflanz.*, **168**, 185–93.

Bourque, D. P. & Wildman, S. G. (1973). Evidence that nuclear genes code for several chloroplast ribosomal proteins. *Biochem. biophys. Res. Commun.*, **50**, 532–7.

Brown, J. S. & Schoch, S. (1982). Comparison of chlorophyll a spectra in wild-type and mutant barley chloroplasts grown under day or intermittent light. *Photosynthesis Research*, vol. 3, pp. 19–30. Martinus Nijhoff/Dr W. Junk Publishers, The Hague, The Netherlands.

Burk, L. G. & Menser, H. A. (1964). A dominant aurea mutation in tobacco. *Tobacco Sci.*, **8**, 101–4.

Burke, J. J., Steinback, K. E. & Arntzen, C. J. (1979). Analysis of the light-harvesting pigment-protein complex of wild type and a chlorophyll *b*-less mutant in barley. *Pl. Physiol.*, **63**, 237–43.

Carlsen, B. (1977). Barley mutants with defects in photosynthetic carbon dioxide fixation. *Carlsberg Res. Commun.*, **42**, 199–209.

Casadoro, G., Hoyer-Hansen, G., Kannangara, C. G. & Gough, S. P. (1983). An analysis of temperature and light sensitivity in *tigrina* mutants of barley. *Carlsberg Res. Commun.*, **48**, 95–129.

Cashmore, A. R. (1979). Reiteration frequency of the gene coding for the small subunit of ribulose-1,5-bisphosphate carboxylase. *Cell*, **17**, 383–8.

Chan, P.-H. & Wildman, S. G. (1972). Chloroplast DNA codes for the primary structure of the large subunit of fraction I protein. *Biochim. biophys. Acta*, **277**, 677–80.

Chen, K., Wildman, S. G. & Smith, H. H. (1977). Chloroplast DNA distribution in parasexual hybrids as shown by polypeptide composition of fraction I protein. *Proc. natn. Acad. Sci. USA*, **74**, 5109–12.

Chollet, R. & Paolillo, D. J. Jr (1972). Greening in a virescent mutant of maize. I. Pigment, ultrastructural, and gas exchange studies. *Z. Pfl. Physiol.*, **68**, 30–44.

Dale, J. E. & Heyes, J. K. (1970). A virescens mutant of *Phaseolus vulgaris*: growth, pigment and plastid character. *New Phytol.*, **69**, 733–42.

Dean, C. & Leech, R. M. (1982). Genome expression during normal leaf development. 2. Direct correlation between ribulose bisphosphate carboxylase content and nuclear ploidy in a polyploid series of wheat. *Pl. Physiol.*, **70**, 1605–8.

Demeter, S., Sagromsky, H. & Faludi-Daniel, A. (1976). Orientation of chlorophyll b in thylakoids of barley chloroplasts. *Photosynthetica (Prague)*, **10**, 193–7.

Ellis, R. J. (1981). Chloroplast proteins: synthesis, transport, and assembly. *A. Rev. Pl. Physiol.*, **32**, 111–37.

Ellis, R. J. (1983). Chloroplast protein synthesis: principles and problems. *Subcell. Biochem.*, **9**, 237–61.

Genge, S., Pilger, D. & Hiller, R. G. (1974). The relationship between chlorophyll *b* and pigment protein complex. II. *Biochim. biophys. Acta*, **347**, 22–30.

Gillham, N. W. (1978). *Organelle Heredity*. Raven Press, New York.

Goodchild, D. J., Highkin, H. R. & Boardman, N. K. (1966). The fine structure of chloroplasts in a barley mutant lacking chlorophyll *b*. *Expl. Cell Res.*, **43**, 684–8.

Gough, S. P. (1972). Defective synthesis of porphyrins in barley plastids caused by mutation in nuclear genes. *Biochim. biophys. Acta*, **286**, 36–54.

Gough, S. P. & Kannangara, C. G. (1979). Biosynthesis of δ-aminolevulinate in greening barley leaves. III. The formation of δ-aminolevulinate in tigrina mutants of barley. *Carlsberg Res. Commun.*, **44**, 403–16.

Gray, J. C. (1980). Maternal inheritance of cytochrome f in interspecific *Nicotiana* hybrids. *Eur. J. Biochem.*, **112**, 39–46.

Gray, J. C., Kung, S. D., Wildman, S. G. & Sheen, S. J. (1974). Origin of *Nicotiana tabacum* L. detected by polypeptide composition of fraction I protein. *Nature*, **252**, 226–7.

Hagemann, R. (1976). Plastid distribution and plastid competition in higher plants and the induction of plastom mutations by nitroso-urea compounds. In *The Genetics and Biogenesis of Chloroplasts and Mitochondria*, ed. T. Bücher, W. Neupert, W. Sebald & S. Werner, pp. 331–8. Elsevier/North-Holland, Amsterdam.

Hagemann, R. (1979). Genetics and molecular biology of plastids of higher plants. *Stadler Symp.*, **11**, 91–116.

Hagemann, R. & Börner, T. (1978). Plastid ribosome-deficient mutants of higher plants as a tool in studying chloroplast biogenesis. In *Chloroplast Development*, ed. G. Akoyunoglou & J. H. Argyroudi-Akoyunoglou, pp. 709–20. Elsevier/North-Holland, Amsterdam.

Hayden, D. B. & Hopkins, W. G. (1977). A second distinct chlorophyll-protein complex in maize mesophyll chloroplasts. *Can. J. Bot.*, **55**, 2525–9.

Heber, U. & Gottschalk, W. (1963). Die Bestimmung des genetisch fixierten Stoffwechselblockes einer Photosynthese-Mutante von *Vicia faba*. *Z. Naturf.*, **18b**, 36–44.

Henningsen, K. W., Nielsen, N. C. & Smillie, R. M. (1974). The effect of nuclear mutations on the assembly of photosynthetic membranes in barley. *Port. Acta biol. ser. A*, **14**, 323–44.

Henningsen, K. W. & Thorne, S. W. (1974). Esterification and spectral shifts of chlorophyll(ide) in wildtype and mutant seedlings developed in darkness. *Physiol. Pl.*, **30**, 82–9.

Henriques, F. & Park, R. B. (1975). Further chemical and morphological characterization of chloroplast membranes from a chlorophyll b-less mutant of *Hordeum vulgare*. *Pl. Physiol.*, **55**, 763–7.

Henriques, F. & Park, R. B. (1976). Development of the photosynthetic unit in lettuce. *Proc. natn. Acad. Sci. USA*, **73**, 4560–4.

Henry, L. E. A., Mikkelsen, J. D. & Moller, B. L. (1983). Pigment and acyl lipid composition of photosystem I and II vesicles and of photosynthetic mutants in barley. *Carlsberg Res. Commun.*, **48**, 131–48.

Herrmann, F. von & Hagemann, R. (1971). Struktur und Funktion der genetischen Information in den Plastiden. III. Genetik, Chlorophylle und Photosynthese-verhalten der Plastommutante 'Mrs Pollock' und der Genmutante 'Cloth of Gold' von *Pelargonium zonale*. *Biochem. Physiol. Pflanz.*, **162**, 390–409.

Herrmann, F. H., Schumann, B., Börner, T. & Knoth, R. (1976). Struktur und Funktion der genetischen Information in den Plastiden. XII. Die plastidalen Lamellarproteine der photosynthesedefekten Plastommutante *en: gil-l* ('Mrs Pollock') und der Genmutante 'Cloth of Gold' von *Pelargonium zonale* Arr. *Photosynthetica*, **10**, 164–71.

Heyes, J. K. & Dale, J. E. (1971). A virescens mutant of *Phaseolus vulgaris*: photosynthesis and metabolic changes during leaf development. *New Phytol.*, **70**, 415–26.

Highkin, H. R. (1950). Chlorophyll studies on barley mutants. *Pl. Physiol.*, **25**, 294–306.

Highkin, H. R., Boardman, N. K. & Goodchild, D. J. (1969). Photosynthetic studies on a pea mutant deficient in chlorophyll. *Pl. Physiol.*, **44**, 1310–20.

Highkin, H. R. & Frenkel, A. W. (1962). Studies of growth and metabolism of a barley mutant lacking chlorophyll *b*. *Pl. Physiol.*, **37**, 814–20.

Hiller, R. G. & Goodchild, D. J. (1981). Thylakoid membrane and pigment organization. In *The Biochemistry of Plants*, vol. 8, *Photosynthesis*, ed. M. D. Hatch & N. K. Boardman, pp. 1–49. Academic Press, New York.

Hiller, R. G., Moller, B. L. & Hoyer-Hansen, G. (1980). Characterization of six putative photosystem I mutants in barley. *Carlsberg Res. Commun.*, **45**, 315–28.

Hirono, Y. & Rédei, G. P. (1963). Multiple allelic control of chlorophyll b level in *Arabidopsis thaliana*. *Nature*, **197**, 1324–5.

Homan, P. H. (1968). Fluorescence properties of chloroplasts from manganese deficient and mutant tobacco. *Biochim. biophys. Acta*, **162**, 545–54.

Homan, P. H. & Schmid, G. H. (1967). Photosynthetic reactions of chloroplasts with unusual structures. *Pl. Physiol.*, **42**, 1619–32.

Hopkins, W. G., German, J. B. & Hayden, D. B. (1980). A light-sensitive mutant in maize (*Zea mays* L.). II. Photosynthetic properties. *Z. Pfl. Physiol.*, **100**, 15–24.

Hopkins, W. G., Hayden, D. B. & Neuffer, M. G. (1980). A light-sensitive mutant in maize (*Zea mays* L.). I. Chlorophyll, chlorophyll–protein and ultrastructural studies. *Z. Pfl. Physiol.*, **99**, 417–26.

Horak, A. & Zalik, S. (1975). Development of photoreductive activity in plastids of a virescens mutant of barley. *Can. J. Bot.*, **53**, 2399–404.

Hoyer-Hansen, G. & Casadoro, G. (1982). Unstable chloroplast ribosomes in the cold-sensitive barley mutant *tigrina-o*[34]. *Carlsberg Res. Commun.*, **47**, 103–18.

Hoyer-Hansen, G., Moller, B. L., Henry, L. E. A. & Casadoro, G. (1982). Thylakoid polypeptide synthesis and assembly in wild-type and mutant barley. In *Cell Function and Differentiation, Part B*, pp. 111–25. Alan R. Liss, Inc., New York.

Hoyer-Hansen, G., Moller, B. L. & Pan, L. C. (1979). Identification of coupling factor subunits in thylakoid polypeptide patterns of wild-type and mutant barley thylakoids using crossed immunoelectrophoresis. *Carlsberg Res. Commun.*, **44**, 337–51.

Huffaker, R. C., Cox, E. L., Kleinkopf, G. E. & Standford, E. H. (1970). Regulation of synthesis of chlorophyll, carotene, ribulose-1,5-diP carboxylase and phosphoribulokinase in a temperature-sensitive chlorophyll mutant of *Medicago sativa*. *Physiologia Pl.*, **23**, 404–11.

Jhamb, S. & Zalik, S. (1973). Soluble and lamellar proteins in seedlings of barley and its virescens mutant in relation to chloroplast development. *Can J. Bot.*, **51**, 2147–54.

Kahn, A., Avivi-Bleiser, N. & Wettstein, D. von (1976). Genetic regulation of chlorophyll synthesis analysed with double mutants in barley. In *Genetics and Biogenesis of Chloroplasts and Mitochondria*, ed. Th. Bücher, W. Neupert, W. Sebald & W. Werner, pp. 119–31. Elsevier/North-Holland, Amsterdam.

Kannangara, C G., Gough, S. P. & Wettstein, D. von (1978). Biosynthesis of δ-aminolevulinate and chlorophyll and its genetic regulation. In *Chloroplast Development*, ed. G. Akoyunoglou & J. H. Argyroudi-Akoyunoglou, pp. 147–60. Elsevier/North-Holland, Amsterdam.

Kawashima, N. & Wildman, S. G. (1972). Studies on fraction I protein. IV. Mode of inheritance of primary structure in relation to whether chloroplast or nuclear DNA contains the code for a chloroplast protein. *Biochim. biophys. Acta*, **262**, 42–9.

Kay, R. E. & Phinney, B. O. (1956). The control of plastid pigment formation by a virescent gene, *pale-yellow-l*, of maize. *Pl. Physiol.*, **31**, 415–20.

Keck, R. W., Dilley, R. A., Allen, C. F. & Biggs, S. (1970*a*). Chloroplast composition and structure differences in a soybean mutant. *Pl. Physiol.*, **46**, 692–8.

Keck, R. W., Dilley, R. A. & Ke, B. (1970). Photochemical characteristics in a soybean mutant. *Pl. Physiol.*, **46**, 699–704.

Kirk, J. T. O. & Tilney-Bassett, R. A. E. (1978). *The Plastids: Their Chemistry, Structure, Growth and Inheritance*, 2nd edn. Elsevier/North-Holland Biomedical Press, Amsterdam.

Kumar, A., Wilson, D. & Cocking, E. C. (1981). Polypeptide composition of fraction I protein of the somatic hybrid between *Petunia parodii* and *Petunia parviflora*. *Biochem. Genet.*, **19**, 255–61.

Kung, S. D. (1976). Tobacco fraction I protein: a unique genetic marker. *Science*, **191**, 429–34.

Kung, S. D., Gray, J. C., Wildman, S. G. & Carlson, P. S. (1975*a*). Polypeptide composition of fraction I protein from parasexual hybrid plants in the genus *Nicotiana*. *Science*, **187**, 353–5.

Kung, S. D., Sakano, K., Gray, J. C. & Wildman, S. G. (1975*b*). The evolution of fraction I protein during the origin of a new species of *Nicotiana*. *J. molec. Evol.*, **7**, 59–64.

Kung, S. D., Sakano, K. & Wildman, S. G. (1974). Multiple peptide composition of the large and small subunits of *Nicotiana* fraction I protein ascertained by fingerprinting and electrofocusing. *Biochim. biophys. Acta*, **365**, 138–47.

Kung, S. D., Thornber, J. P. & Wildman, S. G. (1972). Nuclear DNA codes for the photosystem II chlorophyll–protein of chloroplast membranes. *FEBS Lett.*, **24**, 185–8.

Kwanyuen, P. & Wildman, S. G. (1978). Evidence that genetic information for chloroplast coupling factor I is shared by nuclear and chloroplast DNA. *Biochim. biophys. Acta*, **502**, 269–75.

Kyle, D. J. & Zalik, S. (1982*a*). Development of photochemical activity in relation to pigment and membrane protein accumulation in chloroplasts of barley and its virescens mutant. *Pl. Physiol.*, **69**, 1392–400.

Kyle, D. J. & Zalik, S. (1982*b*). Photosystem II activity, plastoquinone A levels, and fluorescence characterization of a virescens mutant of barley. *Pl. Physiol.*, **70**, 1026–31.

Leto, K. & Arntzen, C. J. (1981). Cation-mediated regulation of excitation energy distribution in chloroplasts lacking organized photosystem II complexes. *Biochim. biophys. Acta*, **637**, 107–17.

Leto, K. J., Keresztes, A. & Arntzen, C. J. (1982). Nuclear involvement in the appearance of a chloroplast-encoded 32000 Dalton thylakoid membrane polypeptide integral to the photosystem II complex. *Pl. Physiol.*, **69**, 1450–8.

Leto, K. L. & Miles, C. D. (1980). Characterization of three photosystem II mutants in *Zea mays* L. lacking a 32000 Dalton lamellar polypeptide. *Pl. Physiol.*, **66**, 18–24.

Machold, O. & Hoyer-Hansen, G. (1976). Polypeptide composition of thylakoids from viridis and xantha mutants in barley. *Carlsberg Res. Commun.*, **41**, 359–66.

Machold, O., Meister, A., Sagromsky, H., Hoyer-Hansen, G. & Wettstein,

D. von (1977). Composition of photosynthetic membranes of wild-type barley and chlorophyll *b*-less mutants. *Photosynthetica*, **11**, 200–6.

Machold, O., Simpson, D. J. & Moller, B. L. (1979). Chlorophyll proteins of thylakoids from wild-type and mutants of barley (*Hordeum vulgare* L.). *Carlsberg Res. Commun.*, **44**, 235–54.

Maclachlan, S. & Zalik, S. (1963). Plastid structure, chlorophyll concentration, and free amino acid composition of a chlorophyll mutant of barley. *Can. J. Bot.*, **41**, 1053–62.

Mascia, P. N. (1978). An analysis of precursors accumulated by several chlorophyll biosynthetic mutants of maize. *Molec. gen. Genet.*, **161**, 237–44.

Mascia, P. N. & Robertson, D. S. (1978). Studies of chloroplast development in four maize mutants defective in chlorophyll biosynthesis. *Planta*, **143**, 207–11.

Mascia, P. N. & Robertson, D. S. (1980). Genetic studies of the chlorophyll biosynthetic mutants of maize. *J. Hered.*, **71**, 19–24.

Melchers, G., Sacristan, M. D. & Holder, A. A. (1978). Somatic hybrid plants of potato and tomato regenerated from fused protoplasts. *Carlsberg Res. Commun.*, **43**, 203–18.

Meyers, S. P., Nichols, S. L., Baer, G. R., Molin, W. T. & Schrader, L. E. (1982). Ploidy effects in isogenic populations of alfalfa. 1. Ribulose-1,5-bis-phosphate carboxylase, soluble protein, chlorophyll, and DNA in leaves. *Pl. Physiol.*, **70**, 1704–9.

Miles, D. (1980). Mutants of higher plants: Maize. In *Methods in Enzymology*, vol. 69, ed. A. San Pietro, pp. 3–23. Academic Press, New York.

Miles, C. D. & Daniel, D. J. (1974). Chloroplast reactions of photosynthetic mutants in *Zea mays*. *Pl. Physiol.*, **53**, 589–95.

Miles, C. D., Markwell, J. P. & Thornber, J. P. (1979). Effect of nuclear mutation in maize on photosynthetic activity and content of chlorophyll–protein complexes. *Pl. Physiol.*, **64**, 690–4.

Miller, K. R. (1980). A chloroplast membrane lacking photosystem I. Changes in unstacked membrane regions. *Biochim. biophys. Acta*, **592**, 143–52.

Miller, K. R., Miller, G. J. & McIntyre, K. R. (1976). The light-harvesting chlorophyll–protein complex of photosystem II. *J. Cell Biol.*, **71**, 624–38.

Miller, R. A. & Zalik, S. (1965). Effect of light quality, light intensity and temperature on pigment accumulation in barley seedlings. *Pl. Physiol.*, **40**, 569–74.

Millerd, A., Goodchild, D. J. & Spencer, D. (1969). Studies on a maize mutant sensitive to low temperature. II. Chloroplast structure, development, and physiology. *Pl. Physiol.*, **44**, 567–83.

Millerd, A. & McWilliam, J. R. (1968). Studies on a maize mutant sensitive to low temperature. I. Influence of temperature and light on the production of chloroplast pigments. *Pl. Physiol.*, **43**, 1967–72.

Moller, B. L., Hoyer-Hansen, G. & Henry, L. E. A. (1982). The use of chloroplast proteins in crop improvement. In *Plant Cell Culture in Crop Improvement*, ed. K. L. Giles & S. K. Sen, pp. 249–57. Plenum Press, New York.

Moller, B. L., Nugent, J. H. A. & Evans, M. C. W. (1981). Electron para-magnetic resonance spectrometry of photosystem I mutants in barley. *Carlsberg Res. Commun.*, **46**, 373–82.

Moller, B. L., Smillie, R. M. & Hoyer-Hansen, G. (1980). A photosystem I mutant in barley (*Hordeum vulgare* L.). *Carlsberg Res. Commun.*, **45**, 87–99.

Newell, E. & Rienits, K. G. (1975). Chlorophyll synthesis in a yellow mutant of wheat. *Aust. J. Pl. Physiol.*, **2**, 543–52.

Nielsen, N. C. (1975). Electrophoretic characterization of membrane proteins during chloroplast development in barley. *Eur. J. Biochem.*, **50**, 611–23.

Nielsen, N. C., Henningsen, K. W. & Smillie, R. M. (1974). Chloroplast membrane proteins in wild-type and mutant barley. In *Proc. 3rd int. Congr. Photosynthesis*, ed. M. Avron, pp. 1603–14. Elsevier/North-Holland, Amsterdam.

Nielsen, N. C., Smillie, R. M., Henningsen, K. W., Wettstein, D. von & French, C. S. (1979). Composition and function of thylakoid membranes from grana-rich and grana-deficient chloroplast mutants of barley. *Pl. Physiol.*, **63**, 174–82.

Nielsen, O. F. (1973). Protochlorophyll(ide) holochrome subunits from a mutant defective in the regulation of protochlorophyll(ide) synthesis. *FEBS Lett.*, **38**, 75–8.

Nielsen, O. F. (1974*a*). Macromolecular physiology of plastids. XII. Tigrina mutants in barley: genetic, spectroscopic and structural characterization. *Hereditas*, **76**, 269–304.

Nielsen, O. F. (1974*b*). Photoconversion and regeneration of active proto-chlorophyll(ide) in mutants defective in the regulation of chlorophyll synthesis. *Archs. Biochem. Biophys.*, **160**, 430–9.

Nielsen, O. F. (1975). Macromolecular physiology of plastids. XIII. The effect of photoinactive protochlorophyllide on the function of proto-chlorophyllide holochrome. *Biochem. Physiol. Pfl.*, **167**, 195–206.

Nielsen, O. F. & Gough, S. P. (1974). Macromolecular physiology of plastids. XI. Carotenes in etiolated tigrina and xantha mutants of barley. *Physiologia Pl.*, **30**, 246–54.

Okabe, K. & Schmid, G. H. (1978). Properties of the tobacco aurea mutant *Su/su* var. Aurea. On photorespiration and on the structure and function relationship in chloroplasts. In *Chloroplast Development*, ed. G. Akoyu-noglou & J. H. Argyroudi-Akoyunoglou, pp. 501–6. Elsevier/North-Holland, Amsterdam.

Okabe, K., Schmid, G. H. & Straub, J. (1977). Genetic characterization and high efficiency photosynthesis of an aurea mutant of tobacco. *Pl. Physiol.*, **60**, 150–6.

Papageorgious, G. (1975). Chlorophyll fluorescence: an intrinsic probe of photosynthesis. In *Bioenergetics of Photosynthesis*, ed. Govindjee, pp. 319–71. Academic Press, New York.

Pettigrew, R., Driscoll, C. J. & Rienits, K. G. (1969). A spontaneous chlorophyll mutant in hexaploid wheat. *Heredity*, **24**, 481–7.

Röbbelen, G. (1957). Untersuchungen an strahlinduzierten Blattfarb-mutanten von *Arabidopsis thaliana* (L.) Heynh. *Z. VererbLehre*, **88**, 189–252.

Sakano, K., Kung, S. D. & Wildman, S. G. (1974). Identification of several chloroplast DNA genes which code for the large subunit of *Nicotiana* fraction I protein. *Molec. gen. Genet.*, **130**, 91–7.

Salin, M. L. & Homan, P. H. (1971). Changes of photorespiratory activity with leaf age. *Pl. Physiol.*, **48**, 193–6.

Sane, P. V. & Zalik, S. (1970). Metabolism of acetate-2-^{14}C, glycine-2-^{14}C, leucine-U-^{14}C, and effect of δ-aminolevulinic acid on chlorophyll synthesis in Gateway barley and its mutant. *Can J. Bot.*, **48**, 1171–8.

Schmid, G. H. (1967*a*). The influence of different light intensities on the growth of the tobacco aurea mutant, *Su/su*. *Planta*, **77**, 77–94.

Schmid, G. H. (1967*b*). Photosynthetic capacity and lamellar structure in various chlorophyll-deficient plants. *J. Microsc.*, **6**, 485–98.

Schmid, G. H. (1971). Origin and properties of mutant plants: Yellow tobacco. *Meth. Enzymol.*, **23**, 171–94.

Schmid, G. H. & Gaffron, H. (1967). Light metabolism and chloroplast structure in chlorophyll deficient tobacco mutants. *J. gen. Physiol.*, **50**, 563–82.

Schmid, G. H. & Gaffron, H. (1971). Fluctuating photosynthetic units in higher plants and fairly constant units in algae. *Photochem. Photobiol.*, **14**, 451–64.

Schmid, G. H., Price, J. M. & Gaffron, H. (1966). Lamellar structure in chlorophyll deficient but normally active chloroplasts. *J. Microsc.*, **5**, 205–12.

Searle, G. F. W., Tredwell, C. J., Barber, J. & Porter, G. (1979). Picosecond time-resolved fluorescence study of chlorophyll organization and excitation energy distribution in chloroplasts from wild-type barley and a mutant lacking chlorophyll *b*. *Biochim. biophys. Acta*, **545**, 496–507.

Sears, B. B. (1980). Elimination of plastids during spermatogenesis and fertilization in the plant kingdom. *Plasmid*, **4**, 233–55.

Simpson, D. J. (1979). Freeze-fracture studies on barley plastid membranes. III. Location of the light-harvesting chlorophyll protein. *Carlsberg Res. Commun.*, **44**, 305–36.

Simpson, D. J. (1982). Freeze-fracture studies on barley plastid membranes. V. *Viridis-n*^{34}, a photosystem I mutant. *Carlsberg Res. Commun.*, **47**, 215–25.

Simpson, D. J., Hoyer-Hansen, G., Chua, N.-H. & Wettstein, D. von (1977). The use of gene mutants in barley to correlate thylakoid polypeptide composition with the structure of the photosynthetic membrane. In *Proc. 4th int. Congr. Photosynthesis*, ed. D. O. Hall, J. Coombs & T. W. Goodwin, pp. 537–48. Biochemical Society Press, London.

Simpson, D. J., Moller, B. L. & Hoyer-Hansen, G. (1978). Freeze-fracture structure and polypeptide composition of thylakoids of wild type and mutant barley plastids. In *Chloroplast Development*, ed. G. Akoyunoglou & J. H. Argyroudi-Akoyunoglou, pp. 507–12. Elsevier/North-Holland, Amsterdam.

Simpson, D. J. & Wettstein, D. von (1980). Macromolecular physiology of plastids. XIV. Viridis mutants in barley: genetic fluoroscopic and ultrastructural characterization. *Carlsberg Res. Commun.*, **45**, 283–314.

Smillie, R. M., Henningsen, K. W., Bain, J. M., Critchley, C., Foster, T. & Wettstein, D. von (1978). Mutants of barley heat-sensitive for chloroplast development. *Carlsberg Res. Commun.*, **43**, 351–64.

Smillie, R. M., Nielsen, N. C., Henningsen, K. W. & Wettstein, D. von (1977*a*). Development of photochemical activity in chloroplast membranes. I. Studies with mutants of barley grown under a single environment. *Aust. J. Pl. Physiol.*, **4**, 415–38.

Smillie, R. M., Nielsen, N. C., Henningsen, K. W. & Wettstein, D. von (1977*b*). Development of photochemical activity in chloroplast membranes. II. Studies with a mutant barley grown under different environments. *Aust. J. Pl. Physiol.*, **4**, 439–49.

Smith, S. M. & Ellis, R. J. (1981). Light-stimulated accumulation of transcripts of nuclear and chloroplast genes for ribulose bisphosphate carboxylase. *J. molec. appl. Genet.*, **1**, 127–37.

Somerville, C. R., Portis, A. R. Jr & Ogren, W. L. (1982). A mutant of *Arabidopsis thaliana* which lacks activation of RuBP carboxylase *in vivo*. *Pl. Physiol.*, **70**, 381–7.

Steinback, K. E., Burke, J. J., Mullet, J. E. & Arntzen, C. J. (1978). The role of the light-harvesting complex in cation-mediated grana formation. In *Chloroplast Development*, ed. G. Akoyunoglou & J. H. Argyroudi-Akoyunoglou, pp. 389–400. Elsevier/North-Holland, Amsterdam.

Stephansen, K. & Zalik, S. (1971). Inheritance and qualitative analysis of pigments in a barley mutant. *Can. J. Bot.*, **49**, 49–51.

Stummann, B. M., Veierskov, B., Jacobsen, S. E. & Henningsen, K. W. (1980). Two temperature-sensitive photosystem II mutants of pea. *Pl. Physiol.*, **49**, 135–40.

Sun, C. N. (1963). The effect of genetic factors on the submicroscopic structure of soybean chloroplasts. *Cytologia*, **28**, 257–63.

Thomson, L. W. & Zalik, S. (1981*a*). Acyl lipids, pigments, and gramine in developing leaves of barley and its virescens mutants. *Pl. Physiol.*, **67**, 646–54.

Thomson, L. W. & Zalik, S. (1981*b*). Acetyl coenzyme A carboxylase activity in developing seedlings and chloroplasts of barley and its virescens mutant. *Pl. Physiol.*, **67**, 655–61.

Thornber, J. P. & Highkin, J. R. (1974). Composition of the photosynthetic apparatus of normal barley leaves and a mutant lacking chlorophyll *b*. *Eur. J. Biochem.*, **41**, 109–16.

Tilney-Bassett, R. A. E. (1973). The control of plastid inheritance in *Pelargonium*. II. *Heredity*, **30**, 1–13.

Tilney-Bassett, R. A. E. (1976). The control of plastid inheritance in *Pelargonium*. IV. *Heredity*, **37**, 95–107.

Tilney-Bassett, R. A. E. & Abdel-Wahab, O. A. L. (1979). Maternal effects and plastid inheritance. In *Maternal Effects in Development*, ed. D. R. Newth & M. Balls, pp. 29–45. Cambridge University Press, London.

Tilney-Bassett, R. A. E. & Abdel-Wahab, O. A. L. (1982). Irregular segregation at the *Pr* locus controlling plastid inheritance in *Pelargonium*: gametophytic lethal or incompatibility system? *Theoret. appl. Genet.*, **62**, 185–91.

Tilney-Bassett, R. A. E. & Birky, C. W. Jr (1981). The mechanism of the mixed inheritance of chloroplast genes in *Pelargonium*. Evidence from gene frequency distributions among the progeny of crosses. *Theoret. appl. Genet.*, **60**, 43–53.

Vaughn, K. C. (1981). Organelle transmission in higher plants: organelle alteration vs physical exclusion. *J. Hered.*, **72**, 335–7.

Vaughn, K. C., De Bonte, L. R. & Wilson, K. G. (1980). Organelle alteration as a mechanism for maternal inheritance. *Science*, **208**, 196–8.

Vaughn, K. C., Kimpel, D. L. & Wilson, K. G. (1981). Control of organelle transmission in *Chlorophytum*. *Curr. Genet.*, **3**, 105–8.

Vernotte, C., Briantais, J. M. & Remy, R. (1976). Light-harvesting pigment protein complex requirement for spill-over changes produced by cations. *Pl. Sci. Lett.*, **6**, 135–41.

Walbot, V. & Coe, E. H. (1979). Nuclear gene *iojap* conditions a programmed change in ribosome-less plastids in *Zea mais*. *Proc. natn. Acad. Sci. USA*, **76**, 2760–4.

Waldron, J. C. & Anderson, J. M. (1979). Chlorophyll–protein complexes from thylakoids of a mutant barley lacking chlorophyll *b*. *Eur. J. Biochem.*, **102**, 357–62.

Walker, G. W. R., Dietrich, J., Miller, R. & Kasha, K. (1963). Recent barley mutants and their linkages. II. Genetic data for further mutants. *Can. J. Genet., Cytol.*, **5**, 200–19.

Weber, C. R. & Weiss, M. G. (1959). Chlorophyll mutant in soybean provides teaching aid. *J. Hered.*, **50**, 53–4.

Weeden, N. F. & Gottlieb, L. D. (1980). The genetics of chloroplast enzymes. *J. Hered.*, **71**, 392–6.

Wettstein, D. von (1974). Structural and regulatory genes for the assembly of photosynthetic membranes. *Biochem. Soc. Trans.*, **2**, 176–9.

Wettstein, D. von, Henningsen, K. W., Boynton, J. E., Kannangara, G. C. & Nielsen, O. F. (1971). The genic control of chloroplast development in barley. In *Autonomy and Biogenesis of Mitochondria and Chloroplasts*, ed. N. K. Boardman, P. Linnane & R. M. Smillie, pp. 205–23. North-Holland, Amsterdam.

Wettstein, D. von, Kahn, A., Nielsen, O. F. & Gough, S. P. (1974). Genetic regulation of chlorophyll synthesis analysed with mutants in barley. *Science*, **184**, 800–2.

Wolf, F. F. (1963). The chloroplast pigments of certain soybean mutants. *Bull. Torrey Bot. Club*, **90**, 139–43.

Wolf, F. T. (1965). Photosynthesis of certain soybean mutants. *Bull. Torrey Bot. Club*, **92**, 99–101.

Zelitch, I. & Day, P. R. (1968). Variation in photorespiration. The effect of genetic differences in photorespiration in net photosynthesis in tobacco. *Pl. Physiol.*, **43**, 1838–44.

R. BROGLIE, G. CORUZZI AND N.-H. CHUA

Differential expression of genes encoding polypeptides involved in C_4 photosynthesis

C_4 plants, such as *Zea mais*, contain two photosynthetically distinct cell types which are arranged in two concentric layers around the vascular bundles. The outer layer comprises *mesophyll* cells and that surrounding the vascular bundles comprises the *bundle sheath* cells. In the cytosol of mesophyll cells, phosphoenolpyruvate (PEP) carboxylase is responsible for the initial fixation of atmospheric CO_2 into C_4-dicarboxylic acids. In maize, the fixed CO_2 is transported, as malate, to the chloroplast of bundle sheath cells where decarboxylation occurs. The CO_2 released is then refixed by ribulose-1,5-bisphosphate (RuBP) carboxylase. Net photosynthetic carbon assimilation then occurs in bundle sheath cells by passage of the fixed CO_2 through the photosynthetic carbon reduction (PCR) or Calvin cycle.

As a result of the specialized leaf anatomy, C_4 plants have the ability to concentrate CO_2 at the site of CO_2 refixation by RuBP carboxylase/oxygenase. These plants exhibit higher rates of net photosynthesis (CO_2 fixed per dm^2 of leaf area per h) than plants utilizing the C_3 pathway. This is because of an inhibition of RuBP oxygenase activity (photorespiration) by the high concentrations of CO_2. This activity in C_3 plants results in the release of fixed CO_2 as a result of the glycolate pathway. For a review of C_4 photosynthesis, see Rathnam & Chollet (1980).

We are interested in studying the regulation of nuclear genes encoding polypeptides involved in C_4 photosynthesis. For this purpose we have isolated mesophyll and bundle sheath cells and analyzed their polypeptide components by both lithium dodecylsulfate–polyacrylamide gel electrophoresis (LDS–PAGE) and immunological techniques. In this chapter we report the results of these experiments and show that several of these proteins are regulated at the level of translatable mRNA.

Materials and methods
Maize seedlings (*Zea mais*, FR 9 × FR 37) were grown at 50% humidity under 14-h day at 27 °C and 10-h night at 25 °C. Seedlings were harvested 7 to 14 days after sowing.

51

Mesophyll protoplasts, prepared from primary leaves according to Kanai & Edwards (1973), were lysed by hypotonic shock in 10 mmol l^{-1} potassium phosphate, 1 mmol l^{-1} phenylmethylsulfonylfluoride. The lysate was centrifuged at $50000 \times g_{max}$ for 10 min to sediment a membrane fraction which contained predominantly thylakoid membranes and leaving in the supernatant all the soluble proteins of the protoplasts. Bundle sheath strands were isolated by the maceration–filtration technique described by Chollet & Ogren (1973). The strands were frozen and then thawed in 10 mmol l^{-1} potassium phosphate, 1 mmol l^{-1} phenylmethylsulfonylfluoride. After homogenization in a Potter homogenizer, the suspension was passed through a 80 μm nylon net and the filtrate centrifuged at $50000 \times g_{max}$ for 10 min. The pellet contained predominantly bundle sheath thylakoid membranes whereas the supernatant contained all the cellular soluble proteins.

Soluble and thylakoid membrane proteins of the two cell types were analyzed by 7.5–15% LDS gradient gels at 4 °C (Delepelaire & Chua, 1981). Samples prepared for electrophoresis were also assayed by double immunodiffusion in 1% agarose gels containing 1% Triton X-100 (Chua & Blomberg, 1979). Rocket immunoelectrophoresis was carried out according to the method of Laurell (1972), except that the samples were subject to electrophoresis through a 0.7 mm layer of 1.5% Lubrol PX (Converse & Papermaster, 1975) to remove sodium dodecyl sulfate (SDS). Crossed immunoelectrophoresis was performed by a modification (Chua & Blomberg, 1979) of the procedure first described by Converse & Papermaster (1975). Preparation of monospecific antibodies to various polypeptides has been described in previous publications (Chua & Blomberg, 1979; Joyard et al., 1982; Grossman et al., 1982). IgG was purified according to Harboe & Ingild (1973).

Total RNA was extracted separately from the two cell types with guanidinium thiocyanate (Chirgwin et al., 1979) and poly(A) RNA was selected by affinity chromatography on a poly(U)–sepharose column (Lindberg & Pearson, 1972). Poly(A) RNAs were translated in a wheat-germ cell-free system (Grossman et al., 1982) and specific products were immunoprecipitated from the translation mixes with monospecific antibodies (Chua & Schmidt, 1978).

Chlorophyll concentrations were measured in 80% acetone extracts (Arnon, 1949) and protein concentrations were determined as described by Lowry et al. (1951), using bovine serum albumin as a standard.

Results

The chlorophyll a/b ratios and the distribution of RuBP carboxylase and PEP carboxylase in purified preparations of mesophyll protoplasts and bundle sheath strands are presented in Table 1. If we assume that the two marker enzymes are cell-type specific, the amount of cross-contamination is estimated to be 7–8%.

Polypeptide compositions of thylakoid membranes and soluble fractions of the two cell types

Analysis of thylakoid membranes isolated from the two cell types by LDS–PAGE reveals several qualitative and quantitative differences in the polypeptide patterns (Fig. 1). The most prominent quantitative deficiency in bundle sheath thylakoids is in the 26–29 kD polypeptides which are subunits of the light-harvesting chlorophyll a/b complex or LHC (Burke, Ditto & Arntzen, 1978; Delepelaire & Chua, 1981). This difference had been noted previously (Bishop, 1974; Genge, Pilger & Hiller, 1974; Kirchanski & Park, 1976) and is consistent with the high chlorophyll a/b ratio of bundle sheath thylakoid membranes (Table 1). In addition to the subunits of the light-harvesting complex, several polypeptide bands in the 30–70 kD region also show quantitative variations between the two thylakoid types. To gain a better understanding of these variations, we have identified some of these bands by crossed immunoelectrophoresis and by their association with chlorophylls to form chlorophyll–protein complexes. Thus, bundle sheath thylakoids are enriched in the α and β subunits of the chloroplast ATPase and the 68 kD apoprotein of the chlorophyll–protein complex I (Chua, Matlin & Bennoun, 1975), but deficient in polypeptides 5 and 6 of the photosystem II reaction center (Chua & Bennoun, 1975; Delepelaire & Chua, 1979; Diner & Wollman, 1980) and in ferredoxin–NADP⁺ reductase (M_r

Table 1. *Chlorophyll a to b ratios and relative distribution of the soluble enzymes RuBP carboxylase and PEP carboxylase in mesophyll and bundle sheath cells. Proteins were determined by quantitative rocket immunoelectrophoresis and expressed as relative units per mg protein*

	Mesophyll	Bundle sheath
Chlorophyll a/b ratio[a]	2.7	4.8
Large subunit of RuBP carboxylase[b]	7.6	100
Small subunit of RuBP carboxylase[b]	2.4	100
PEP carboxylase[b]	100	8.2

[a] Average of five determinations.
[b] Average of three determinations.

36000). Other quantitative differences are also evident in the low M_r region (<20000) but the functions of these polypeptides are unknown. Finally, mesophyll thylakoid membranes contain a 95 kD polypeptide of unidentified function, which appears to be absent from the bundle sheath thylakoid membranes.

Several laboratories have reported the differential localization of enzyme activities in mesophyll and bundle sheath cells of maize (Hatch & Kagawa, 1973; Rathnam & Edwards, 1975). Since most of these enzymes are soluble, we compared total soluble proteins of the two cell types by LDS–PAGE. Figure 1 shows that mesophyll soluble fractions have polypeptide com-

Fig. 1. Comparison of thylakoid membrane polypeptides and total soluble proteins of mesophyll (M) and bundle sheath (B) cells. Samples were analyzed in 7.5–15% LDS gradient gel at 4 °C. (1) M thylakoid membranes, 12 µg chlorophyll; (2) B thylakoid membranes, 12 µg chlorophyll; (3) M soluble proteins, 35 µg; (4) B soluble protein, 40 µg; (5) partially purified PEP carboxylase (Hague & Sims, 1980), 4 µg; and (6) partially purified RuBP carboxylase (Chua & Schmidt, 1978), 25 µg. PEPC, PEP carboxylase; L and S, large and small subunit of ribulose bisphosphate carboxylase (RuBPcase); FNR = ferredoxin–NADP+ reductase; Cyt f = cytochrome f; CPI = chlorophyll–protein complex I.

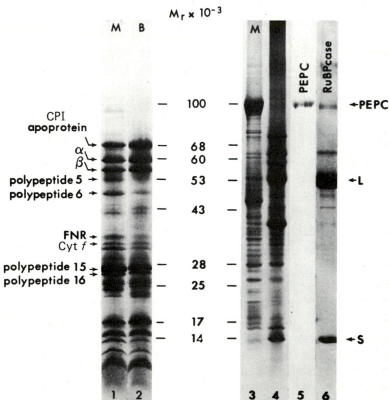

positions which differ dramatically from those of bundle sheath soluble fractions. The major soluble polypeptides of one cell type are not represented in the other cell type. For example, the most prominent soluble polypeptide from mesophyll cells has a $M_r = 100\,000$ (Fig. 1, lane 3). This band, which corresponds to the polypeptide subunit of PEP carboxylase (Hague & Sims, 1980) (Fig. 1, lane 5) is present only in mesophyll protoplasts and accounts for 20–30% of the total soluble protein. Similarly, the major soluble polypeptides of bundle sheath cells (Fig. 1, lane 4) are not detected in mesophyll cells (Fig. 1, lane 3). The predominant polypeptides of the bundle sheath soluble fraction have M_rs of $53\,000$ and $14\,000$ and correspond to the large and small subunits, respectively, of maize RuBP carboxylase (Fig. 1, lane 6).

In addition to RuBP carboxylase activity, the activities of several other Calvin cycle enzymes have been detected only in bundle sheath strands (Hatch & Kagawa, 1973; Rathnam & Edwards, 1975). To see if this differential localization of enzymatic activity is reflected at the protein level, we analyzed soluble proteins of mesophyll and bundle sheath cells by double immuno-diffusion using monospecific antibodies to some of the enzymes. Figure 2(d–f) shows that the low activities of RuBP carboxylase, ribulose-5-phosphate kinase, and sedoheptulose-1,7-bisphosphatase in mesophyll cells can be accounted for by the low levels of the enzyme proteins. Similarly, the lack of PEP carboxylase activity in bundle sheath cells is the result of the absence of this protein in these cells (Fig. 2c).

We also analyzed thylakoid membrane polypeptides from mesophyll and bundle sheath cells by double immunodiffusion. Two thylakoid membrane polypeptide components of photosystem II (PSII) reaction centers, designated 5 and 6 (Chua & Bennoun, 1975; Delepelaire & Chua, 1979; Diner & Wollman, 1980), have been shown by LDS–PAGE to be deficient in bundle sheath thylakoids (Fig. 1 B). Figure 2a shows that these antigens present in mesophyll thylakoids give a strong, single precipitation line with anti-5 IgG and with anti-6 IgG, whereas bundle sheath thylakoids give only a faint band in both cases. In contrast, thylakoids from both cell types react equally with antibodies to cytochrome f and plastocyanin (Fig. 2b).

Unexpected results were obtained in our crossed immunoelectrophoresis experiments with antibodies against ferredoxin–NADP$^+$ reductase (FNR), a protein involved in PSI function. Thylakoid membranes from mesophyll cells produce a distinct immunoprecipitin arc (Fig. 3) and the cross-reacting thylakoid polypeptide has a M_r of 36 kD. In contrast, thylakoid membranes from bundle sheath cells give only a weak reaction suggesting a deficiency of FNR in this thylakoid type (data not shown).

Since FNR is a peripheral protein, there is the possibility that it is dislodged more easily from the bundle sheath thylakoid membranes than mesophyll thylakoid membranes during cell fractionation. Accordingly, we analyzed the soluble fractions of both cell types by double immunodiffusion against anti-FNR IgG. Figure 2(c and f) show that FNR is indeed present in the soluble fractions but its concentration in the bundle sheath cells is still lower than that in the mesophyll cells.

Fig. 2. Immunodiffusion of thylakoid membranes and soluble proteins of mesophyll (M) and bundle sheath (B) cells with various antisera. In (a) and (b) the antigen wells contained thylakoid membranes whereas in (c) to (f) the antigen wells contained total soluble proteins from M (40 μg) and B (35 μg) cells. (a) 40 μl of anti-5 IgG (41 mg ml^{-1}) and 40 μl of anti-6 IgG (40 mg ml^{-1}); (b) 50 μl of anti-cyt.f IgG (11.4 mg ml^{-1}) and 50 μl of anti-PC IgG (32 mg ml^{-1}); (c) 45 μl of anti-PEPC IgG (8 mg ml^{-1}) and 45 μl of anti-FNR IgG (43 mg ml^{-1}); (d) 50 μl of anti-L IgG (23 mg ml^{-1}) and 50 μl of anti-S IgG (50 mg ml^{-1}); (e) 50 μl of anti-FBPase IgG (45 mg ml^{-1}) and 50 μl of anti-SBPase IgG (25 mg ml^{-1}); (f) 50 μl of anti-R-5-P kinase IgG (16 mg ml^{-1}) and 50 μl of anti-FNR IgG (43 mg ml^{-1}). cyt f, cytochrome f; PC, plastocyanin; PEPC, PEP carboxylase; FNR, ferredoxin–NADP$^+$ reductase; L and S, large and small subunit of RuBP carboxylase; FBPase, fructose-1,6-bisphosphatase; SBPase, sedoheptulose-1,7-bisphosphatase; R-5-P kinase, ribulose-5-phosphate kinase.

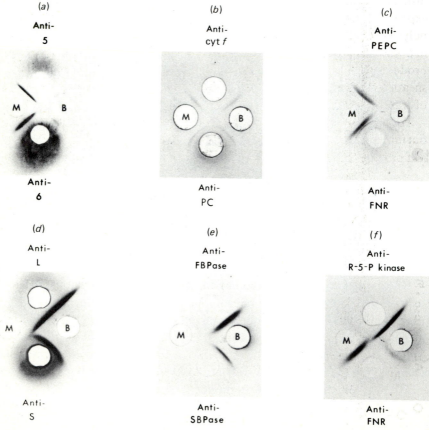

A more precise method for quantifying the level of polypeptides in both cell types is by rocket immunoelectrophoresis. An example of this type of analysis is shown in Fig. 4a. By comparing the area of the immunoprecipitin arc produced versus the amount of protein, it is possible to calculate the levels of antigen present in each cell type. Thus, Fig. 4 shows that the level of FNR in bundle sheath thylakoids and the soluble fraction (data not shown) is only 44% of the level present in mesophyll thylakoids and soluble fractions. In contrast, quantitative rocket immunoelectrophoresis using anti-PEP carboxylase IgG or anti-RuBP carboxylase IgG show that the cross-contamination of our cell-type preparations is only 5–7 % (Table 1).

Translation in vitro *of poly(A) RNAs and identification of products by immunoprecipitation*

So far, our electrophoretic and immunological analyses reveal that many thylakoid membrane polypeptides and soluble proteins in maize leaves are cell-type specific. It is possible that the differential protein distribution could be generated by post-translational events, e.g. differential rates of protein turnover. Alternatively, it could be a manifestation of differential expression of genes encoding these proteins. To examine the second possibility, poly(A)-RNA was isolated from both cell types and translated into protein in a wheat-germ cell-free system. Figure 5 shows that, *in vitro*, the translation products of mesophyll cells are dramatically different from those of bundle sheath cells, demonstrating that the two cell types have different populations of translatable mRNA.

To follow the synthesis of specific proteins in the system *in vitro* we carried out immunoprecipitation experiments with monospecific antibodies to several leaf proteins. Figure 5 shows that maize RuBP carboxylase small subunit and

Fig. 3. Crossed immunoelectrophoresis of mesophyll (M) thylakoid membranes against antibodies to ferredoxin–NADP⁺ reductase (FNR). M thylakoid membranes (15 μg chlorophyll) were resolved in a 7.5–15% LDS gradient gel at 4 °C. The antibody (AG) gel contained 0.243 mg IgG cm⁻². Arrows indicate the position of FNR.

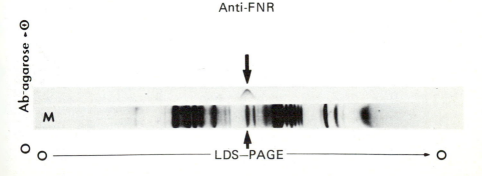

the light-harvesting chlorophyll *a/b* complex (LHC) polypeptides (15 and 16) are synthesized as larger precursors. This finding is consistent with the fact that the polypeptides are products of cytosolic polysomes and are imported post-translationally into the chloroplasts (Chua *et al.*, 1980; Ellis, 1981). The precursors of the LHC polypeptides are 4000 Daltons larger than their respective mature forms. In contrast, the precursor to the small subunit (S)

Fig. 4. Rocket immunoelectrophoresis of mesophyll (M) and bundle sheath (B) thylakoid membranes against antibodies to FNR. In the upper panel (*a*), wells (1) to (8) contain 1 to 8 μg chlorophyll, respectively, of M thylakoid membranes and well (9) contains 5 μg chlorophyll of B thylakoid membranes. The IgG concentration was 0.14 mg cm^{-2}. Lower panel (*b*) shows the linear relationship between the areas of precipitin peaks and the amounts of M thylakoid membranes used for immunoelectrophoresis.

(*a*)   Rocket Immunoelectrophoresis of
FNR

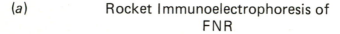
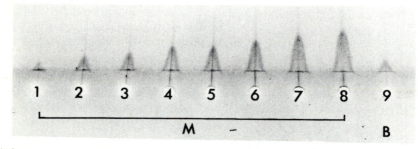

(*b*)

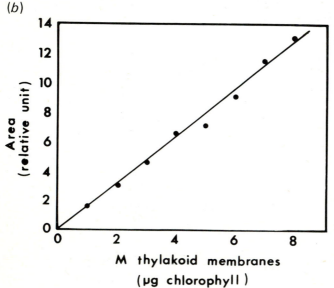

is larger than the mature form by only 2500 Daltons. Thus, the precursor peptide (transit sequence) in this case is smaller than those of other small subunit precursors investigated thus far (cf. Chua *et al.*, 1980; Ellis, 1981). The PEP carboxylase subunit is also synthesized in the wheat-germ system but the product *in vitro* has an electrophoretic mobility identical to that of

Fig. 5. Immunoprecipitation of mesophyll and bundle sheath translation products using monospecific antibodies to the carboxylase small subunit (S); phosphoenolpyruvate carboxylase (PEPC) and the chlorophyll *a/b* light-harvesting complex (polypeptides 15 and 16). pS is the precursor to the small subunit and p15 and p16 are the precursors to polypeptides 15 and 16, respectively.

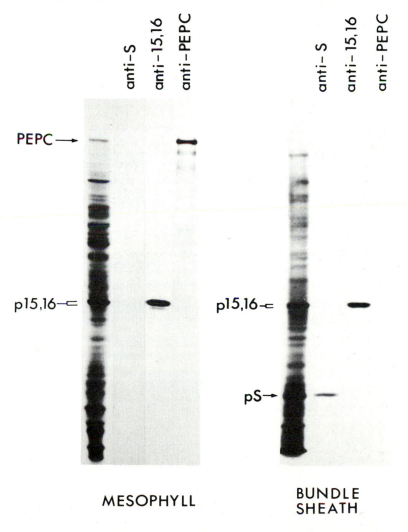

the authentic polypeptide. This finding is not unexpected since PEP carboxylase is a cytosolic enzyme (Perrot *et al.*, 1981). Similar results have been reported previously by Sims & Hague (1981). Comparison of the immunoprecipitated translation products *in vitro* from the two cell types shows that mRNA encoding PEP carboxylase is present in mesophyll but not bundle sheath cells (Fig. 5). The converse is true for the mRNA encoding the carboxylase small subunit (Fig. 5). The mRNAs for the LHC polypeptides are found in both cell types, but the level in the mesophyll cells is higher (Fig. 5). These results provide direct evidence that the differential distribution of PEP carboxylase, the LHC polypeptides, and the carboxylase small subunit are the result of different levels of translatable mRNAs in the two cell types.

Discussion

Previous investigations on the thylakoid membranes of mesophyll and bundle sheath cells of C_4 plants have focused primarily on their photosynthetic electron transport capacities. While the mesophyll thylakoids can perform both PSI and PSII reactions at rates equivalent to those in C_3 plants, the bundle sheath thylakoids contain only 20–30% of the normal PSII activity (Bishop, 1974; Horváth *et al.*, 1978). Electrophoretic analysis of the two types of thylakoid membranes showed a diminution of the LHC polypeptides in bundle sheath thylakoids consistent with their high chlorophyll a/b ratio (Bishop, 1974; Genge *et al.*, 1974; Kirchanski & Park, 1976). However, no biochemical correlates at the membrane polypeptide level have been obtained so far to account for the low PSII activity. In this chapter, we show by immunological methods that two membrane polypeptides (polypeptide 5, $M_r = 51000$ D and polypeptide 6, $M_r = 47000$ D) known to be associated with PSII reaction centers (Chua & Bennoun, 1975; Delepelaire & Chua, 1979; Diner & Wollman, 1980) are deficient in bundle sheath thylakoids. Thus, the low PSII activity is the result not of the synthesis of inactive membrane polypeptides but rather of reduced amounts of reaction center polypeptides 5 and 6 in the thylakoid membranes. PSI and PSII complexes have been isolated from detergent extracts of thylakoid membranes and each complex contains several polypeptide subunits in strict stoichiometry (Mullet, Burke & Arntzen, 1980; Mullet & Arntzen, 1981). Polypeptide subunits of the same complex are affected to the same extent in mutants with lesions in PSI or PSII activity (Girard *et al.*, 1980). Thus, some of the polypeptides of mesophyll thylakoids which are absent or severely deficient in bundle sheath thylakoids are most likely subunits of the PSII complex. On the other hand, bundle sheath thylakoids are enriched in the 68 kD polypeptide of PSI reaction center (Fig. 2, lane 6) (Chua *et al.*, 1975) and 4–5 polypeptides in the 18–22 kD range (Fig. 2, cf. lanes 5 and 6). The latter polypeptides probably correspond to subunits of the PSI peripheral chlorophyll proteins described by Mullet *et al.* (1980).

A surprising finding is the deficiency of FNR in bundle sheath chloroplasts. This enzyme is involved in the terminal step of NADP$^+$ photoreduction using electrons generated from water oxidation by PSII. Since the amount of PSII in bundle sheath chloroplasts is only 10–20% that of mesophyll chloroplasts, it is perhaps not unexpected that a reduced amount of FNR would suffice for maintaining a low level of NADP$^+$ photoreduction.

While maize mesophyll and bundle sheath cells have been shown to contain different soluble enzymatic activities (cf. Edwards & Huber, 1981), the molecular events responsible for such differences are poorly understood. Electrophoretic analyses of soluble proteins show that the major components of one cell type are not found in the other; in fact, the two cell types have strikingly different patterns of soluble protein. These results provide preliminary evidence that the differences in enzymatic activities can be accounted for largely by differential protein distribution. This conclusion is reinforced by results obtained from double immunodiffusion assays using monospecific antibodies to several chloroplast proteins. We found that PEP carboxylase is restricted primarily to the mesophyll cells, whereas RuBP carboxylase, ribulose-5-phosphate kinase, fructose-1,6-bisphosphatase, and sedoheptulose-1,7-bisphosphatase are confined mainly to the bundle sheath cells. Thus, at least for these five enzymes, we have established unequivocally that differential distribution of the enzyme proteins, rather than the presence of inactive enzymes or cell-type specific inhibitors, is responsible for differences in the enzymatic activity. These results confirm and extend previous work (Huber, Hall & Edwards, 1976; Link, Coen & Bogorad, 1978; Perrot *et al.*, 1981).

With the exception of the large subunit of RuBP carboxylase, most of the chloroplast and cytoplasmic soluble proteins in plant cells are products of nuclear genes and are translated on 80S ribosomes (cf. Ellis, 1981). The nuclear contribution to thylakoid membrane biogenesis is also quite substantial, accounting for 70–80% of the thylakoid membrane polypeptides (Chua & Gillham, 1977). Translations of poly(A)-RNAs from the two cell types in a wheat-germ cell-free system produce different patterns of newly synthesized polypeptides. These findings strongly suggest that qualitative and quantitative differences in translatable poly(A)-RNAs are responsible for the differential distribution of nuclear-encoded proteins in the two cell types. Examination of the synthesis *in vitro* of specific proteins by poly(A)-RNAs corroborate this notion. Link *et al.* (1978) reported that the absence of the large subunit of RuBP carboxylase in mesophyll chloroplasts results from the absence of mRNAs encoding this subunit. Our results show that translatable mRNA encoding the small subunit of RuBP carboxylase is not detectable in mesophyll cells which lack this polypeptide. Similar results were obtained with PEP carboxylase and the LHC polypeptides. We are at present identifying cDNA clones encoding proteins that are differentially distributed in bundle sheath and mesophyll cells. The identified clones will be used as probes to determine whether the developmentally regulated genes are controlled at the transcriptional or post-transcriptional level.

Polypeptides 5 and 6 are products of chloroplast protein synthesis (Chua & Gillham, 1977) and are likely to be encoded by the organelle genome. In contrast to the large subunit of RuBP carboxylase (Huber *et al.*, 1976; Link *et al.*, 1978), these two polypeptides are present in mesophyll chloroplasts but greatly diminished in bundle sheath chloroplasts. We are investigating whether chloroplast genes encoding these polypeptides are regulated at the transcriptional level.

This work was supported in part by USDA CRGO Grant #8100294. Dr Richard Broglie and Dr Gloria Coruzzi are recipients of NIH Post-doctoral Fellowships 5F32GM07446 and 1F32GM07776, respectively. We thank Dr Don Hague for sending us a preparation of partially purified PEP carboxylase and Dr Bob B. Buchanan for antibodies to sedoheptulose-1,7-bisphosphatase. We also thank Sara Trillo for excellent technical assistance.

References

Arnon, D. I. (1949). Copper enzymes in isolated chloroplasts. Polyphenol oxidase in *Beta vulgaris. Pl. Physiol.*, **24**, 1–5.

Bishop, D. G. (1974). Lamellar structure and composition of chloroplasts in relation to photosynthetic electron transfer. *Photochem. Photobiol.*, **20**, 281–99.

Burke, J. J., Ditto, C. L. & Arntzen, C. J. (1978). Involvement of the light-harvesting complex in cation regulation of excitation energy distribution in chloroplasts. *Archs Biochem. Biophys.*, **187**, 252–63.

Chirgwin, J. M., Przybyla, A. E., MacDonald, R. J. & Rutter, W. J. (1979). Isolation of biologically active ribonucleic acid from sources enriched in ribonuclease. *Biochemistry*, **18**, 5294–304.

Chollet, R. & Ogren, W. L. (1973). Photosynthetic carbon metabolism in isolated maize bundle sheath strands. *Pl. Physiol.*, **51**, 787–92.

Chua, N.-H. & Bennoun, P. (1975). Thylakoid membrane polypeptides of *Chlamydomonas reinhardtii*: Wild-type and mutant strains deficient in photosystem II reaction center. *Proc. natn. Acad. Sci. USA*, **73**, 2175–9.

Chua, N.-H. & Blomberg, F. (1979). Immunochemical studies of thylakoid membrane polypeptides from spinach and *Chlamydomonas reinhardtii*. *J. biol. Chem.*, **254**, 216–23.

Chua, N.-H. & Gillham, N. W. (1977). The sites of synthesis of the principal thylakoid membrane polypeptides in *Chlamydomonas reinhardtii*. *J. Cell Biol.*, **74**, 441–52.

Chua, N.-H., Grossman, A. R., Bartlett, S. G. & Schmidt, G. W. (1980). Synthesis, transport and assembly of chloroplast proteins. *Mosbach Colloqu.*, **31**, 113–18.

Chua, N.-H., Matlin, K. & Bennoun, P. (1975). A chlorophyll–protein complex lacking in photosystem I mutants of *Chlamydomonas reinhardtii*. *J. Cell Biol.*, **678**, 361–77.

Chua, N.-H. & Schmidt, G. W. (1978). Post-translational transport into intact chloroplasts of a precursor to the small subunit of ribulose-1,5-bisphosphate carboxylase. *Proc. natn. Acad. Sci. USA*, **75**, 6110–14.

Converse, C. A. & Papermaster, D. S. (1975). Membrane protein analysis by two-dimensional immunoelectrophoresis. *Science*, **189**, 469–72.

Delepelaire, P. & Chua, N.-H. (1979). Lithium dodecyl sulfate/polyacrylamide gel electrophoresis of thylakoid membranes at 4 °C: Characterization of two additional chlorophyll a–protein complexes. *Proc. natn. Acad. Sci. USA*, **76**, 111–15.

Delepelaire, P. & Chua, N.-H. (1981). Electrophoretic purification of chlorophyll a/b–protein complexes from *Chlamydomonas reinhardtii* and spinach and analysis of their polypeptide compositions. *J. biol. Chem.*, **256**, 9300–7.

Diner, B. A. & Wollman, F.-A. (1980). Isolation of highly active photosystem II particles from a mutant of *Chlamydomonas reinhardtii*. *Eur. J. Biochem.*, **110**, 521–6.

Edwards, G. E. & Huber, S. C. (1981). The C4 pathway. In *The Biochemistry of Plants. A Comprehensive Treatise*, vol. 8, ed. M. D. Hatch & N. K. Boardman, pp. 238–82. Academic Press, New York & London.

Ellis, R. J. (1981). Chloroplast proteins: synthesis, transport and assembly. *A. Rev. Pl. Physiol.*, **32**, 111–37.

Genge, S., Pilger, D. & Hiller, R. G. (1974). The relationship between chlorophyll *b* and pigment–protein complex II. *Biochim. biophys. Acta*, **347**, 22–30.

Girard, J., Chua, N.-H., Bennoun, B., Schmidt, G. & Delosme, M. (1980). Studies on mutants deficient in the photosystem I reaction centers in *Chlamydomonas reinhardtii*. *Curr. Genet.*, **2**, 215–21.

Grossman, A. R., Bartlett, S. G., Schmidt, G. W., Mullet, J. E. & Chua, N.-H. (1982). Optimal conditions for post-translational uptake of proteins by isolated chloroplasts. *In vitro* synthesis and transport of plastocyanin, ferredoxin–NADP$^+$ oxidoreductase and fructose-1,6-bisphosphatase. *J. biol. Chem.*, **257**, 1558–63.

Hague, D. R. & Sims, T. L. (1980). Evidence for light-stimulated synthesis of phosphoenolpyruvate carboxylase in leaves of maize. *Pl. Physiol.*, **66**, 505–9.

Harboe, N. & Ingild, A. (1973). Immunization, isolation of immunoglobulins, estimation of antibody titre. *Scand. J. Immunol.*, **2**, *Suppl.* 1, 161–4.

Hatch, M. D. & Kagawa, T. (1973). Enzymes and functional capacities of mesophyll chloroplasts from plants with C_4-pathway photosynthesis. *Archs. Biochem. Biophys.*, **159**, 842–53.

Horváth, G., Droppa, M., Mustárdy, L. A. & Faludi-Dániel, A. (1978). Functional characteristics of intact chloroplasts isolated from mesophyll protoplasts and bundle sheath cells of maize. *Planta*, **141**, 239–44.

Huber, S. C., Hall, T. C. & Edwards, G. E. (1976). Differential localization of fraction I protein between chloroplast types. *Pl. Physiol.*, **57**, 730–3.

Joyard, J., Grossman, A. R., Bartlett, S. G., Douce, R. & Chua, N.-H. (1982). Characterization of envelope membrane polypeptides from spinach chloroplasts. *J. biol. Chem.*, **257**, 1095–101.

Kanai, R. & Edwards, G. E. (1973). Separation of mesophyll protoplasts and bundle sheath cells from maize leaves for photosynthetic studies. *Pl. Physiol.*, **51**, 1133–7.

Kirchanski, S. J. & Park, R. B. (1976). Comparative studies of the thylakoid proteins of mesophyll and bundle sheath plastids of *Zea mays*. *Pl. Physiol.*, **58**, 345–9.

Laurell, C.-B. (1972). Electroimmunoassay. *Scand. J. clin. Lab. Invest.*, **29**, *Suppl.* 124, 21–37.

Lindberg, U. & Pearson, T. (1972). Isolation of mRNA from KB-cells by affinity chromatography on polyacridylic acid covalently linked to sepharose. *Eur. J. Biochem.*, **31**, 246–54.

Link, G., Coen, D. M. & Bogorad, L. (1978). Differential expression of the gene for the large subunit of ribulose bisphosphate carboxylase in maize leaf cell types. *Cell*, **15**, 725–31.

Lowry, O. H., Rosebrough, N. J., Farr, A. L. & Randall, R. J. (1951). Protein measurement with the folin phenol reagent. *J. biol. Chem.*, **193**, 265–75.

Mullet, J. E., Burke, J. J. & Arntzen, C. J. (1980). A developmental study of photosystem I peripheral chlorophyll proteins. *Pl. Physiol.*, **65**, 814–22.

Mullet, J. E. & Arntzen, C. J. (1981). Identification of a 32 kilodalton polypeptide as a herbicide receptor protein in photosystem II. *Biochim. biophys. Acta*, **635**, 236–48.

Perrot, C., Vidal, J., Burlet, A. & Gadal, P. (1981). On the cellular localization of phosphoenolpyruvate carboxylase in *Sorghum* leaves. *Planta*, **151**, 226–31.

Rathnam, C. K. M. & Chollet, R. (1980). Photosynthetic carbon metabolism in C_4 plants and C_3–C_4 intermediate species. *Prog. Phytochem.*, **6**, 1–48.

Rathnam, C. K. M. & Edwards, G. E. (1975). Intracellular localization of certain photosynthetic enzymes in bundle sheath cells of plants possessing the C_4 pathways of photosynthesis. *Archs Biochem. Biophys.*, **171**, 214–25.

Sims, T. L. & Hague, D. R. (1981). Light-stimulated increase of translatable mRNA for phosphoenolpyruvate carboxylase in leaves of maize. *J. biol. Chem.*, **256**, 8252–5.

PART II

The chloroplast genetic system

G. S. P. GROOT

Chloroplast DNA of higher plants

The biosynthesis of chloroplasts requires the coordinated expression of two distinct genetic systems: the usual, chromosomal, one located in the nucleus and the other, cytoplasmic, one present in the plastid itself. Like mitochondria, chloroplasts maintain control over the synthesis of some of their constituent polypeptides via their own DNA, RNA and protein synthesizing activity.

Early indications of the existence of a separate chloroplast genetic system stem from the study of non-Mendelian mutants of the chloroplast phenotype in *Pelargonium* (Bauer, 1909). Since the discovery of a separate DNA in chloroplasts, now 20 years ago (Ris & Plaut, 1962), a wealth of information on the physical and genetic properties of different chloroplast (cp) DNAs has been obtained (see recent reviews by Edelman, 1981; Herrmann & Possingham, 1981; Bohnert, Crouse & Schmitt, 1982; Whitfeld & Bottomley, 1983). Restriction fragment mapping, cloning and expression of characterized fragments and DNA sequence analysis have made possible, as in other fields, a detailed analysis of chloroplast genes.

In this review I will summarize the results obtained up to August 1983 and stress the many similarities of the genetic organization of the cp-DNAs in a variety of higher plants, but with some notable exceptions. I realize at the same time that the increase in our knowledge will be rapid and that, as in the case of mitochondrial DNAs (Grivell, 1981), surprises are likely to occur.

General properties

cp-DNAs of higher plants are covalently closed circular molecules with a density of 1.697–1.699 g cm^{-3} corresponding to a G+C content of 37–38 mol percent. In Table 1 I have summarized the physical properties of cp-DNAs from different plants, restricting myself to those species for which a restriction fragment map is available. In general the size as measured by electron microscopy is about 45 μm. Although the length measurements of such large molecules are intrinsically inaccurate there is a good agreement with the molecular weight measurements as determined by the sum of the restriction fragments: 150–155 kbp. There are two notable exceptions known

67

at present. One, represented by the cp-DNA of *Spirodela oligorhiza* which is considerably larger than most of the other cp-DNAs and one, represented by the cp-DNAs of *Pisum sativum* and *Vicia faba*, which is substantially smaller (*vide infra*).

Electron microscopical analysis and restriction fragment mapping have shown that the organization of most of the cp-DNAs is strikingly similar. Two regions, the large and small single copy regions, are separated from each other by two identical inverted repeat regions. An example of the evidence for this anatomy is given in Fig. 1. This figure shows the separation of the restriction fragments of *Spirodela oligorhiza* cp-DNA digested with the enzymes Bam HI, Sac I and Xho I respectively. Several fragments, e.g. Fragment H in the Bam digest, Fragment C in the Xho digest or Fragment I in the Sac digest appear to be produced in a stoichiometry twice that of other bands. Similarly when cp-DNA is denatured and the single-stranded molecules are allowed to renature for a short period, snap-back, dumb-bell shaped molecules

Table 1. *Physical properties of cp-DNAs of higher plants*

Species	(Ref.)	Size μm (EM)	Molecular weight kbp (RE)	Density g cm^{-3}
Atriplex triangularis	(a)	—	152	—
Brassica napus	(l)	—	150	—
Cucumis sativa	(a)	—	155	—
Lycopersicon sp.	(m)	—	149	—
Nicotiana tabacum	(b)	48	160	1.697
Oenothera sp.*	(c)	45	160	1.697
Petunia hybrida	(d)	44	151	1.699
Phaseolus vulgaris	(n)	—	150	—
Sinapsis alba	(f)	44	154	1.697
Solanum sp.	(m)	—	158	—
Spinacia oleracea	(g)	46	154	1.697
Triticum aestivum	(h)	—	135	—
Vigna radiata	(e)	—	150	—
Zea mais	(i)	44	151	1.698
Spirodela oligorhiza	(j)	54	182	1.698
Pisum sativum	(e)	—	120	—
Vicia faba	(k)	35	121	1.696

EM = electron microscopical analysis. RE = restriction endonuclease analysis.
* Some small insertions/deletions have been reported for different species.
References: (a) Palmer, 1982; (b) Jurgenson & Bourque, 1980; (c) Gordon *et al.*, 1981; (d) Bovenberg, Kool & Nijkamp, 1981; (e) Palmer & Thompson, 1981; (f) Link *et al.*, 1981; (g) Herrmann, Whitfeld & Bottomley, 1980; (h) Bowman *et al.*, 1981; (i) Bedbrook & Bogorad, 1976; (j) van Ee, Vos & Planta, 1980; (k) Koller & Delius, 1980; (l) Vedel & Mathieu, 1983; (m) Palmer & Zamir, 1982; (n) Palmer, 1983.

Fig. 1. Agarose gel of restriction fragments of *Spirodela oligorhiza* cp-DNA digested with the endonucleases Bam HI, Sac I and Xho I. Nomenclature of the fragments according to van Ee *et al.*, 1980.

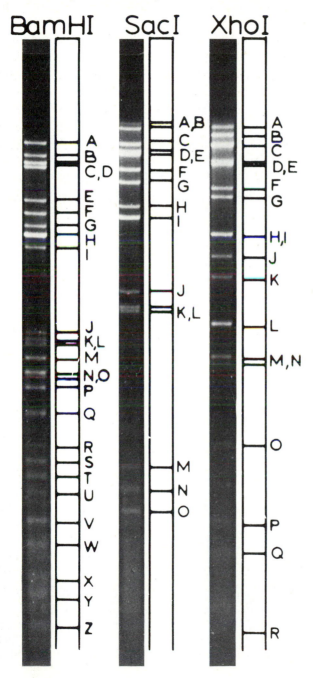

are observed under the electron microscope (Bedbrook, Kolodner & Bogorad, 1979b).

The cp-DNAs of *Pisum sativum* and *Vicia faba*, however, are composed completely of non-repeated DNA and miss at least one of the two repeated regions. The coding capacity of the inverted repeat region and the consequences for these two species are discussed later.

Genes on chloroplast DNA

With modern biochemical techniques the localization of genes on cp-DNA is relatively easy for genes which produce stable or abundant transcripts, e.g. the ribosomal RNAs and transfer RNAs. The localization of genes coding for proteins is more difficult. *A priori*, one does not even know which of the several hundred chloroplast proteins are chloroplast gene products. In principle, it might be possible – although there is no evidence for this – that chloroplast gene products function outside the organelle. In practice, however, the most likely candidates for cp-DNA gene products are chloroplast translation products. Those presently known are summarized in Table 2, together with the stable RNA transcripts.

In general, genes can be localized on cp-DNA by hybridizing characterized RNAs (either rRNAs, tRNAs or abundant mRNAs), appropriately labelled,

Table 2. *Chloroplast transcription and translation products*

A. *Involved in expression of genetic information*
   (1) Large (23S) ribosomal RNA (*a*)
   (2) Small (16S) ribosomal RNA (*a*)
   (3) 5S and 4.5S ribosomal RNA (*b*)
   (4) 35–40 transfer RNAs (*c*)
   (5) 11 ribosomal proteins (*d*)
   (6) 2 elongation factors (*e*)

B. *Involved in chloroplast function*
   (1) Large subunit of ribulose-1,5-bisphosphate carboxylase (*f*)
   (2) α, β and ϵ subunits of CF_1 (*g*)
   (3) 1(2) subunit(s) of chloroplast CF_0 (*h*)
   (4) 3 subunits of PSI (*i*)
   (5) '32 kD' shielding protein (binds herbicides) (*j*)
   (6) Cytochrome *f*, *b*-6 and *b*-559 (*k*, *l*)
   (7) CP_I apoprotein (*l*)

References: (*a*) Tewari & Wildman, 1968; (*b*) Dyer, Bowman & Payne, 1977; (*c*) Tewari & Wildman, 1970; (*d*) Eneas-Filho, Hartley & Mache, 1981; (*e*) Tiboni, diPasquale & Ciferri, 1978; (*f*) Blair & Ellis, 1973; (*g*) Mendiola-Morgenthaler, Morgenthaler & Price, 1976; (*h*) Sebald, Hoppe & Wachter, 1979; (*i*) Neuchushtai *et al.*, 1981; (*j*) Edelman & Reisfeld, 1978; (*k*) Doherty & Gray, 1979; (*l*) Zielinski & Price, 1980.

to Southern blots of restriction fragments (Southern, 1975). The complementary approach is to identify, mostly with the help of specific antibodies, protein products obtained either (1) in a coupled transcription–translation system from *E. coli* programmed with suitably cloned cp-DNA fragments or (2) by translation of purified mRNAs in a heterologous translation system. The latter approach must be followed by establishing the transcriptional origin of the mRNA. Finally, DNA sequence analysis can be used not only to obtain information on the structural part of the gene, but also to study 5′- and 3′-flanking base sequences such as transcriptional start and stop sequences and processing sites.

Genes for ribosomal RNAs

The ribosomal RNAs of chloroplasts of higher plants consist of four species: the usual 23S, 16S and 5S rRNA and a typical 4.5S rRNA, which shows base sequence homology with the 3′ end of 23S prokaryotic rRNA (see Crouse *et al.*, this volume). By hybridization and electron microscopical studies it was shown that these genes are located in the inverted repeat segment of cp-DNA in the order 5′-16S–spacer–23S–spacer–4.5S–spacer–5S-3′. As an example Fig. 2 shows the organization of the rRNA genes in *Spirodela oligorhiza*. In all cp-DNAs of higher plants studied so far a similar organization is found. Between different plants, differences in the length of the spacers are observed which are real, but whose significance is unknown.

Recently Kössel and coworkers (Schwarz & Kössel, 1980; Edwards & Kössel, 1981; Koch, Edwards & Kössel, 1981) and Sugiura *et al.* (Takaiwa & Sugiura, 1982*a*; Tohdoh & Sugiura, 1982) have reported an extensive nucleotide sequence analysis of the ribosomal RNA operon of *Zea mais* and *Nicotiana*. Both rRNA genes are more than 92% homologous. Comparison of the primary structure of the corresponding RNAs suggests a secondary structure that is remarkably similar to that proposed for *E. coli* 23S rRNA.

Genes for transfer RNAs

Saturation hybridization experiments have shown that chloroplasts contain 30–40 different tRNA genes (Tewari & Wildman, 1970; Haff & Bogorad, 1976; Steinmetz & Weil, 1976). Driesel and coworkers (Driesel *et*

Fig. 2. Organization of the rRNA genes on cp-DNA of *Spirodela oligorhiza*. (From Keus *et al.*, 1983.)

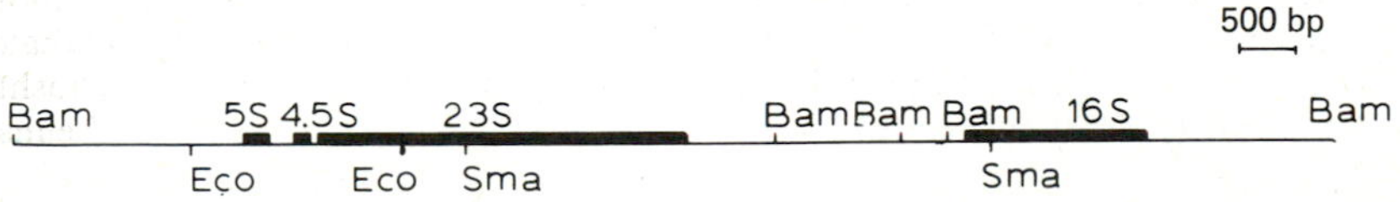

al., 1979) have determined the physical localization of different tRNA species. Their results show that most genes are scattered over the plastid genome although a limited clustering cannot be ruled out. A very similar organization of the genes for tRNAs has been reported for *Zea mais* (Weil *et al.*, 1981) and for *Spirodela oligorhiza* (Groot & van Harten-Loosbroek, 1981).

From the sequence data reported by Kössel's group (Koch *et al.*, 1981), it appears that the spacer region between 16S and 23S rRNAs of *Zea mais* cp-DNA contains the genes for tRNA^{ile} and tRNA^{ala}, with the remarkable feature that both contain large intervening sequences of 949 and 806 base-pairs, respectively. Sugiura's group found similar results in *Nicotiana tabacum*, the intervening sequences being about 100–200 base-pairs shorter (Takaiwa & Sugiura, 1982b). Recent evidence from Weil's laboratory shows that tRNAs of the predicted nucleotide sequence are present in *Zea mais* chloroplasts (Weil *et al.*, 1982). This means that chloroplasts can carry out the splicing reaction required to produce the mature tRNAs from a primary transcript. The existence of split genes in a DNA which is otherwise strikingly prokaryotic in character is remarkable although not completely unexpected (cf. Allet & Rochaix, 1979). In this respect it should also be noted that tRNA genes sequences thus far do not code for the 3'-CCA as is the case in prokaryotes (Steinmetz, Gubbins & Bogorad, 1982).

Genes for proteins

Coen *et al.* (1978) have reported that transcription of a cloned cp-DNA fragment of *Zea mais* followed by translation *in vitro* of the produced transcript leads to the synthesis of the large subunit of ribulose-1,5-bisphosphate carboxylase. This for the first time was direct evidence for the protein coding capacity of cp-DNA. Bottomley & Whitfeld (1979), using a coupled transcription–translation system, obtained similar results. From these data, the gene for the large subunit (LSU) could be localized on the restriction fragment maps (Bedbrook *et al.*, 1979a). Subsequently DNA sequence analysis confirmed this location (McIntosh, Poulsen & Bogorad, 1980).

The gene for the '32 kD' herbicide binding protein (Steinback *et al.*, 1981) (also known as peak D or photogene 32) has been mapped by hybridizing a 14S chloroplast mRNA shown to produce the 35 kD precursor of the '32 kD' protein when translated in a heterologous translation system, to restriction fragments of cp-DNA (Bedbrook *et al.*, 1978; Driesel, Speirs & Bohnert, 1980). The position on the restriction map (see Fig. 3) is similar in *Zea mais*, *Spinacia* and *Sinapsis* (Link, Cohen & Bogorad, 1981). The gene has been sequenced by Zurawski *et al.* (1982a).

Both genes are under transcriptional control: the LSU gene is transcribed in *Zea mais* only in the bundle sheath cells and not in the mesophyll cells (Bedbrook *et al.*, 1978). The '32 kD' protein gene is transcribed only in the light (Link *et al.*, 1978).

Recently Westoff *et al.* (1981) have located, in an elegant series of experiments, the genes for the α, β and ϵ subunits of the chloroplast F_1-ATPase. They purified specific mRNAs for these subunits using immobilized cp-DNA fragments of known map position. Translation of these RNAs produces polypeptides which could be immunologically identified as the α, β and ϵ subunits. In this way they established that the β and ϵ genes are located close together near the gene for the LSU. The gene for subunit α is located some 40 kbp away. This organization makes the coordinated expression of these genes as in the *unc* operon of *E. coli* impossible, and requires a different mode of regulation of the synthesis of the ATPase complex. A coordinated expression is plausible for the genes for β and ϵ, since they are likely to be part of a common transcriptional unit (see also Howe *et al.*, 1982). Sequence analysis even shows that both the genes partly overlap (Krebbers *et al.*, 1982; Shinozaki & Sugiura, 1982; Zurawski, Bottomley & Whitfeld, 1982*b*). This observation also suggests that synthesis of different quantities of the two subunits might be controlled at the level of translation, in a way analogous to the regulation of the expression of the lysis enzyme in bacteriophage MS2 (Kastelein *et al.*, 1982).

Herrmann's group has located, using a similar approach, the genes coding for the subunits I and III of the membrane part of the proton-translocating ATPase (Alt *et al.*, 1983*b*), the subunits of the cytochrome *b*-6/*f* complex (Alt *et al.*, 1983*a*) and the subunits of the photosystems (Herrmann *et al.*, 1983). All these data are summarized in a schematic physical map of *Spinacia* chloroplast DNA presented in Fig. 3.

The published DNA sequence data enable us to investigate the 'prokaryotic' character of cp-DNA in more detail. In the case of the proton-translocating ATPase, a direct comparison between the deduced amino-acid sequence of chloroplast subunits and their *E. coli* counterparts is possible, since the nucleotide sequence of the *unc* operon of *E. coli* is known. It appears that the homology is high for the β subunit (67%, Zurawski *et al.*, 1982*b*) and for the α subunit (55%, Deno, Shinozaki & Sugiura, 1983) and much lower (26%) for the ϵ subunit (Zurawski *et al.*, 1982*b*). The high homology for the α and β subunit is in line with their role in the enzymatic activity of the ATPase complex.

The strong homology between the deduced amino-acid sequence of chloroplast proteins and the corresponding *E. coli* proteins has enabled Subramanian, Steinmetz & Bogorad (1983) and Sugita & Sugiura (1983) to identify the genes coding for the chloroplast counterparts of the *E. coli* ribosomal proteins S4 and S19 by comparison only. In both cases the homology is 50–55%.

Furthermore one can compare the putative regulatory signals found in the 5′ and 3′ flanking areas of the genes with those found in bacteria. About 10 nucleotides upstream from the ATG start codon of the LSU gene a pentanucleotide sequence is found which is complementary to the 3′ end of 16S rRNA, and is thought to play a crucial role in initiation of translation (McIntosh *et al.*, 1980; Zurawski *et al.*, 1981).

In the vicinity of the transcription start of the LSU gene in *Spinacia*

Fig. 3. Physical map of *Spinacia* cp-DNA. The inverted repeat regions are indicated by the thick bars. 16S, 23S, 4.5S and 5S are the rRNA genes. The black dots are tRNA genes, α, β and ϵ are subunits of the chloroplast ATPase complex, LSU is the gene for the large subunit of ribulose-1,5-bisphosphate carboxylase and 32 kD is the gene for the light-induced, herbicide-binding protein. I and III are subunits of the membrane part of the proton-translocating ATPase; cyt. *f*, cyt. *b*-6 and pp4 are cytochrome *f*, cytochrome *b*-6 and subunit 4 of the cytochrome *b*-6/*f* complex; PS_{II}-1 and PS_{II}-2 are subunits of 1 and 2 of photosystem II, and AP700 is the apoprotein of P700. The arrows indicate the direction of transcription. (Redrawn from Herrmann *et al.*, 1983.)

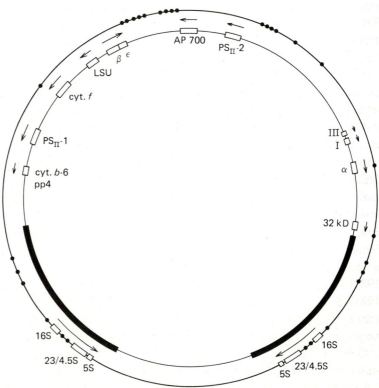

sequences homologous to the *E. coli* TATAAT (Pribnow) box and '-35' sequence are found (Zurawski *et al.*, 1981). However, less homologous sequences could be found in the corresponding part of the *Zea mais* gene (McIntosh *et al.*, 1980). Similarly no obvious prokaryotic promoter sites could be found in the 5' leading sequence of the *Zea mais* 16S rRNA gene, although several stretches of nucleotides in that region could be protected against DNAase I degradation by binding to the heterologous RNA polymerase from *E. coli* (Schwarz *et al.*, 1981).

At the 3' terminus of the LSU gene, both in *Zea mais* and *Spinacia*, an extensive stem-loop structure can be formed, although at different distances from the translation stop codon, that might be involved as a transcriptional stop signal. Most of these statements should be interpreted with care since they are based on DNA sequence data alone, while the exact transcriptional start and stop points are unknown.

Although it is obviously too early to draw firm conclusions, there seems to be an indication that the organization of cp-DNA regulatory sequences does resemble that in their prokaryotic counterparts. In line with this idea is the finding (Gatenby, Castleton & Saul, 1981) that chloroplast genes are properly expressed when introduced into bacteria. However, it is also clear that cp-DNA is not completely prokaryotic in character, e.g. the split structure of several tRNA genes. Studies with the homologous RNA polymerase are required to settle these questions.

It appears that the base sequences of the structural genes discussed so far are well conserved during evolution. This makes the use of heterologous probes possible in determining the positions of the corresponding genes on cp-DNAs of different plants. Using this method, Link (1981) could establish the position of the genes for the LSU and '32 kD' protein on the physical map of *Sinapsis* cp-DNA. Palmer (1982) has shown that these genes occupy similar positions on *Atriplex* and *Cucumis* cp-DNA. We have recently established that the α, β and ϵ genes on *Spirodela* cp-DNA are located at map positions corresponding to those on *Spinacia* cp-DNA. Moreover using cloned *Spirodela* cp-DNA fragments we could show that *Spirodela* and *Petunia* cp-DNA share a common sequence arrangement around the entire molecule (de Heij *et al.*, 1983). Similar findings are announced by Palmer (1982) for *Spinacia*, *Cucumis* and *Atriplex* and reported by Fluhr & Edelman (1981) for *Nicotiana* and *Spinacia*. It seems therefore very likely that cp-DNAs from most higher plants have a similar base sequence arrangement and that the general structure for these cp-DNAs can be summarized as shown in Fig. 3. Genes for rRNAs and tRNAs are located in similar positions and a common arrangement of protein genes is found in most cases.

A detailed comparison of organization of a large number of cp-DNAs by

Palmer & Thompson (1981, 1982) shows that, compared to *Spinacia* cp-DNA, a 50 kb area, spanning the position of the α gene to the LSU gene, is inverted in *Vigna*. *Zea*, in its turn, differs from *Spinacia* by a 10 kb inversion around the α gene. They attribute the overall stability of the cp-DNA organization to the presence of the inverted repeat. The absence of this repeat in *Pisum*, *Vicia* and other members of the Leguminosae family leads to a scrambling of the informational order, compared with the other cp-DNAs (Palmer & Thompson, 1981).

The presence of an inverted repeat causes cp-DNA in *Phaseolus* to exist in two orientations (Palmer, 1983). Although not shown for other cp-DNAs, it is likely that this is a common property, suggesting that recombination does occur within chloroplasts.

Further studies are required to confirm this current picture. In the near future an extended gene map can be expected. Moreover, the comparison between different cp-DNA base sequences possibly involved in regulatory phenomena will be particularly interesting.

References

Allet, B. & Rochaix, J.-D. (1979). Structure analysis at the ends of the intervening DNA sequences in the chloroplast 23S ribosomal genes of *C. reinhardtii*. *Cell*, **18**, 55–60.

Alt, J., Westhoff, P., Sears, B. B., Nelson, N., Hurt, E., Hauska, G. & Herrmann, R. G. (1983*a*). Genes and transcripts for the polypeptides of the cytochrome b_6/f complex from spinach thylakoid membranes. *EMBO J.*, **2**, 979–86.

Alt, J., Winter, P., Sebald, W., Moser, J. G., Schedel, R., Westhoff, P. & Herrmann, R. G. (1983*b*). Localization and nucleotide sequence of the gene for the ATP synthase proteolipid subunit on the spinach plastid chromosome. *Curr. Genet.*, **7**, 129–38.

Bauer, E. (1909). Das Wesen und die Erblichkeitsverhältnisse der 'Varietates albomarginateae hort' von *Pelargonium zonale*. *Z. VererbLehre*, **1**, 333–51.

Bedbrook, J. R. & Bogorad, L. (1976). Endonuclease recognition sites mapped on *Zea mays* chloroplast DNA. *Proc. natn. Acad. Sci. USA*, **73**, 4309–13.

Bedbrook, J. R., Coen, D. M., Beaton, A. R., Bogorad, L. & Rich, A. (1979*a*). Location of the single gene for the large subunit of ribulose-bisphosphate carboxylase on the maize chloroplast chromosome. *J. biol. Chem.*, **254**, 905–10.

Bedbrook, J. R., Kolodner, R. & Bogorad, L. (1979*b*). *Zea mays* chloroplast ribosomal RNA genes are part of a 22000 base pair inverted repeat. *Cell*, **11**, 739–49.

Bedbrook, J. R., Link, G., Coen, D. M., Bogorad, L. & Rich, A. (1978). Maize plastid gene expressed during photoregulated development. *Proc. natn. Acad. Sci. USA*, **75**, 3060–4.

Blair, G. E. & Ellis, R. J. (1973). Protein synthesis in chloroplasts. I. Light-driven synthesis of the large subunit of Fraction I protein by isolated pea chloroplasts. *Biochim. biophys. Acta*, **319**, 223–34.

Bohnert, H. J., Crouse, E. J. & Schmitt, J. (1982). Organization and expression of plastid genomes. In *Encyclopedia of Plant Physiology: Nucleic Acids and Proteins*, vol. 14B, ed. D. Boulter & B. Parthier, pp. 475–530. Springer Verlag, Berlin.

Bottomley, W. & Whitfeld, P. R. (1979). Cell-free transcription and translation of total spinach chloroplast DNA. *Eur. J. Biochem.*, **93**, 31–9.

Bovenberg, W. A., Kool, A. H. & Nijkamp, H. J. J. (1981). Isolation, characterization and restriction endonuclease mapping of the *Petunia hybrida* chloroplast DNA. *Nucleic Acids Res.*, **9**, 503–17.

Bowman, C. M., Koller, B., Delius, H. & Dyer, T. A. (1981). A physical map of wheat chloroplast DNA showing the location of the structural genes for the ribosomal RNAs and the large subunit of ribulose-1,5-bisphosphate carboxylase. *Molec. gen. Genet.*, **183**, 93–101.

Coen, D. M., Bedbrook, J. R., Bogorad, L. & Rich, A. (1978). Maize chloroplast DNA fragment encoding the large subunit of ribulose-bisphosphate carboxylase. *Proc. natn. Acad. Sci. USA*, **74**, 5487–91.

de Heij, J. T., Lustig, H., Moeskops, D.-J. M., Bovenberg, W. A., Bisanz, C. & Groot, G. S. P. (1983). Chloroplast DNAs of *Spinacia*, *Petunia* and *Spirodela* have a similar gene organization. *Curr. Genet.*, **7**, 1–6.

Deno, H., Shinozaki, K. & Sugiura, M. (1983). Nucleotide sequence of tobacco chloroplast gene for the α-subunit of proton-translocating ATPase. *Nucleic Acids Res.*, **11**, 2185–93.

Doherty, A. & Gray, J. C. (1979). Synthesis of cytochrome *f* by isolated pea chloroplasts. *Eur. J. Biochem.*, **98**, 87–92.

Driesel, A. M., Crouse, E. J., Gordon, K., Bohnert, H. J., Herrmann, R. G., Steinmetz, A., Mubumbila, M., Keller, M., Burkard, G. & Weil, J. H. (1979). Fractionation and identification of spinach chloroplast transfer RNAs and mapping of their genes on the restriction map of chloroplast DNA. *Gene*, **6**, 285–306.

Driesel, A. J., Speirs, J. & Bohnert, H. J. (1980). Spinach chloroplast mRNA for a 32000 dalton polypeptide. Size and localization on the physical map of the chloroplast DNA. *Biochim. biophys. Acta*, **610**, 297–310.

Dyer, T. A., Bowman, C. M. & Payne, P. I. (1977). The low-molecular-weight RNAs of plant ribosomes: their structure, function and evolution. In *Nucleic Acids and Protein Synthesis in Plants*, ed. L. Bogorad & J. H. Weil, pp. 121–3. Plenum Press, New York and London.

Edelman, M. (1981). Nucleic acids of chloroplasts and mitochondria. In *The Biochemistry of Plants*, vol. 6, ed. A. Marcus, pp. 249–301. Academic Press, New York.

Edelman, M. & Reisfeld, A. (1978). Characterization, translation and control of the 32000 dalton chloroplast membrane in *Spirodela*. In *Chloroplast Development*, ed. G. Akoyunoglou & J. H. Argyroudi-Akoyunoglou, pp. 641–53. Elsevier/North-Holland, Amsterdam.

Edwards, K. & Kössel, H. (1981). The rRNA operon from *Zea mays* chloroplasts: nucleotide sequence of 23S rDNA and its homology with *E. coli* 23S rDNA. *Nucleic Acids Res.*, **9**, 2853–69.

Eneas-Filho, J., Hartley, M. R. & Mache, R. (1981). Pea chloroplast ribosomal proteins: characterization and site of synthesis. *Molec. gen. Genet.*, **184**, 484–8.

Fluhr, R. & Edelman, M. (1981). Conservation of sequence arrangement among higher plant chloroplast DNAs: molecular cross hybridization among the Solanaceae and between *Nicotiana* and *Spinacia*. *Nucleic Acids Res.*, **9**, 6841–53.

Gatenby, A. A., Castleton, J. A. & Saul, M. W. (1981). Expression in *E. coli* of maize and wheat chloroplast genes for large subunit of ribulose bisphosphate carboxylase. *Nature*, **291**, 117–21.

Gordon, K. H. J., Crouse, E. J., Bohnert, H. J. & Herrmann, R. G. (1981). Restriction endonuclease cleavage site map of chloroplast DNA from *Oenothera parviflora* (Eunoenothera plastome IV). *Theoret. appl. Genet.*, **59**, 281–96.

Grivell, L. A. (1981). Mitochondrial genes at Cold Spring Harbor. *Curr. Genet.*, **4**, 167–71.

Groot, G. S. P. & van Harten-Loosbroek, N. (1981). Physical mapping of 4S RNA genes on chloroplast DNA of *Spirodela oligorhiza*. *Curr. Genet.*, **4**, 187–90.

Haff, L. A. & Bogorad, L. (1976). Hybridization of maize chloroplast DNA with transfer RNA. *Biochemistry*, **15**, 4105–9.

Herrmann, R. G. & Possingham, J. V. (1981). Plastid DNA – the plastome. In *Results and Problems in Cell Differentiation*, vol. 10, ed. J. Reinert, pp. 45–96. Springer Verlag, Berlin.

Herrmann, R. G., Westhoff, P., Alt, J., Winter, P., Tittgen, J., Bisanz, C., Sears, B. B., Nelson, N., Hurt, E., Hauska, G., Viebrock, A. & Sebald, W. (1983). Identification and characterization of genes for polypeptides of the thylakoid membrane. In *Structure and Function of Plant Genomes*, ed. O. Ciferri, pp. 143–54. Academic Press, New York.

Herrmann, R. G., Whitfeld, P. R. & Bottomley, W. (1980). Construction of a SalI/PstI restriction map of spinach chloroplast DNA using low-gelling-temperature-agarose electrophoresis. *Gene*, **8**, 179–91.

Howe, C. J., Bowman, C. M., Dyer, T. A. & Gray, J. C. (1982). Localization of wheat chloroplast genes for the beta and epsilon subunits of ATP synthase. *Molec. gen. Genet.*, **186**, 525–30.

Jurgenson, J. E. & Bourque, D. P. (1980). Mapping of rRNA genes in an inverted repeat in *Nicotiana tabacum* chloroplast DNA. *Nucleic Acids Res.*, **8**, 3505–16.

Kastelein, R. A., Remaut, E., Fiers, W. & van Duin, J. (1982). Lysis gene expression of RNA phage MS 2 depends on a frameshift during translation of the overlapping coat protein gene. *Nature*, **295**, 35–41.

Keus, R. J. A., Roovers, D. J., van Heerikhuizen, H. & Groot, G. S. P. (1983). Molecular cloning and characterization of the chloroplast ribosomal RNA genes from *Spirodela oligorhiza*. *Curr. Genet.*, **7**, 7–12.

Koch, W., Edwards, K. & Kössel, H. (1981). Sequencing of the 16S–23S spacer in a ribosomal RNA operon of *Zea mays* chloroplast DNA reveals two split tRNA genes. *Cell*, **25**, 203–13.

Koller, B. & Delius, H. (1980). *Vicia faba* chloroplast DNA has only one set of ribosomal RNA genes as shown by partial denaturation mapping and R-loop analysis. *Molec. gen. Genet.*, **178**, 261–9.

Krebbers, E. T., Larrinua, I. M., McIntosh, L. & Bogorad, L. (1982). The maize chloroplast genes for the β and ϵ subunits of the photosynthetic coupling factor CF_1 are fused. *Nucleic Acids Res.*, **10**, 4985–5002.

Link, G. (1981). Cloning and mapping of the chloroplast DNA sequences for two messenger RNAs from mustard (*Sinapis alba* L.). *Nucleic Acids Res.*, **9**, 3681–94.

Link, G., Chambers, S. E., Thompson, J. A. & Falk, H. (1981). Size and physical organization of chloroplast DNA from mustard (*Sinapis alba* L.). *Molec. gen. Genet.*, **181**, 454–7.

Link, G., Coen, D. M. & Bogorad, L. (1978). Differential expression of the gene for the large subunit of ribulose bisphosphate carboxylase in maize leaf cell types. *Cell*, **15**, 725–31.

McIntosh, L., Poulsen, C. & Bogorad, L. (1980). Chloroplast gene sequence for the large subunit of ribulose bisphosphate carboxylase of maize. *Nature*, **288**, 556–60.

Mendiola-Morgenthaler, L. R., Morgenthaler, J. J. & Price, C. A. (1976). Synthesis of coupling factor CF_1 protein by isolated spinach chloroplasts. *FEBS Letts.*, **62**, 96–100.

Nechushtai, R., Nelson, N., Mattoo, A. K. & Edelman, M. (1981). Site of synthesis of subunits to photosystem I reaction center and the proton-ATPase in *Spirodela. FEBS Lett.*, **125**, 115–18.

Palmer, J. D. (1982). Physical and gene mapping of chloroplast DNA from *Atriplex triangularis* and *Cucumis sativa. Nucleic Acids Res.*, **10**, 1593–605.

Palmer, J. D. (1983). Chloroplast DNA exists in two orientations. *Nature*, **301**, 92–3.

Palmer, J. D. & Thompson, W. F. (1981). Rearrangements in the chloroplast genomes of mung bean and pea. *Proc. natn. Acad. Sci. USA*, **78**, 5533–7.

Palmer, J. D. & Thompson, W. F. (1982). Chloroplast DNA rearrangements are more frequent when a large inverted repeat sequence is lost. *Cell*, **29**, 537–50.

Palmer, J. D. & Zamir, D. (1982). Chloroplast DNA evolution and phylogenetic relationships in *Lycopersicon. Proc. natn. Acad. Sci. USA*, **79**, 5006–10.

Ris, H. & Plaut, W. (1962). Ultrastructure of DNA-containing areas in the chloroplast of *Chlamydomonas. J. Cell Biol.*, **13**, 383–91.

Schwarz, Z. & Kössel, H. (1980). The primary structure of 16S rDNA from *Zea mays* chloroplast is homologous to *E. coli* 16S rRNA. *Nature*, **283**, 739–42.

Schwarz, Z., Kössel, H., Schwarz, E. & Bogorad, L. (1981). A gene coding for $tRNA^{val}$ is located near the 5′ terminus of 16S rRNA gene in *Zea mays. Proc. natn. Acad. Sci. USA*, **78**, 4748–52.

Sebald, W., Hoppe, J. & Wachter, E. (1979). Amino acid sequence of the ATPase proteolipid from mitochondria, chloroplasts and bacteria (wild type and mutants). In *Function and Molecular Aspects of Biomembrane Transport*, ed. E. Quagliariello, E. Palmieri, S. Papa & M. Klingenberg, pp. 63–74. Elsevier/North-Holland, Amsterdam.

Shinozaki, K. & Sugiura, M. (1982). Sequence of the intercistronic region between the LSU and coupling factor β subunit gene. *Nucleic Acids Res.*, **10**, 4923–34.

Southern, E. (1975). Detection of specific sequences among DNA fragments separated by gel electrophoresis. *J. molec. Biol.*, **98**, 503–17.

Steinback, K. E., McIntosh, L., Bogorad, L. & Arntzen, C. J. (1981). Identification of the triazine receptor protein as a chloroplast gene product. *Proc. natn. Acad. Sci. USA*, **78**, 7463–7.

Steinmetz, A., Gubbins, E. J. & Bogorad, L. (1982). The anticodon of the maize chloroplast gene for $tRNA_{UAA}^{Leu}$ is split by a large intron. *Nucleic Acids Res.*, **10**, 3027–37.

Steinmetz, A. & Weil, J. H. (1976). Hybridization of bean chloroplast transfer RNAs to chloroplast DNA. *Biochim. Biophys. Acta*, **454**, 429–35.

Subramanian, A., Steinmetz, A. & Bogorad, L. (1983). Maize chloroplast DNA encodes a protein sequence homologous to the bacterial ribosome assembly protein S4. *Nucleic Acids Res.*, **15**, 5277–87.

Sugita, M. & Sugiura, M. (1983). A putative gene of tobacco chloroplast coding for a ribosomal protein similar to *E. coli* ribosomal protein S19. *Nucleic Acids Res.*, **11**, 1913–18.

Takaiwa, F. & Sugiura, M. (1982a). The complete nucleotide sequence of a 23-S rRNA gene from tobacco chloroplasts. *Eur. J. Biochem.*, **124**, 13–19.

Takaiwa, F. & Sugiura, M. (1982b). Nucleotide sequence of the 16S–23S spacer region in an rRNA gene cluster from tobacco chloroplast DNA. *Nucleic Acids Res.*, **10**, 2665–76.

Tewari, K. K. & Wildman, S. G. (1968). Function of chloroplast DNA. I. Hybridization studies involving nuclear and chloroplast DNA with RNA from cytoplasmic (80S) and chloroplast (70S) ribosomes. *Proc. natn. Acad. Sci. USA*, **59**, 569–76.

Tewari, K. K. & Wildman, S. G. (1970). Information content in the chloroplast DNA. In *Control of Organelle Development*, ed. P. L. Miller, pp. 147–9. Cambridge University Press, Cambridge.

Tiboni, O., diPasquale, G. & Ciferri, O. (1978). Purification, characterization and site of synthesis of chloroplast elongation factors. In *Chloroplast Development*, ed. G. Akoyunoglou & J. H. Argyroudi-Akoyunoglou, pp. 675–8. Elsevier/North-Holland, Amsterdam.

Tohdoh, N. & Sugiura, M. (1982). The complete nucleotide sequence of a 16S ribosomal RNA gene from tobacco chloroplasts. *Gene*, **17**, 213–18.

van Ee, J. H., Vos, Y. J. & Planta, R. J. (1980). Physical map of chloroplast DNA of *Spirodela oligorrhiza*; analysis by the restriction endonucleases PstI, XhoI and SacI. *Gene*, **12**, 191–200.

Vedel, F. & Mathieu, C. (1983). Physical and gene mapping of chloroplast DNA from normal and cytoplasmic male-sterile (radish cytoplasm) lines of *Brassica napus*. *Curr. Genet.*, **7**, 13–20.

Weil, J. H., Guillemaut, P., Burkard, G., Canaday, J., Mubumbila, M., Osorio, M. L., Keller, M., Gloeckler, R., Steinmetz, A., Keith, G., Heiser, D. & Crouse, E. J. (1981). Comparative studies on chloroplast transfer RNAs: tRNA sequences and tRNA gene localization in the rDNA units.

In *Photosynthesis, Proc. 5th int. Congr. Photosynthesis*, ed. G. Akoyunoglou, pp. 777–86. Balaban International Science Services, Philadelphia.

Weil, J. H., Mubumbila, M., Kuntz, M., Keller, M., Steinmetz, A., Crouse, E. J., Burkard, G., Guillemaut, P., Selden, R., McIntosh, L., Bogorad, L., Löfelhardt, W., Mucke, H. & Bohnert, H. J. (1982). Gene mapping studies and sequence determination on chloroplast tRNAs from various photosynthetic organisms. In *Abstract Special FEBS Meeting in Arolla*, p. 53.

Westhoff, P., Nelson, N., Bünneman, H. & Herrmann, R. G. (1981). Localization of genes for coupling factor subunits on the spinach plastid chromosome. *Curr. Genet.*, **4**, 109–20.

Whitfeld, P. R. & Bottomley, W. (1983). Organization and structure of chloroplast genes. *A. Rev. Pl. Physiol.*, **34**, 279–327.

Zielinski, R. E. & Price, C. A. (1980). Synthesis of thylakoid membrane proteins by chloroplasts isolated from spinach. *J. Cell Biol.*, **85**, 435–45.

Zurawski, G., Bohnert, H. J., Whitfeld, P. R. & Bottomley, W. (1982*a*). Nucleotide sequence of the gene for the 32 kD thylakoid membrane protein from *S. oleracea* and *N. debuyi* predicts a totally conserved primary translation product of M_r 38950. *Proc. natn. Acad. Sci. USA*, **79**, 7699–703.

Zurawski, G., Bottomley, W. & Whitfeld, P. (1982*b*). Structures of the genes for the β and ϵ subunits of spinach chloroplast ATPase indicate a dicistronic mRNA and an overlapping translation stop/start signal. *Proc. natn. Acad. Sci. USA*, **79**, 6260–4.

Zurawski, G., Perrot, B., Bottomley, W. & Whitfeld, P. R. (1981). The structure of the gene for the large subunit of ribulose-1,5-bisphosphate carboxylase from spinach chloroplast DNA. *Nucleic Acids Res.*, **9**, 3251–70.

E. J. CROUSE, H. J. BOHNERT AND
J. M. SCHMITT

Chloroplast RNA synthesis

Structural and functional aspects of the organization and expression of plastid genomes have been extensively studied during the past several years. Chloroplasts contain multiple copies of a circular DNA molecule. Physical maps of the restriction endonuclease cleavage sites on the chloroplast genome from a variety of plant species have been constructed. The rRNA genes, most of the tRNA genes and several genes coding for proteins have been located on defined segments of these maps. The nucleotide sequences of some of these genes have been determined.

In comparison, relatively little is known about the mechanisms of chloroplast gene transcription, their control elements, or regulation of gene expression either in different plastid types or at different stages of chloroplast development. Initially, studies on transcription of chloroplast DNA (cp-DNA) were performed mainly on two different levels. First, reports on chloroplast RNA (cp-RNA) synthesis *in vivo* concerned the study of developmental events. Secondly, investigations on RNA synthesis *in organello*, using intact chloroplasts, were performed to measure the transcription capacity of cp-DNA. Precise methods for distinguishing a single mRNA among the many RNAs in a cell have not been fully developed. Some information on the transcription of specific genes has been obtained for only the most abundantly transcribed RNAs. Consequently it has been difficult clearly to relate identified RNA or protein components, and phenotypic changes associated with them, to chloroplast genes and their transcription and translation.

Recently, more detailed studies on the mechanisms, control and co-ordination of chloroplast gene expression have become possible using a combination of different methods *in vitro*. These general techniques have been adapted and optimized to chloroplast research (Edelman, Hallick & Chua, 1982). With the introduction of recombinant DNA technology, specific transcripts of segments of the chloroplast genome can be studied. Coupled transcription–translations of cloned cp-DNA fragments are presently being carried out in heterologous systems *in vitro*. It is anticipated that even more

83

information will be obtained by using homologous transcription and transcription–translation systems derived from chloroplasts.

By reference to selected papers, this review article focuses on cp-RNAs and cp-RNA synthesis. Emphasis is put on the expression of chloroplast genes or gene families and components of the transcription machinery, as well as the available experimental systems which may further our understanding of control mechanisms on a molecular basis. Review articles concerning these and related subjects are as follows: Whitfeld (1977), Kirk & Tilney-Bassett (1978), Becker (1979), Bedbrook & Kolodner (1979), Hall (1979), Leaver (1979), Herrmann & Possingham (1980), Wollgiehn & Parthier (1980), Dyer & Leaver (1981), Bartlett, Boynton & Gillham (1981), Edelman (1981), Ellis (1981), Gillham & Boynton (1981), Guilfoyle (1981), Rochaix (1981), Steinback (1981), Bohnert *et al.* (1982), Bottomley & Bohnert (1982), Buetow (1982), Delihas & Andersen (1982), Dyer (1982), Gray & Doolittle (1982), Grierson (1982), Wallace (1982), Bogorad *et al.* (1983), Ellis (1983), Kozak (1983), Whitfeld & Bottomley (1983), Hallick (1984), Rochaix (1984) and Groot (this volume). Control of transcription is reviewed elsewhere in this book (see Bennett *et al.*, Edelman *et al.*, Gallagher *et al.* and Tobin *et al.*, this volume).

Approaches for studying chloroplast RNA synthesis

Synthesis of cp-RNA has been studied using systems of different complexity. Studies *in vivo* of plants or unicellular organisms, experiments *in organello* using intact, functional chloroplasts and systems *in vitro* involving purified components or broken chloroplasts have been performed. The merits and limitations of these approaches will briefly be reviewed.

Studies in vivo

The earliest attempts to characterize transcription in chloroplasts looked at the situation *in vivo*. Whole organisms were incubated with radioactive precursors of RNA. Their appearance in different cellular compartments was monitored by autoradiography (for a review see Kirk & Tilney-Bassett, 1978) or the radioactivity of isolated and fractionated RNA was measured (Ingle, 1968; Detchon & Possingham, 1973; Hartley & Ellis, 1973; Munsche & Wollgiehn, 1973; Grierson & Loening, 1974; Heizmann, 1974; Miller & McMahon, 1974; Scott, 1976). These studies were the first to establish that transcription occurs in chloroplasts. In addition, the amount of cp-DNA transcribed at a given developmental state and/or under a given set of experimental conditions was measured by hybridization of cp-RNA to cp-DNA or to fragments of cp-DNA (Rawson, 1975; Chelm & Hallick, 1976; Rawson & Boerma, 1976; Howell & Walker, 1977; Chelm, Gray & Hallick,

1978; Chelm, Hallick & Gray, 1979; Matsuda & Surzycki, 1980; Oishi, Sumnicht & Tewari, 1981; Rawson *et al.*, 1981; Dix & Rawson, 1983; see also pp. 104–9). For example, by using saturating amounts of total pea (*Pisum sativum*) chloroplast [³H]RNA hybridized to pea cp-DNA, transcripts from 50% of the DNA, i.e. RNA molecules equivalent to a complete single strand of the cp-DNA, were found to be transcribed *in vivo* from light-grown plants (Oishi *et al.*, 1981). In this study, a reduction of about 2–3% of the total RNA hybridization was observed when cp-RNA from dark-grown plants was hybridized to cp-DNA. Only 0.15–0.2% of the transcripts from pea cp-DNA were found to contain poly(A) tracts. The conclusion that cp-RNA contains little or no poly(A) stretches had earlier been reached by Wheeler & Hartley (1975). In contrast, Haff & Bogorad (1976*b*) reported 6% of total poly(A)⁺-RNA from maize (*Zea mais*) seedlings hybridizing to cp-DNA. Milner, Hershberger & Buetow (1979) showed that 7.7% of the total cell poly(A)⁺-RNA from fully green *Euglena* hybridized to *Euglena* cp-DNA; however, poly(A) was not detected in the mature mRNAs prepared from chloroplast polysomes.

Problems with studies *in vivo* are two-fold. First, the contribution of other cellular compartments in the process of synthesis is not clear. The intracellular precursor pools and turn-over rates, as well as their compartmentalization and utilization by different biochemical pathways, are difficult to assess. Secondly, identification of the RNA products is critically dependent on cell fractionation. Cross-contamination between the chloroplast and nucleo-cytosolic compartments was not uncommon in early studies (see Kirk, 1971).

RNA synthesis in organello

Some of the problems encountered *in vivo* have been overcome by using isolated intact chloroplasts and light as the energy source (Hartley & Ellis, 1973; Bohnert, Schmitt & Herrmann, 1974). Intactness, i.e. the presence of an envelope, can conveniently be monitored by phase contrast microscopy (Kahn & von Wettstein, 1961) or interference contrast microscopy (Schmitt & Herrmann, 1977). Microscopic criteria do not, however, signify physiological competence of the organelles, which is best assessed by measuring rates of photosynthesis (see below). Systems *in organello* require radioactively labelled nucleosides or inorganic phosphate as precursors of RNA synthesis. These are taken up by the organelle and converted into ribonucleoside triphosphates which can be used for the synthesis of RNA. In this way, contaminating nuclear RNA-synthesizing activities can be elegantly circumvented, since nucleoside triphosphates do not cross the chloroplast envelope at appreciable rates (Heber & Heldt, 1981).

Systems *in organello* have the drawback of being prone to artifacts resulting from damage to the organelles during isolation or incubation. In our hands, the rate of CO_2 fixation dropped to approximately 30% of the initial rate after 2 h of incubation for RNA synthesis at 15 °C. Transcription also proceeds under high salt conditions (200 mmol l^{-1} KCl) which inhibit CO_2 fixation. The use of studies *in organello* may also be limited to studying elongation of RNA chains as RNA chain initiation has not yet been demonstrated (Bennett & Milewska, 1976).

However, this approach has allowed the identification of several proteins which are coded by the chloroplast genome. Isolated chloroplasts have been extensively used for studying both protein synthesis *in organello* (Blair & Ellis, 1973; Bottomley, Spencer & Whitfeld, 1974; for reviews see Ellis, 1977, 1981; Bottomley & Bohnert, 1982) and transport of cytosolically synthesized chloroplast proteins (Chua & Schmidt, 1978; Highfield & Ellis, 1978; Grossman, Bartlett & Chua, 1980; for a recent review see Ellis, 1983). In view of the lack of experimental proof that mRNAs from the cytosol are transported into chloroplasts and then translated, it is believed that all proteins made inside the organelle are encoded by the cp-DNA, and all proteins, or their precursors, transported across the envelope into the chloroplast are encoded in other compartments (Ellis, Smith & Barraclough, 1980; Ellis, 1981, 1983).

Systems in vitro

In this article, the term *in vitro* is used to describe studies either on chloroplasts lacking an intact envelope or on subchloroplast systems. Since no membrane barrier shields the components of the transcription machinery, these systems are open to a broad range of experimental manipulation. They can be probed or primed with heterologous components and thus fully reconstituted systems with well-characterized components should soon be available (see pp. 109–13).

As an intermediate step towards a reconstituted system *in vitro*, chloroplast transcription complexes have been prepared. Such complexes have been isolated, after the lysis of chloroplasts, from *Euglena* (Hallick *et al.*, 1976; Schiemann, Wollgiehn & Parthier, 1977) and from spinach (*Spinacia oleracea*) (Briat, Laulhere & Mache, 1979; Blanc, Briat & Laulhere, 1981; Briat *et al.*, 1982*b*). The complex consists of RNA polymerase still associated with other uncharacterized proteins and at least part of the chloroplast genome. The complex from *Euglena* was shown to initiate RNA transcripts *in vitro* at a few sites on the endogenous template (Rushlow *et al.*, 1980) leading to transcripts larger than marker 5S rRNA. The analysis of RNase digestion products was consistent with selective rather than random initiation. When the transcripts were hybridized back to filter-immobilized *Euglena* cp-DNA fragments, those fragments carrying the rRNA genes became predominantly labelled (Fig. 1).

Transcription of cloned cp-DNA fragments has also been studied using *Escherichia coli* RNA polymerase holoenzyme (Zech, Hartley & Bohnert, 1981; Koller, Delius & Dyer, 1982; Briat *et al.*, 1982*a*) and eukaryotic RNA polymerase III (Gruissem *et al.*, 1982). The usefulness of the prokaryotic system, however, is limited. For example, it could be shown that under the experimental conditions used, the bacterial enzyme initiates transcripts from an established, sequenced promoter (Zurawski *et al.*, 1982*a*), but in both directions (H. J. Bohnert & M. Zech, unpublished). The use of homologous transcriptional components will be reviewed later (pp. 109–13).

The use of inhibitors

Inhibitors have been used in all three types of experiments described above. An observed reduction in RNA synthesis does not necessarily mean

Fig. 1. Hybridization of radioactively labelled RNA, synthesized *in vitro* from a transcriptional complex isolated from *Euglena* chloroplasts, to filter-immobilized Eco-RI-generated *Euglena* cp-DNA fragments.

Hybridization occurs exclusively with DNA fragments harboring rRNA genes (from Rushlow *et al.*, 1980; with permission). Fragment nomenclature (*left*) and sizes in kbp (*right*) are indicated.

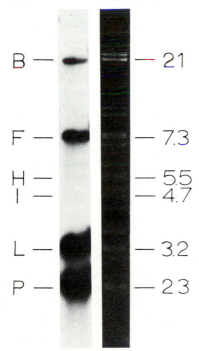

B —	— 21
F —	— 7.3
H —	— 5.5
I —	— 4.7
L —	— 3.2
P —	— 2.3

that the primary cellular target is RNA synthesis proper. In Table 1, the effects of some inhibitors on RNA synthesis and on the reduction of phosphoglyceric acid (PGA) *in organello* are compared. While actinomycin D, chromomycin and distamycin inhibit RNA synthesis, but not the complex process of PGA reduction, ethidium bromide in low concentrations inhibits both processes. It is concluded that in this case the effect on RNA synthesis is probably indirect, because of the interference with other organellar reactions. This view is corroborated by the fact that ethidium bromide at similar concentrations ($5 \mu mol \, l^{-1}$) does not inhibit RNA synthesis promoted by isolated transcription complexes using nucleoside triphosphates as precursors (Hallick *et al.*, 1976; Briat *et al.*, 1979; Briat & Mache, 1980; see also Galling, 1982 for a review). Likewise, rifampicin inhibited *Euglena* cp-RNA synthesis *in vivo* but not the RNA polymerase/DNA transcription complexes from *Euglena* and spinach (Hallick *et al.*, 1976; Briat *et al.*, 1979) or the isolated RNA polymerase from maize chloroplasts (Bottomley, Smith & Bogorad, 1971). For similar observations on inhibitors of translation see Ellis (1976, 1977, 1982).

Synthesis of chloroplast RNAs
Synthesis of rRNAs

The 70S ribosomes of chloroplasts contain three (*Euglena*), four (higher plants) or five (*Chlamydomonas*) RNA molecules which differ in size and nucleotide sequence (Whitfeld, 1977; Dyer & Leaver, 1981; Edelman, 1981; Rochaix, 1981; Bohnert *et al.*, 1982; Buetow, 1982; Dyer, 1982;

Table 1. *Inhibition of RNA synthesis and phosphoglyceric acid (PGA) reduction in isolated intact chloroplasts*

	Concentration ($\mu g \, ml^{-1}$)	RNA synthesis (%)	PGA reduction (%)
Isolated chloroplasts		100	100
+Actinomycin D	20	53	100
+Chromomycin	2	27	n.d.
+Chromomycin	20	n.d.	100
+Distamycin	2	22	100
+Ethidium bromide	10	n.d.	22
+Ethidium bromide	2	10	n.d.

Isolation of chloroplasts and measurement of RNA synthesis as described by Bohnert *et al.* (1974). PGA reduction was determined with an oxygen electrode using a substrate concentration of $2 \, mmol \, l^{-1}$. Inhibitors were added about 2 min after the start of the reaction by illumination. Uninhibited reaction rates were 50 pmol [^{3}H]uridine $\times mg^{-1}$ chlorophyll $\times h^{-1}$ and 87 μmol PGA reduced $\times mg^{-1}$ chlorophyll $\times h^{-1}$; n.d. = not determined.

Whitfeld & Bottomley, 1983). The 30S ribosomal subunit particle contains 16S rRNA. In higher plants, the 50S ribosomal subunit contains 23S and 5S rRNAs, as well as 4.5S rRNA (Whitfeld *et al.*, 1978). In the algae *Chlamydomonas* and *Euglena*, no counterpart to the higher plant 4.5S rRNA has been demonstrated in addition to the 23S and 5S rRNAs (Gray & Hallick, 1979; Rochaix, 1981). Instead, the large ribosomal subunit of *Chlamydomonas* contains two different small rRNAs of 3S and 7S, respectively (Rochaix & Malnoë, 1978; Rochaix, 1981). Their genes have been mapped close to the 5′ end of the 23S rRNA gene. The 3S and 7S rRNAs have not been detected in any other chloroplast ribosome. The precise sizes of these rRNAs have been determined from rDNA and rRNA sequencing data (listed in Table 2). The 23S rRNA is about 2800 to 2900 nucleotides long, the 16S rRNA is about 1500 nucleotides and the 5S rRNA is about 120 nucleotides. The 5S rRNA varies in length even in the same plant (see Table 2 and Erdmann *et al.*, 1983). The 4.5S rRNA from various higher plants and from the fern *Dryopteris acuminata* are about 63 to 106 nucleotides in length, depending on the species (Table 2). Furthermore, the 4.5S rRNA has even been found to have variable length in the same plant, e.g. in *Lemna minor*, where related molecules of about 63, 96 and 103 nucleotides have been seen. The function of 4.5S rRNA in chloroplasts is unclear. As inferred from the location of its gene adjacent to the 3′ end of the 23S rRNA gene, the 4.5S rRNA might once have been contiguous with the 23S rRNA which is known to contain hidden nicks of unknown function (Whitfeld, 1977). A sequence comparison between a tobacco (*Nicotiana tabacum*) chloroplast 4.5S rRNA (Takaiwa & Sugiura, 1980*a*, 1980*b*) and the *E. coli* 23S rRNA sequence showed homologies at the 3′ end of the 23S rRNA (MacKay, 1981). Regions of homology were also observed by Machatt *et al.* (1981), who found that the 4.5S rRNA sequences of tobacco and wheat showed common short sequence stretches not only with the bacterial 23S rRNA, but also with the 3′ end of eukaryotic 28S rRNA, and that a highly similar secondary structure can be constructed for all three types of rRNAs. A short sequence at the 3′ end of both the 4.5S rRNAs from wheat (Wildeman & Nazar, 1980) and tobacco (Takaiwa & Sugiura, 1980*a*) may form a double-stranded stem with the 5′ end of the *Euglena* 23S rRNA (Orozco *et al.*, 1980*a*, 1980*b*). This secondary structure is also possible between the 3′ end of the 4.5S rRNA and the 5′ end of the 23S rRNA of maize (Edwards *et al.*, 1981). Although the 4.5S rRNA may be a breakdown product, it is also possible that it represents an end-product of the chloroplast 23S rRNA maturation (Hartley, 1979; Hartley & Head, 1979) which is entrapped in the assembled ribosome. It can also not be excluded that the function fulfilled by the uninterrupted bacterial 23S rRNA has been split into two genes (or has been retained as two chloroplast genes) during evolution.

Table 2. *Size of some chloroplast rRNAs as revealed from DNA or RNA sequencing data*

Type of rRNA	Gene	Plant source	Size of rRNA (nucleotides)	Intron	Reference
23S	23S rDNA	Tobacco	2804	No	Takaiwa & Sugiura, 1982a
		Maize	2890	No	Edwards & Kössel, 1981
		Chlamydomonas		0.87 kbp[a]	Rochaix & Malnoë, 1978; Rochaix, 1981
16S	16S rDNA	Tobacco	1486	No	Tohdoh & Sugiura, 1982
		Maize	1491	No	Schwarz & Kössel, 1980
		Euglena	1491	No	Graf, Roux & Stutz, 1982
		Chlamydomonas	1475	No	Dron, Rahire & Rochaix, 1982b
7S	7S rDNA	*Chlamydomonas*	282	No	Rochaix & Darlix, 1982
5S	5S rDNA	Spinach	121	—	Delihas *et al.*, 1981
			122	—	Pieler *et al.*, 1982
		Tobacco	119	—	Dyer & Bowman, 1979;
			120	No	Takaiwa & Sugiura, 1980a;
			121	—	1981
		Broad bean	122	—	Dyer & Bowman, 1979
		Dwarf bean	120	—	Dyer & Bowman, 1979
		Maize	122	No	Dyer & Bedbrook, 1980
		Lemna minor	119	—	Dyer & Bowman, 1979
			121	—	
		Spirodela oligorhiza	120	No	Keus *et al.*, 1983b
		Dryopteris acuminata	119	—	Takaiwa & Sugiura, 1982b
			120	—	
			122	—	
		Euglena	*ca* 121	No	El-Gewely *et al.*, 1984
4.5S	4.5S rDNA	Spinach	106	—	Kumagai *et al.*, 1982
		Tobacco	101	—	Bowman & Dyer, 1979
			103	No	Takaiwa & Sugiura, 1980a, c
		Broad bean	72	—	Bowman & Dyer, 1979
		Dwarf bean	103	—	Bowman & Dyer, 1979
		Maize	95	No	Edwards *et al.*, 1981
		Wheat	96	—	Wildeman & Nazar, 1980
		Lemna minor	*ca* 63	—	Bowman & Dyer, 1979
			96	—	
			103	—	
		Spirodela oligorhiza	102	No	Keus *et al.*, 1983b
		Dryopteris acuminata	103	—	Takaiwa, Kusuda & Sugiura, 1982
3S	3S rDNA	*Chlamydomonas*	47	No	Rochaix & Darlix, 1982

[a] Determined by electron microscopy and by hybridization of purified rRNA to various rDN restriction fragments.

Chloroplasts can be isolated after labelling *in vivo* and the RNAs extracted from them. After a short pulse of [^{32}P]orthophosphate, RNA from chloroplast preparations of young tobacco or spinach leaves was fractionated into size classes on polyacrylamide gels (Hartley & Ellis, 1973; Munsche & Wollgiehn, 1973; Mache *et al.*, 1978). Two prominent peaks of radioactivity were observed. The labelled RNAs were slightly larger in molecular mass, 1.25 and 0.65 MD, respectively, than the 23S and 16S rRNA peaks, approximately 1.08 and 0.56 MD, respectively, as detected in an optical density profile. There was a possibility that the radioactivity might be the result of contaminating cytosolic rRNAs (25S and 18S rRNAs) which have peaks approximately comparable in size to those of the labelled RNA (Whitfeld, 1977; Dyer & Leaver, 1981). When the short pulse-labelling period was followed by a longer time of incubation with unlabelled phosphate, the radioactivity could be chased towards the position of the 23S and 16S rRNAs. While the initial synthesis was inhibited by actinomycin D applied together with the [^{32}P]orthophosphate, the maturation step was not inhibited. Similar labelling experiments were also reported using other higher plants (Posner & Rosner, 1975) and algae, including *Euglena* (Carritt & Eisenstadt, 1973; Heizmann, 1974; Scott, 1976; Wollgiehn & Parthier, 1979) and *Chlamydomonas* (Miller & McMahon, 1974).

Further information on the mode of rRNA synthesis was obtained by monitoring RNA synthesis *in organello* (Hartley & Ellis, 1973; Bohnert *et al.*, 1974). After the cp-RNA products were isolated, purified and separated on low percentage polyacrylamide gels, the optical density profiles showed the two abundant stable large RNAs and the 4S RNA peak. In addition, several minor peaks could be discerned, giving molecular masses of 1.25, 0.65, and 0.45 MD (corresponding to approximately 25S, 18S, and 14S, respectively).

The radioactivity profile and the optical density profile do not coincide (Bohnert *et al.*, 1974; Bohnert, Driesel & Herrmann, 1976). The fit is best after a short time of labelling, e.g. for 20 min (Fig. 2). With longer incubation times, as more radioactivity has accumulated, a main peak of radioactivity was observed, corresponding to a molecular mass of approximately 2.7 MD (Fig. 2). This region was increasingly labelled for several hours at 15 °C while the incorporation into other RNA populations was decreased or eventually stopped (Hartley, Head & Gardiner, 1977; Hartley & Head, 1979; Bohnert *et al.*, 1976, 1977). Apparently, isolated chloroplasts process the primary transcript only for a short time whereas the incorporation of radioactivity into the 2.7 MD transcript proceeds for longer. It is not yet known whether this incorporation results from the initiation of new transcripts or the elongation of existing ones. The latter possibility is made more likely because Bennett & Milewska (1976) did not find that initiation occurred.

Fig. 2. Distribution of radioactivity in chloroplast RNA after labelling *in organello* of chloroplasts with [³H]uridine in the light.

Chloroplasts were labelled for 15, 25, 45, and 70 min, respectively (radioactivity distributions from bottom to top); the RNAs were isolated and separated on 2.4% cylindrical polyacrylamide gels which were cut into slices for counting. Some molecular weights and the positions of 23S, 16S, and 4S RNAs are indicated.

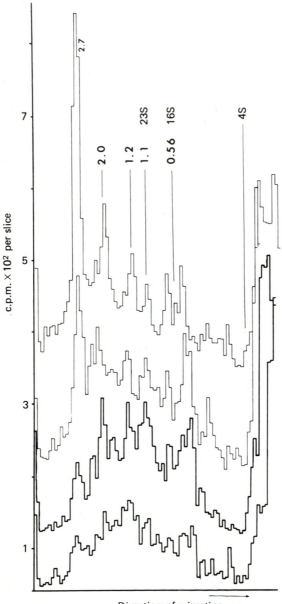

In addition, two peaks of radioactivity which were intermediary between the 2.7 MD and the 1.25 MD optical density peaks could be discerned. Radioactivity was also found at positions corresponding to the 1.25 MD and the 0.65 MD peaks, which were considered to be the immediate precursors of the two large rRNAs. By RNA/DNA hybridization the 2.7 MD RNA was shown to be a transcript of cp-DNA. By saturation–hybridization of this transcript to cp-DNA, it could be calculated that two copies were present per chromosome (Bohnert *et al.*, 1976; Schmitt *et al.*, 1981). Upon addition of unlabelled, purified chloroplast 23S and 16S rRNAs as competitors in the hybridization reaction, it was shown that the 2.7 MD RNA contains one copy of both 23S and 16S RNAs (Hartley *et al.*, 1977; Hartley & Head, 1979; Bohnert *et al.*, 1977). From the same type of experiment it was also suggested that small rRNAs, such as the 4.5S and 5S, could be present in this 2.7 MD transcript of the rDNA unit.

Additional radioactively labelled material, migrating in the gel at a position corresponding to a molecular mass of about 0.45 MD, was not easily explained. This 0.45 MD RNA showed labelling kinetics similar to those of the 2.7 MD RNA and not like the precursor (p) 23S rRNA. When unlabelled 0.45 MD RNA, as well as 16S and 23S rRNA, was used to compete with the 2.7 MD RNA labelled *in organello*, the reduction in the plateau-hybridization revealed the presence of the 0.45 MD component(s) in the 2.7 MD RNA. The 0.45 MD size class represents, however, a heterogeneous population of RNA molecules. It normally contains some specific breakdown products of the 23S rRNA and mRNA (see below; Rosner *et al.*, 1975; Speirs & Grierson, 1978; Driesel, Speirs & Bohnert, 1980).

The transcription-processing model (Fig. 3) of the chloroplast rDNA unit could be deduced only by combining time-dependent RNA synthesis *in*

Fig. 3. Processing pathway of higher plant chloroplast rRNAs (according to Bohnert *et al.*, 1977; Hartley *et al.*, 1977; Hartley & Head, 1979; Zenke *et al.*, 1982).

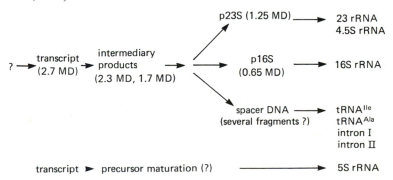

organello with saturation- and competition-hybridization experiments, and with recent work involving hybridization of the transcript to filter-immobilized DNA or DNA fragments. Results vary as to whether or which of the small rRNAs are part of the 2.7 MD transcript. Isotopic label added to isolated chloroplasts does not immediately appear in the 4.5S rRNA (Hartley, 1979). It was suggested that the 4.5S rRNA persisted as part of the p23S rRNA. However, the 5S rRNA is transcribed as such in isolated organelles (Hartley, 1979). Even after 20 min labelling in the light, taking into account the lag-phase of 10 min, the position of 5S rRNA in gels shows radioactivity (Fig. 4). This points either to initial rapid processing, if the sequence for 5S RNA appears

Fig. 4(*a–b*). Synthesis of low molecular weight RNA *in organello* and *in vitro* and processing of transcripts.

The RNA was separated on a 7% polyacrylamide gel. Track 1, RNA synthesized *in organello* for 20 min; track 2, same amount of RNA as in (1), incubated for 5 min at 20 °C with 0.5 units of *E. coli* RNAase III; track 3, as in (2), incubated for 15 min; track 4, as in (1), incubated for 15 min at 20 °C with a protein extract from chloroplasts; track 5, *E. coli* rRNA used as marker. (*a*) Photograph of the stained gel; (*b*) autoradiograph of the gel shown in (*a*).

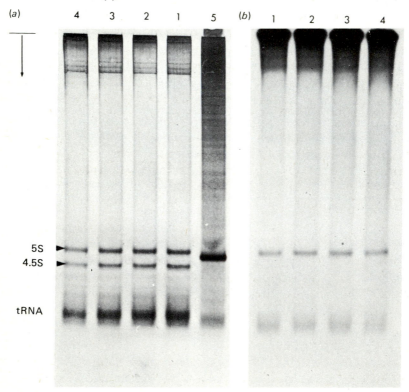

first in 2.7 MD RNA, or to independent transcription. Our attempts to process 2.7 MD RNA *in vitro* using either *E. coli* RNAase III or chloroplast extracts have not demonstrated a significant increase in the amount of labelled 5S rRNA (Fig. 4). The existence of an independent transcriptional unit for the 5S rRNA gene is supported by the finding of a promoter-like sequence between the 4.5S and the 5S rRNA genes (Dyer & Bedbrook, 1980; Takaiwa & Sugiura, 1980*a*; Keus *et al.*, 1983*b*; see pp. 109–13). In spinach, labelling kinetics in the presence of chloramphenicol also support this notion, since the immediate precursors of 23S and 16S rRNAs (p23S and p16S, respectively) accumulate, while the 5S rRNA appears to be unaffected (Hartley, 1979).

Transcription *in vitro* of a tobacco cp-DNA fragment containing the start of the 16S rRNA gene and an upstream gene for tRNA^{Val} revealed that these genes are co-transcribed (Tohdoh, Shinozaki & Sugiura, 1981). In contrast, the spinach gene for tRNA^{Val} appears not to be co-transcribed with the 16S rRNA gene (Briat *et al.*, 1982*a*). The extent of the maize rDNA transcriptional unit was determined by Zenke *et al.* (1982), using small DNA probes which contain the genes for the sequence: (5′)-tRNA^{Val}–16S rRNA–tRNA^{Ile} (including intron)–tRNA^{Ala} (including intron)–23S rRNA–4.5S rRNA–5S rRNA (3′) (Schwarz & Kössel, 1979; Schwarz *et al.*, 1981*b*; Koch, Edwards & Kössel, 1981; Edwards & Kössel, 1981). These probes were hybridized to filter-immobilized cp-RNA from young leaves. The results (Fig. 5) showed that a transcript encompassing the 16S rRNA to the 4.5S rRNA is present in a high molecular weight RNA. The transcript contains neither the tRNA^{Val} sequence nor the 5S rRNA sequence. The introns which split the two tRNA genes appear to be excised intact as 12S–14S RNA. In tobacco chloroplasts, an 8.2 kb RNA species, which is about the same size as the spinach 2.7 MD rRNA precursor, was shown to be the common precursor of both tRNA^{Ile} and tRNA^{Ala} and at least both 23S and 16S rRNAs (Takaiwa & Sugiura, 1982*c*). Recently, the *Euglena* chloroplast rDNA unit, which is about 2 kbp smaller than higher plant rDNA units, was shown to be transcribed as a 6.0 kb RNA (Dix & Rawson, 1983) which is about the size of the 1.8 MD transcript presumed to be the initial common precursor of the 16S and 23S rRNA sequences (Wollgiehn & Parthier, 1979). Chloroplasts of the brown alga *Pylaiella littoralis* are known to synthesize a similar size RNA precursor (1.88 MD) which included the nucleotide sequences of at least the two large rRNAs (Loiseaux, Rozier & Dalmon, 1980).

Synthesis of tRNAs

Chloroplast tRNAs, which are different from those of the cytosol or the mitochondria (for reviews see Barnett, Schwartzbach & Hecker, 1978;

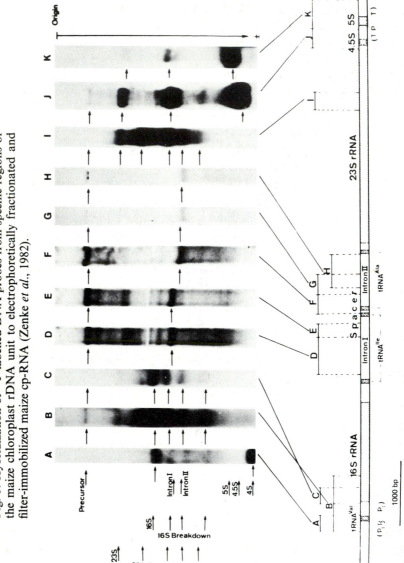

Fig. 5. Hybridization of ^{32}P-labelled DNA probes from specific regions of the maize chloroplast rDNA unit to electrophoretically fractionated and filter-immobilized maize cp-RNA (Zenke et al., 1982).

Weil, 1979), are gene products of the chloroplast genome (reviewed in Bohnert *et al.*, 1982; Weil & Parthier, 1982). *Euglena* contains 23–9 tRNA genes as indicated by plateau-hybridization studies (Gruol & Haselkorn, 1976; McCrea & Hershberger, 1976; Schwartzbach, Hecker & Barnett, 1976). A similar number of chloroplast tRNA genes was reported for maize and *Chlamydomonas* (Haff & Bogorad, 1976*a*; Malnoë & Rochaix, 1978; Meeker & Tewari, 1980) while spinach and pea cp-DNAs have about 40 tRNA genes (Meeker & Tewari, 1982). Pseudogenes for tRNAs have been found in the repeated region of *Euglena* cp-DNA (Orozco *et al.*, 1980*b*; Miyata *et al.*, 1982; El-Gewely, Helling & Dibbits, 1984).

The chloroplast 4S RNAs from various plant species have been fractionated, e.g. by two-dimensional gel electrophoresis (Burkard *et al.*, 1982), and the individual tRNAs identified by aminoacylation *in vitro* using either chloroplast or *E. coli* aminoacyl-tRNA synthetases and ^{3}H-labelled amino acids (Table 3). This has led to the identification of 27 spinach and 27 common bean (*Phaseolus vulgaris*) tRNA species specific for 16 amino acids, 20 tobacco (*Nicotiana tabacum*) tRNAs for 14 amino acids, 19 soybean (*Glycine max*) tRNAs for 14 amino acids, 28 pea tRNAs for 20 amino acids, 26 maize tRNAs for 17 amino acids, 24 broad bean tRNAs for 17 amino acids, 11 or 12 wheat tRNAs for 6 amino acids, and 23 *Euglena* tRNAs for 18 amino acids. The genes for most of the identified tRNAs have been located on the physical map of the cp-DNA from spinach, tobacco, common bean, pea, broad bean, maize, and *Euglena* (for references see Table 3). The identification of tRNAs specific for some amino acids, e.g. Asp, Cys, Gln and Glu, is difficult to achieve because of technical problems. Aminoacylation of pea chloroplast tRNAs and hybridization of these tRNAs to pea cp-DNA showed that the pea chloroplast genome contains tRNA genes for each of the 20 amino acids (Meeker & Tewari, 1980). Thus the chloroplast contains a complete set of tRNAs. The primary structures of a number of chloroplast tRNAs (Dyer, 1982; Gauss & Sprinzl, 1983*a*) and tRNA genes (Gauss & Sprinzl, 1983*b*) from spinach, tobacco, common bean, maize, *Euglena* and *Scenedesmus* have been determined (Table 3). In general, a very high degree of sequence homology exists among the same chloroplast tRNA isoacceptors from different higher plants (Mubumbila *et al.*, 1980). Homologies between chloroplast tRNAs and among chloroplast and prokaryotic tRNAs having the same anticodon have also been found (Karabin & Hallick, 1983; Steinmetz *et al.*, 1983; Gillham, Boynton & Harris, 1984).

Our knowledge concerning the existence of chloroplast tRNAs, the location of tRNA genes on the physical maps of cp-DNAs, and the primary structure of tRNAs or tRNA genes, contrasts sharply with our ignorance of tRNA synthesis. Studies on the synthesis of tRNAs or 4S RNAs have been

Table 3. *Chloroplast tRNAs and tRNA genes*

tRNAs accepting	Plant source	No. of isoacceptors identified[a,b]	tRNAs sequenced[c,d]	tRNA genes mapped on cp-DNA[a]	tRNA genes sequenced[c]	Reference for sequencing data
Ala	Spinach	1		1		
	Tobacco	1		2	trnA-UGC 710 bp intron	Takaiwa & Sugiura, 1982c
	Common bean	1		2		
	Soybean	1				
	Pea	1		1		
	Broad bean	1		1		
	Maize	1		2	trnA-UGC 806 bp intron	Koch et al., 1981
	Euglena	1		2 or 3	trnA-UGC	Graf, Kössel & Stutz, 1980 Orozco et al., 1980b
Arg	Spinach	2		1		
	Tobacco	2		3	trnR-UCU	Sugiura et al., 1983
	Common bean	2		2		
	Pea	1		1		
	Broad bean	2		2		
	Maize	1		2		
	Euglena	2		1	trnR-ACG	Orozco & Hallick, 1982b
Asn	Spinach	1		1		
	Tobacco			2	trnN-GUU	Kato et al., 1981
	Common bean	2		2		
	Pea	2		1		
	Broad bean	1		1		
	Maize	2		4		
	Euglena	1		1	trnN-GUU	Orozco & Hallick, 1982b
Asp	Tobacco			1		
	Soybean	1				
	Pea	1		1		
	Broad bean	2		2		
	Maize	1		1		
Cys	Pea	1				

Table 3 (*cont.*)

RNAs accepting	Plant source	No. of iso-acceptors identified[a,b]	tRNAs sequenced[c,d]	tRNA genes mapped on cp-DNA[a]	tRNA genes sequenced[c]	Reference for sequencing data
Glu	Pea	1				
	Wheat	1				
	Euglena	1		1	*trn*E-UUC	Hollingsworth & Hallick, 1982
Gln	Pea	1				
	Euglena	1		1	*trn*Q-UUG	Karabin & Hallick, 1983
Gly	Spinach	2		1		
	Tobacco	1		2	*trn*G-UCC	Sugiura *et al.*, 1983
	Common bean	2		1		
	Soybean	1				
	Pea	1		1		
	Broad bean	1		1 or 2		
	Maize	2		4		
	Euglena	1		2	*trn*G-GCC	Karabin & Hallick, 1983
					*trn*G-UCC	Hollingsworth & Hallick, 1982
His	Spinach	1		1	*trn*H-GUG	W. Bottomley pers. comm.
	Tobacco	1		1	*trn*H-GUG	Sugiura *et al.*, 1983
	Common bean	1		1		
	Soybean	1			*trn*H-GUG	Spielmann & Stutz, 1983
	Pea	1		1		
	Broad bean	1		1		
	Cotton	1				
	Maize	1		2	*trn*H-GUG	Schwarz *et al.*, 1981*a*
	Euglena	1			*trn*H-GUG	Hollingsworth & Hallick, 1982
	Spinach	2	tRNA$_{GAU}^{Ile}$ tRNA$_{XAU}^{Ile}$	4	*trn*I-CAU	Guillemaut & Weil, 1982 Francis & Dudock, 1982 Kashdan & Dudock, 1982*b*
	Tobacco	2		4	*trn*I-GAU 707 bp intron	Takaiwa & Sugiura, 1982*c*

Table 3 (cont.)

tRNAs accepting	Plant source	No. of iso-acceptors identified[a,b]	tRNAs sequenced[c,d]	tRNA genes mapped on cp-DNA[a]	tRNA genes sequenced[c]	Reference for sequencing data
	Common bean	2		4		
	Soybean	2				
	Pea	2		2		
	Broad bean	2		2		
	Cotton	2				
	Maize	2	tRNA$_{GAU}^{Ile}$	4	trnI-GAU 949 bp intron	Koch et al., 1981 Guillemaut & Weil 1982
	Euglena	2		2 or 3	trnI-GAU	Graf et al., 1980 Orozco et al., 1980
Leu	Spinach	3	tRNA$_{UAG}^{Leu}$	3 to 5		Canaday et al., 19?
	Tobacco	3		4	trnL-UAG	M. Sugiura, pers. c
	Common bean	3	tRNA$_{U*AA}^{Leu}$ tRNA$_{CmAA}^{Leu}$ tRNA$_{UAm7G}^{Leu}$	4		Canaday et al., 19? Osorio-Almeida et al., 1980
	Soybean	3				
	Pea	3		3		
	Broad bean	3		3 to 5		
	Cotton	3				
	Maize	3		4	trnL-UAA 458 bp intron / trnL-CAA	Steinmetz, Gubbin? Bogorad, 1982 / Steinmetz et al., 1983
	Wheat	3				
	Euglena	2		2 or 3	trnL-UAG	Orozco & Hallick 1982b
Lys	Spinach	1				
	Tobacco	1		1		
	Common bean	1		1		
	Soybean	1				
	Pea	1		1		
	Broad bean	1		1		
	Cotton	1				
	Maize	1		1		
	Euglena	1		1		
Met	Spinach	3	tRNA$_{CAU}^{Met}$(f) tRNA$_{CAU}^{Met}$(m)	2 or 3		Calagan et al., 19? Pirtle et al., 1981

Table 3 (*cont.*)

tRNAs accepting	Plant source	No. of iso-acceptors identified[a,b]	tRNAs sequenced[c,d]	tRNA genes mapped on cp-DNA[a]	tRNA genes sequenced[c]	Reference for sequencing data
	Tobacco			2	*trn*M-CAU(m)	Deno *et al.*, 1982
					*trn*M-CAU(f)	Sugiura *et al.*, 1983
	Common bean	2	tRNA$_{CAU}^{Met}$(f)	1		Canaday, Guillemaut & Weil, 1980*b*
	Soybean	2				
	Pea	1		1	*trn*M-CAU	Zurawski (see Karabin & Hallick, 1983)
	Broad bean	2		1		
	Cotton	2				
	Maize	2		2	*trn*M-CAU(m)	Steinmetz *et al.*, 1983
	Wheat	1				
	Euglena	2		2	*trn*M-CAU(f)	Karabin & Hallick, 1983
					*trn*M-CAU(m)	Hollingsworth & Hallick, 1982
	Scenedesmus	2	tRNA$_{CAU}^{Met}$(f)			McCoy & Jones, 1980
			tRNA$_{CAU}^{Met}$(m)			Jones, 1980
he	Spinach	1	tRNA$_{GAA}^{Phe}$	1		Canaday *et al.*, 1980*a*
	Tobacco	1		2		
	Common bean	2	tRNA$_{GAA}^{Phe}$	2		Guillemaut & Keith, 1977
						Canaday *et al.*, 1980*a*
	Soybean	1				
	Pea	2				A. Steinmetz, pers. comm.
	Broad bean	1		1	*trn*F-GAA	
	Cotton	2				
	Barley	2				
	Maize	1		1	*trn*F-GAA	Steinmetz *et al.*, 1983
	Wheat	1				
	Euglena	1	tRNA$_{GAA}^{Phe}$	1	*trn*F-GAA	Chang *et al.*, 1976; R. B. Hallick, pers. comm.
o	Spinach	1	tRNA$_{U*GG}^{Pro}$	1		Francis *et al.*, 1982
	Tobacco	1		2	*trn*P-UGA	M. Sugiura, pers. comm.
	Common bean	1		2		
	Pea	1		1		
	Broad bean	1		1		
	Maize	1		1		
	Euglend	1				

Table 3 (*cont.*)

tRNAs accepting	Plant source	No. of iso-acceptors identified[a,b]	tRNAs sequenced[c,d]	tRNA genes mapped on cp-DNA[a]	tRNA genes sequenced[c]	Reference for sequencing data
Ser	Spinach	3		2		
	Tobacco	1		3	trnS-GCU	M. Sugiura, pers. c
	Common bean	3		3		
	Soybean	1				
	Pea	3		3		
	Broad bean	2		1		
	Maize	3		3	trnS-GGA	Steinmetz et al., 1983
	Wheat	3			trnS-UGA	L. Bogorad, pers.
	Euglena	1		1	trnS-GCU	Karabin & Hallick 1983
Thr	Spinach	2	tRNA$_{GGU}^{Thr}$	1	trnT-GGU	Kashdan & Dudoc 1982a Kashdan et al., 19
	Tobacco	2		2		
	Common bean	1		2		
	Soybean	1				
	Pea	1		2		
	Broad bean	1		1	trnT-GGU	A. Steinmetz, pers comm.
	Maize	2		2	trnT-UGU	Steinmetz et al., 1983
	Euglena	2		1	trnT-UGU	Karabin & Hallick 1983
Trp	Spinach	1	tRNA$_{CCA}^{Trp}$	1		Canaday et al., 19
	Tobacco	1		1	trnW-CCA	M. Sugiura, pers.
	Common bean	1		1		
	Soybean	1				
	Pea	1		1		
	Broad bean	1		1		
	Cotton	2				
	Maize	1		1		
	Euglena	1		1	trnW-CCA	Hollingsworth & Hallick, 1982
	Chlamydomonas	1				
Tyr	Spinach	1		1	trnY-GUA	W. Bottomley, pe comm.
	Tobacco	1		1	trnY-GUA	M. Sugiura, pers.

Table 3 (*cont.*)

tRNAs accepting	Plant source	No. of iso-acceptors identified[a,b]	tRNAs sequenced[c,d]	tRNA genes mapped on cp-DNA[a]	tRNA genes sequenced[c]	Reference for sequencing data
	Common bean	1		1		
	Soybean	2				
	Pea	1		1		
	Broad bean	1		1	trnY-GUA	A. Steinmetz, pers. comm.
	Maize	1		1		
	Wheat	1				
	Euglena	1		1	trnY-GUA	Hollingsworth & Hallick, 1982
Val	Spinach	2	tRNA$_{\mathrm{U*AC}}^{\mathrm{Val}}$	3	trnV-GAC	Sprouse et al., 1981 / Briat et al., 1982a
	Tobacco	2		3	trnV-GAC / trnV-UAC 571 bp intron	Tohdoh et al., 1981 / Deno et al., 1982
	Common bean	2		2		
	Soybean	1				
	Pea	2		1		
	Broad bean	1		1		
	Cotton	1				
	Maize	1		2	trnV-GAC	Schwarz et al., 1981b
	Wheat	2 or 3			trnV-UAC 603 bp intron	L. Bogorad, pers. comm.
	Spirodela oligorhiza				trnV-GAC	Keus et al., 1983a
	Euglena	1		1	trnV-UAC	Orozco & Hallick, 1982b

Bohnert *et al.* (1979), Driesel *et al.* (1979), Keller *et al.* (1980), Briat *et al.* (1982a), El-Gewely *al.* (1982), Kuntz *et al.* (1982), Mubumbila *et al.* (1983), Selden *et al.* (1983), Hallick (1984), Bergmann, P. Seyer, G. Burkard, & J. H. Weil, personal communication; E. J. Crouse, . Mubumbila, B. M. Stummann, G. Bookjans, J. H. Weil & K. W. Henningsen, personal mmunication; M. Mubumbila, E. J. Crouse & J. H. Weil, personal communication.
Barnett, Pennington & Fairfield (1969), Leis & Keller (1970), Reger *et al.* (1970), Guderian, lliam & Gordon (1972), Merrick & Dure (1972), Guillemaut *et al.* (1973), Hiatt & Snyder 973), Preddie *et al.* (1973), Karwowska, Gózdzicka-Józefiak & Augustyniak (1979), Meeker Tewari (1980), Mubumbila *et al.* (1980), Swamy & Pillay (1982), Gózdzicka-Józefiak & gustyniak (1983).
Gauss & Sprinzl (1983a, 1983b); see references, last column.
tRNA$_{\mathrm{U*AA}}^{\mathrm{Leu}}$ = a leucine tRNA with the anticodon 5'-U*AA; U* is a derivative of uridine.

reported for *Euglena* (Barnett *et al.*, 1969; Reger *et al.*, 1970; Parthier & Krauspe, 1974). Synthesis *in vivo* was also measured in higher plants (Burkard, Vaultier & Weil, 1972; Merrick & Dure, 1972). In these cases a light-induced increase of chloroplast-specific tRNAs was observed. However, during cotton (*Gossypium arboreum*) cotyledon development, increasing amounts of chloroplast tRNAs were observed even in the dark (Brantner & Dure, 1975). Isolated intact chloroplasts from spinach are capable of light-driven synthesis of 4S RNAs *in organello* (Bohnert *et al.*, 1977; Hartley, 1979). The products have not been characterized further.

Little or no specific information is yet available about precursors of tRNAs or the possibility of polycistronic transcription of those tRNAs whose genes have been mapped close together (Driesel *et al.*, 1979; Kuntz *et al.*, 1982; Orozco & Hallick, 1982*a*, 1982*b*; Hollingsworth & Hallick, 1982; Karabin & Hallick, 1983; Mubumbila *et al.*, 1983; Selden *et al.*, 1983). Several genes for tRNAs in *Euglena* are located on the same strand of the DNA, separated by only a few base-pairs. Their transcription as a polycistronic RNA is very likely (Hollingsworth & Hallick, 1982; Karabin & Hallick, 1983). In tobacco, common bean, pea, broad bean, maize and *Euglena*, genes for tRNA^{Ile} and tRNA^{Ala} are located in the spacer DNA sequence between the 16S and the 23S rRNA genes (see references in Table 3). At present, only a gene for tRNA^{Ile} has been mapped in the spacer region of the spinach rDNA units (Bohnert *et al.*, 1979). The primary transcript of some rDNA units has been shown to contain both rRNAs and the spacer tRNAs which are subsequently processed to the mature molecules (see pp. 88–95).

Synthesis of mRNAs

The pea chloroplast genome (120 kbp, Palmer & Thompson, 1981; or 135 kbp, Chu, Oishi & Tewari, 1981) contains one set of rRNA genes (Chu *et al.*, 1981; Palmer & Thompson, 1981) and about 40 tRNA genes (Meeker & Tewari, 1980). These rRNA and tRNA genes occupy about 10% (*ca* 8 kbp and *ca* 4 kbp, respectively) of the coding capacity of the cp-DNA. The remainder of this genome could code for 70–80 proteins, assuming an average size protein is about 40 kD (equivalent to a gene of about 1.2–1.5 kbp, which would include translated and non-translated regions; cf. Table 4).

Transcripts to be translated into proteins amount to only 2–3% of the total cp-RNA. Several groups have tried to measure the extent and complexity of chloroplast transcription by hybridization experiments using total cp-RNA or leaf RNA as probes. In pea, the amount of RNA which can be hybridized to filter-bound cp-DNA is equivalent to the complexity of one strand of the DNA (Oishi *et al.*, 1981). Similarly, Zech *et al.* (1981) have shown that all the fragments generated by Sal I endonuclease cleavage of spinach cp-DNA hybridized, although to varying amounts, with cp-RNA labelled *in vitro* (Fig. 6). Similar results have been obtained for *Chlamydomonas* cp-DNA; however, no RNA sequences were detected on three small dA+dT-rich fragments (Matsuda & Surzycki, 1980). Howell & Walker (1977) showed that, depending on developmental and cell cycle conditions, between 39 and 60% of the double-stranded *Chlamydomonas* cp-DNA was transcribed. These results permit the conclusion that these cp-DNAs do not contain any major transcriptionally silent regions. In contrast, 22 to 57% of one strand of *Euglena* cp-DNA has been found to be transcribed, depending on the growth

Fig. 6. Binding of *E. coli* RNA polymerase holoenzyme to spinach chloroplast DNA fragments compared with amounts of transcripts from regions of the cp-DNA present *in vivo*.

The map shows (inner circle) the cleavage sites of the restriction endonucleases Sal I, Sma I and the location of the Hind III fragment 1 (HF 1), and Bam HI fragments 3 and 5 (BF 3 and BF 5, respectively). The location and direction of transcription of the rDNA genes, of the *rbc*L (LSU) gene and the location of the *psb*A (32 kD) gene are shown. The hatched areas show DNA fragments with strong binding of the bacterial enzyme. The strongest binding is observed with Sal I fragment F or some of its subfragments. The outer circle shows DNA fragments for which transcripts of high molecular weight have been found *in vivo* (Zech *et al.*, 1981).

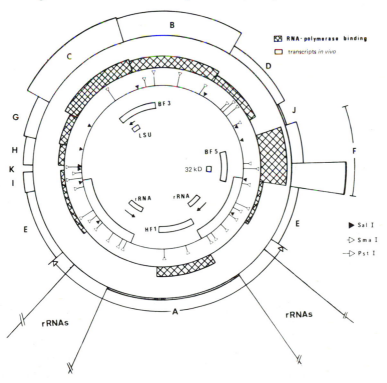

conditions and the developmental state of the plastids (Rawson, 1975; Chelm & Hallick, 1976; Rawson & Boerma, 1976; Chelm *et al.*, 1978, 1979; Rawson *et al.*, 1981). In addition, these studies revealed that a large fraction of the transcripts are present in dark-grown cells and throughout the development of the chloroplast. Furthermore, variation in the abundance of the RNA derived from certain regions of the cp-DNA molecule occurred during chloroplast development. The difference in the total amount of cp-DNA transcribed in higher plants and *Euglena* can be partly explained by the possibility of symmetrical transcription in some regions of the cp-DNA (Rawson *et al.*, 1981) and by the low dG + dC content of *Euglena* cp-DNA (28 % ; Schmitt *et al.*, 1981) as compared to higher plant cp-DNA (about 37 % ; reviewed by Bohnert *et al.*, 1982). In *Euglena*, about 30 % of the chloroplast genome has less than 12 % dG + dC which might be too low to contain genes (Schmitt *et al.*, 1981).

There is no proof that the RNAs which hybridize to various cp-DNA fragments are mRNAs unless they can be translated into protein. Only a small number of chloroplast mRNAs have been characterized via their translational products, either in isolated plastids or in heterologous translation systems *in vitro* derived from wheat germ, reticulocyte lysate or *E. coli* (Hartley, Wheeler & Ellis, 1975; Sagher *et al.*, 1976; Bedbrook *et al.*, 1978; Bogorad *et al.*, 1978; Malnoë *et al.*, 1979; Sano, Spaeth & Burton, 1979; Driesel *et al.*, 1980; Westhoff *et al.*, 1981; Camerino *et al.*, 1982*a*; Watanabe & Price, 1982; Oishi & Tewari, 1983; for a review see Bottomley & Bohnert, 1982). Specific mRNAs can also be isolated using the hybrid selection-method by which a mRNA is annealed to a purified DNA fragment from which it was transcribed, non-complementary mRNAs are removed from double-stranded hybrid, and the purified mRNA then released from the hybrid and subsequently translated *in vitro* (Driesel *et al.*, 1980; Westhoff *et al.*, 1981). Alternatively, and more elegantly, a coupled or linked transcription/translation system has been used both to map and to identify a product as being derived from a specific cp-DNA fragment (Coen *et al.*, 1977; Bottomley & Whitfeld, 1979; Andrews & Rawson, 1982). Some genes or gene products have also been identified by biochemical and immunochemical methods, or by genetic complementation, i.e. by introduction of a chloroplast gene of presumed function into *E. coli* cells which contain a non-functional corresponding gene (Tiboni *et al.*, 1981). Recently, a protein-synthesizing system has been prepared from spinach chloroplasts which is capable of translating spinach chloroplast mRNAs and heterologous mRNAs (Camerino, Savi & Ciferri, 1982*b*).

Among the genes for mRNAs which have been localized on various chloroplast genomes are the genes for LSU, a 32 kD thylakoid membrane protein, EF-Tu, two ribosomal proteins, the alpha, beta, epsilon and

Table 4. *Properties of some chloroplast mRNAs as deduced from DNA sequencing data*

mRNA for	Gene	Plant source	Size of mRNA; nucleotides	Size of protein product	Introns	Reference
Stromal polypeptides						
LSU	*rbc*L	Spinach	1690 ∓ 3^a	52760 D 475 amino acids	No	Zurawski *et al.*, 1981
		Tobacco	1756	52936 D 477 amino acids	No	Shinozaki & Sugiura 1982*a*
		Maize		52682 D 475 amino acids	No	McIntosh, Poulsen & Bogorad, 1980 Poulsen, 1981
		Chlamy-domonas	*ca* 1600	475 amino acids	No	Dron *et al.*, 1982*a*, 1983
EF-Tu	*tuf*A	*Euglena*	*ca* 1950	45011 D 408 amino acids	Yes	Montandon & Stutz, 1983
30S ribosomal protein 4	*rps*4	Maize		*ca* 23 500 D 201 amino acids	No	Subramanian, Steinmetz & Bogorad, 1983
30S ribosomal protein 19	*rps*19	Tobacco		10443 D 92 amino acids	No	Sugita & Sugiura, 1983
Thylakoid membrane polypeptides						
CF$_1$ alpha subunit	*atp*A	Tobacco		55446 D 507 amino acids	No	Deno, Shirozaki & Sugiura, 1983
CF$_1$ beta subunit	*atp*B	Spinach		53874 D 498 amino acids	No	Zurawski *et al.*, 1982*b*
		Tobacco		498 amino acids	No	Shinozaki *et al.*, 1983
		Maize		54042 D 498 amino acids	No	Krebbers *et al.*, 1982
CF$_1$ epsilon subunit	*atp*E	Spinach		14702 D 134 amino acids	No	Zurawski *et al.*, 1982*b*
		Tobacco		133 amino acids	No	Shinozaki *et al.*, 1983
		Maize		15218 D 139 amino acids	No	Krebbers *et al.*, 1982
CF$_0$ subunit III (proteolipid)	*atp*H	Spinach		7968 D 81 amino acids	No	Alt *et al.*, 1983*b*

Table 4 (*cont.*)

mRNA for	Gene	Plant source	Size of mRNA nucleotides	Size of protein product	Introns	Reference
PS II '32 kD' protein	*psbA*	Wheat		*ca* 8000 D 81 amino acids	No	Howe *et al.*, 1983*a*
		Spinach	*ca* 1250	38 950 D 353 amino acids	No	Zurawski *et al.*, 1982*a*
		Nicotiana debneyi	*ca* 1250	38 950 D 353 amino acids	No	Zurawski *et al.*, 1982*a*
		Soybean		38 904 D 353 amino acids	No	Spielmann & Stut 1983

a Synthesized as a precursor (Langridge, 1981).

proton-translocating subunits of the chloroplast ATP-synthase, three of the components of the cytochrome *b6/f* complex and for the P700 chlorophyll *a* apoprotein (Link *et al.*, 1978; Malnoë *et al.*, 1979; Driesel *et al.*, 1980; Whitfeld & Bottomley, 1980; Westhoff *et al.*, 1981; Link, 1981*b*; Dron *et al.*, 1982*a*; Howe *et al.*, 1982; Link, 1982; van Ee *et al.*, 1982; Stiegler *et al.*, 1982; Alt *et al.*, 1983*b*; Howe *et al.*, 1983*a*, 1983*b*; Shinozaki & Sugiura, 1982*a*, 1982*b*; Westhoff *et al.*, 1983; also see references listed in Table 4). Some of these genes have been sequenced (see Table 4).

Polycistronic transcription of some chloroplast genes coding for proteins has been suggested (Zurawski, Bottomley & Whitfeld, 1982*b*; Alt *et al.*, 1983*a*; Howe *et al.*, 1983*b*). The genes for the β and ϵ subunits of maize and tobacco chloroplast ATPase were shown to be co-transcribed as 2.2-kbp and 2.7-kbp polycistronic mRNAs, respectively (cf. Jolly *et al.*, 1981; Krebbers *et al.*, 1982; Shinozaki *et al.*, 1983). Additional mRNAs for other polypeptides will certainly be isolated and characterized in the near future. Likely candidates may be found in the proteins which are identified as products of protein synthesis by isolated organelles (e.g. Eneas-Filho, Hartley & Mache, 1981; for reviews see Bohnert *et al.*, 1982; Bottomley & Bohnert, 1982).

The mRNAs for the LSU and the 32 kD thylakoid membrane protein are the most abundant mRNA species in chloroplasts (Hartley *et al.*, 1975; Rosner *et al.*, 1975; Bedbrook *et al.*, 1978; Link *et al.*, 1978; Driesel *et al.*, 1980; Link, 1981*a*). Neither RNA is polyadenylated (Rosner *et al.*, 1975;

Wheeler & Hartley, 1975; Sagher *et al.*, 1976; Howell *et al.*, 1977). The LSU mRNA sediments in non-denaturing sucrose gradients at a position of 18–20S. The mRNA for the 32 kD protein sediments at 13–14S in sucrose gradients (Sagher *et al.*, 1976; Speirs & Grierson, 1978; Driesel *et al.*, 1980; Edelman & Reisfeld, 1980). Sequences coding for LSU and the 32 kD protein in these genes on spinach cp-DNA are 1425 nucleotides (LSU) and 1068 nucleotides (32 kD) long (Zurawski *et al.*, 1981, 1982*a*). However, the total length of the transcribed regions is 1690 ± 3 (LSU) and about 1230 nucleotides (32 kD protein). In addition, transcripts of larger size for LSU (approximately 2400 and 2600 nucleotides) have been identified in legume species (Palmer *et al.*, 1982). Large transcripts for other mRNAs have also been found in *Euglena* and barley (Dix & Rawson, 1983; Poulsen, 1983).

Possible ribosome binding sites (Shine & Dalgarno, 1974), which appear to play an important role in the prokaryotic-type of translational initiation (Gold *et al.*, 1981), have been identified in some, but not all, chloroplast genes sequenced so far (see references in Table 4; for reviews see Bohnert *et al.*, 1982; Whitfeld & Bottomley, 1983).

Components of the transcriptional machinery

RNA synthesis requires a DNA template, the four ribonucleoside triphosphates, divalent cations, and the DNA-dependent RNA polymerase to catalyze the reaction. Bona fide transcription of genes requires certain protein factors which are at least transiently part of the DNA/RNA-polymerase complex. These factors are involved in the recognition of certain base sequences upstream in the gene sequence. The known constituents of the chloroplast system are compared to those of the better-characterized components of the prokaryotic and eukaryotic nuclear transcriptional machineries. Review articles on this subject are available (Chambon, 1975; Losick & Chamberlin, 1976; Whitfeld, 1977; Rosenberg & Court, 1979; Siebenlist, Simpson & Gilbert, 1980; Wollgiehn & Parthier, 1980; Guilfoyle, 1981; Dyer & Leaver, 1981; Wollgiehn, 1982; Hawley & McClure, 1983).

RNA polymerase

In 1964 it was shown that chloroplasts can synthesize RNA *in vitro*. This synthesis was inhibited by actinomycin D (Kirk, 1964). Subsequently, several attempts were made to isolate the enzymes. This proved to be difficult since the chloroplast polymerase appears to be tightly associated with the thylakoid membrane. It can, at least partially, be solubilized by treatment with the nonionic detergent Triton X-100 (Joussaume, 1973; Hallick *et al.*, 1976; Schiemann *et al.*, 1977, 1978; Briat *et al.*, 1979, 1982*b*; Ness & Woolhouse, 1980; Tewari & Goel, 1983), digitonin (Schiemann *et al.*, 1977), EDTA

(Bottomley *et al.*, 1971; Kidd & Bogorad, 1980) or low salt (Bennett & Ellis, 1973). The isolated enzyme complex still contains at least some cp-DNA in addition to several protein species whose stoichiometry is not easy to assess. In Table 5, properties of cp-RNA polymerases from three species are compared to those of prokaryotic and eukaryotic enzymes. The chloroplast DNA-dependent RNA polymerases from *Euglena* (Hallick *et al.*, 1976), spinach (Briat *et al.*, 1979) and maize (Smith & Bogorad, 1974) exhibit features which clearly distinguish them from the *E. coli* enzyme (e.g. subunit complexity, salt requirement and sensitivity to rifampicin: Surzycki, 1969; Jolly & Bogorad, 1980). The enzyme activity is increased by light (Apel & Bogorad, 1976). Pronounced similarities exist with the eukaryotic nuclear RNA polymerase I or A. It is, however, already known that nuclear RNA polymerases I, e.g. those from yeast and mammals, differ in several aspects (Guilfoyle, 1981). Thus, it still has to be shown whether one should speak of a single type of enzyme in all chloroplasts or whether different types may exist.

Specificity factors

The sigma factor of bacterial RNA polymerase is an absolute requirement for specific initiation of RNA chains at proper sites of the *E. coli* chromosome (see below). As the RNA chain is being elongated, sigma dissociates from the complex. Without the sigma factor, the core enzyme transcribes DNA at random. Comparable protein factors for initiating proper transcription of different gene families have been found among the subunits of eukaryotic polymerases.

Another factor, rho, functions in the termination of RNA synthesis in *E. coli*. Rho is multimeric, and dissociates the transcript while the polymerase remains bound to the template in an inactive state. No counterpart to the rho factor has yet been demonstrated in either eukaryotes or chloroplasts.

Since the chloroplast RNA polymerase molecule cannot be isolated easily in an active and stable form, not much is known about specificity factors. One report (Surzycki & Shellenbarger, 1976) demonstrated a chloroplast sigma-type protein, which has a molecular mass of 39 kD. This factor stimulated the core enzyme of *E. coli* and the isolated chloroplast RNA polymerase from *Chlamydomonas*, whereas the activity of two nuclear RNA polymerases from *Chlamydomonas* was not changed. In contrast, no effect of the *E. coli* sigma factor on purified RNA polymerase from maize chloroplasts could be found, using cloned cp-DNA fragments as template (Jolly & Bogorad, 1980). These authors isolated a factor, S, with a molecular mass of 27 kD, which changed the specificity of their chloroplast RNA polymerase from preference for denatured templates to selective transcription

... chloroplast and prokaryotic enzymes

	Chloroplast enzymes from			Prokaryotic	Eukaryotic polymerase classes		
	Maize	Euglena	Spinach	E. coli	I	II	III
Transcription of	Total DNA	Total DNA	Total DNA	Total DNA	rRNA	mRNA (hnRNA)	5SrRNA tRNA
Subunit composition	Complex. 14	Transcription complex	Transcription complex. approx. 10	5	7–9	11–14	10–12
Optimal ionic strength	$Me^+ > 100$ mmol l^{-1} inhibits	—	$Me^+ > 100$ mmol l^{-1} inhibits	100–250 mmol l^{-1}	50 mmol l^{-1}	200–600 mmol l^{-1}	200–600 mmol l^{-1}
Mg^{2+} optimum	10–40 mmol l^{-1}	<5 mmol l^{-1}	20 mmol l^{-1}	20 mmol l^{-1}	5–10 mmol l^{-1}	5–10 mmol l^{-1}	5–10 mmol l^{-1}
Mn^{2+} optimum	8 mmol l^{-1}	5 mmol l^{-1}	5 mmol l^{-1}	0	1–2 mmol l^{-1}	1–2 mmol l^{-1}	1–2 mmol l^{-1}
$(NH_4)_2SO_4$	—	—	Inhibits at 50 mmol l^{-1}	—	50 mmol l^{-1}	100 mmol l^{-1}	50–200 mmol l^{-1}
Temperature opt.	48 °C	—	37–40 °C	37 °C	—	—	—
Opt. template	Denatured DNA	—	Denatured DNA	Some T5 promoters	Nat./denat. DNA	Denat. DNA	Nat./denat. DNA
Rifampicin (20 µg ml^{-1})	—	No effect	No effect	Initiation inhibited	No effect	No effect	No effect
Rifamycin SV (200µg ml^{-1})	No effect	—	—	—	—	—	—
α-Amanitin	No effect	No effect	—	No effect	No effect	50% (0.02 µg ml^{-1})	50% (15 µg ml^{-1})
Streptolydigin	—	No effect	—	Elongation inhibited	—	—	—
Actinomycin D	—	50% (3.5 µg ml^{-1})	14% (3.5 µg ml^{-1})	0% (20 µg ml^{-1})	—	—	—
Ethidium bromide	—	86% (2 mmol l^{-1})	96% (2 mmol l^{-1})	—	—	—	—
Specificity factors	S (27000 D) preference for ccc–cp-DNA	In complex	—	Sigma (90000 D)	Subunits of the enzyme	Subunits of the enzyme	Subunits of the enzyme

Me$^+$ = monovalent metal ion. Opt. = optimum. hnRNA = heterogeneous nuclear RNA. Nat. = native, denat. = denatured. ccc = covalently closed circular.

of supercoiled maize cp-DNA inserts in the plasmid pMB9. It was not reported if the S factor had the same effect on pMB9 without insert. The ratio by which the cloned DNA was recognized more efficiently $(+S/-S = 5.1)$ points to a real change in specificity. The S factor also showed an effect when T4 DNA was used in combination with the chloroplast enzyme, although the stimulation $(+S/-S = 3.6)$ was less pronounced. The chloroplast rRNA gene region was chosen to test the specificity of the factor S. DNA fragments preceding the 16S rRNA gene, and a DNA fragment containing the 5S rRNA gene and sequences downstream were especially well transcribed. The S factor is present in the crude polymerase preparation and was separated from the enzyme by ion-exchange chromatography.

Control sequences flanking chloroplast genes

Bona fide transcription requires the recognition by the polymerase of specific nucleotide sequences at the beginning and at the end of a given gene or operon. In prokaryotes the RNA polymerase holoenzyme binds to DNA stretches known as '−10' (Pribnow-box) and '−35' regions (for a review see Bujard, 1980). A consensus sequence was compiled from a comparison of many such sites (see Rosenberg & Court, 1979; Siebenlist *et al.*, 1980; Hawley & McClure, 1983).

Chloroplast DNA from spinach has been probed with *E. coli* RNA polymerase under stringent conditions and found to contain several strong binding sites (Fig. 6) distributed over the unique-copy regions of the molecule (Zech *et al.*, 1981). The strongest binding of polymerase was observed at the *psb*A site. No strong binding of polymerase was found in the region of the rRNA genes. Even at a high polymerase/DNA ratio of 80, no binding was observed (Zech *et al.*, 1981) to the region of the 16S rRNA gene (Crouse *et al.*, 1978), whereas at a ratio of 10–15 the *psb*A binding site was saturated (Fig. 6).

The extensive work on characterization of chloroplast genes (for some recent reviews see Bohnert *et al.*, 1982; Buetow, 1982; Whitfeld & Bottomley, 1983; Groot, this volume), and the growing number of gene sequences, now permits an examination of putative chloroplast promoters and a comparison with the *E. coli* consensus sequence (Table 6). It should, however, be kept in mind that sequence homology is no proof for a promoter function of the chloroplast sequence in question. Supporting evidence is required for this. The 5′ end of the transcript should for instance be located only a few base-pairs downstream from the putative promoter site. This situation has been observed for several chloroplast genes where the start nucleotide of the mRNA is near the Pribnow-box (Table 6).

Several promoter-like sequences have been found by sequencing of tRNA genes (Table 6). At least in two cases in maize (Steinmetz *et al.*, 1983), these putative promoters are similar to the T7A2 promoter (Bujard, 1980). From studies of the 5′ end of the *Euglena* strain B 16S rDNA region, sequences similar to the consensus sequence of *E. coli* are known (El-Gewely *et al.*, 1984; R. Helling, personal communication). In accordance with the results mentioned above, putative promoter sequences are not that easily recognizable close to the chloroplast rDNA in maize. Since the start nucleotide for the primary transcript is not known, sequences upstream of the 5′ end of the 16S rRNA gene have to be analyzed. In maize cp-DNA three positions around nucleotides -420 (F1), -330 (F2) and -130 (F3) from the start of the 16S rRNA gene have been found to bind to *E. coli* RNA polymerase (Schwarz *et al.*, 1981*b*). Sites F1 and F3 exhibit sequence homologies (Table 6) with *E. coli rrn* promoters (Rosenberg & Court, 1979) within the region of close contact with the *E. coli* RNA polymerase. A sequence similar to the maize F2 putative promoter has been found upstream from the 5′ end of the spinach, tobacco and *Spirodela* 16S rRNA genes (see Table 6). In the region separating the chloroplast 4.5S and 5S rRNA genes, which is sequenced in maize (Dyer & Bedbrook, 1980), tobacco (Takaiwa & Sugiura, 1980*a*) and *Spirodela* (Keus *et al.*, 1983*b*), such a sequence is again found which might act as a transcription start.

In bacterial systems, another type of termination exists in addition to the aforementioned rho-dependent mechanism. It requires a specific site on the template and is factor-independent. This termination sequence is located downstream relative to the protein termination codon, where the denatured DNA may form a short double-stranded stem of variable length and a loop which is also variable in size. This structure appears to be recognized by the approaching RNA polymerase for termination, although a certain degree of readthrough occurs. Such a possible terminator was found downstream of the 3′ end of several chloroplast genes. Some examples are shown in Fig. 7 for the terminator structures of the *rbc*L gene of spinach (Zurawski *et al.*, 1981) and of maize (Zurawski *et al.*, 1981, as deduced from the published sequence by McIntosh *et al.*, 1980) and in the spinach and *Nicotiana debneyi psb*A genes (Zurawski *et al.*, 1982*a*). Hairpin structures in the 3′ flanking regions of some chloroplast tRNA genes have also been found (e.g. Steinmetz *et al.*, 1983).

Outlook

Genes coding for the constituents of chloroplasts are dispersed among the nuclear, chloroplast and, possibly, mitochondrial genomes. Cooperation between these genetic systems is essential for the biogenesis and function of the organelle. Many multimeric chloroplast structural elements

Table 6. *Putative promoter sequences for some chloroplast genes*

Gene(s)	Source	'−35 region'[a]		'−10 region'[a]		Reference
trnV-GAC	Spinach	(TTGAGT)[a]	17b	(TAGCAT)[a]	(33b)	Briat et al., 1982a
16S rDNA		TTGACG	18b	TATATT	4b	
trnV-GAC	Tobacco	TTGAGT	17b	TAGGAT	5–11b	Tohdoh et al., 1981
16S rDNA		(TTGACG)[a]	18b	(TATATT)[a]	(121b)	
trnV-GAC	Maize F1[b]	TTGACA	21b	TATTTG	(105b)	Schwarz et al., 1981b
16S rDNA	F2[b]	TTGCGT	17b	TAGGAT	(31b)	
	F3[b]	TTGACG	18b	TATACT	(124b)	
trnV-GAC	Spirodela	TTGAGT	17b	TAGGAT	(32b)	Keus et al., 1983a
16S rDNA		TTGACG	18b	TATATT	4b	
		TTGCTG	18b	TAATAT	(97b)	
16S rDNA	Chlamydomonas	(TTGACA)	17b	(TAAATT)[a]	(80b)	Dron et al., 1982b
5S rDNA	Tobacco	TTGGGG	19b	TATGCT	(32b)	Takaiwa & Sugiura, 1980a
5S rDNA	Spirodela	TTGGGG	19b	TATGCT	(36b)	Keus et al., 1983b
trnN-GUU	Tobacco	(TGAATG)[a,c]	20b	TATAAT	(152b)	Kato et al., 1981
		TTGGGA	11b	TATAAT	(152b)	
trnH-GUG	Maize	TGAATG[c]	18b	TTAGCT	4–7b	Schwarz et al., 1981a; Steinmetz et al., 1983
trnL-UAA	Maize	TTCAAA	18b	TAAATT	(35b)	Steinmetz et al., 1982, 1983
trnL-CAA	Maize	TTGTCA	18b	TTCGAT	(38b)	Steinmetz et al., 1983
		TTCCAT	17b	TATCAT	(29b)	
trnM-CAU	Tobacco	TTGCTT	18b	TATAAT	(47b)	Deno et al., 1982
trnM-CAU	Maize	TTGCTT	17b	TATAAT	(38b)	Steinmetz et al., 1983
trnF-GAA	Maize	TTGACA	18b	TAAGAT	(15b)	Steinmetz et al., 1983
trnS-GGA	Maize	TTCACT	17b	TAAGAT	(6b)	Steinmetz et al., 1983
trnT-UGU	Maize	TTTAGA	13b	TAAGAT	(75b)	Steinmetz et al., 1983
		TAATA	15b	TAAGAT	(62b)	
trnV-UAC	Tobacco	TTGACA	17b	TAAAAT	(17b)	Deno et al., 1982
rbcL	Spinach	TTGCGC	18b	TACAAT	4b	Zurawski et al., 1981; Whitfeld & Bottomley, 1983
rbcL	Tobacco	TTGCGC	18b	TACAAT	4b	Shinozaki & Sugiura, 1982a

Gene	Organism	'−35 region'	nucleotides between	'−10 region'	nucleotides following	Reference
rbcL	Maize	TTGATA	9b	TATCAT	34b	McIntosh et al., 1980; Jolly et al., 1981
		TTGATA	20b	TTAGAT	23b	
		TTGATA	21b	TAGATT	22b	
rbcL	Pea	TTGCGC	18b	TAGAAT	2b	Whitfeld & Bottomley, 1983
rbcL	Chlamydomonas	TTTACA	20b	TATAAT	6b	Dron et al., 1982a
rps4	Maize	TTGCTA	21b	TAATAT	(191b)	Subramanian et al., 1983
		TTGTGT	15b	TACTCT	(109b)	
		TTGAAT	14b	TAATGT	(59b)	
		TTGTGT	15b	TAAAAT	(15b)	
rps19	Tobacco	TTGGAA	23b	TATAGT	(103b)	Sugita & Sugiura, 1983; M. Sugiura, pers. comm.
atpA	Tobacco	(TTGAAC)[a]	17b	(TACCAT)[a]	(102b)	Deno et al., 1983
atpB	Spinach	TTGACA	21b	TATCCT	3b	Zurawski et al., 1982b; Whitfeld & Bottomley, 1983
atpB	Tobacco	TAGATA	17b	TATAAT	4–5b	Shinozaki & Sugiura, 1982b
atpB	Pea	TTGACA	21b	AATCCT	3b	Whitfeld & Bottomley, 1983
atpB	Maize	TTGACA	18b	TAGTAT	2–6b	Krebbers et al., 1982
psbA	Spinach	TTGACA	18b	TATACT	4b	Zurawski et al., 1982a
psbA	N. debneyi	TTGACA	18b	TATACT	4b	Zurawski et al., 1982a
psbA	Soybean	TTGACA	18b	TATACT	(87b)	Spielmann & Stutz, 1983
Plastidic consensus[a]		TTGaNa	9–23b	TAtaaT	2–34b	Rosenberg & Court, 1979; Siebenlist et al., 1980; Hawley & McClure, 1983
E. coli consensus[a]		TTGaca	16–19b	TAtaaT	7–10b	

Numbers between the '−35 region' and the '−10 region' (the Pribnow box) indicate the number of nucleotides separating these regions. Numbers following the '−10 region' give the number of nucleotides between the '−10 region' and the assumed start nucleotide of transcription or, as shown in parentheses, the start of the mature transcript.

[a] Sequence is not indicated as a putative promoter region in the reference.
[b] F1, F2 and F3 represent binding sites of E. coli polymerase (holoenzyme) on maize chloroplast DNA.
[c] Note similarity to the '−35 region' (GATTGAATGTAT) of bacteriophage λc in (Rosenberg & Court, 1979).
[a] Upper-case letters indicate that a base appears more frequently in that position of the promoter regions than do bases indicated by lower-case letters.

Fig. 7. Proposed terminator structures of some *rbc*L and *psb*A genes.

The number of nucleotides is counted from the protein termination codon (McIntosh *et al.*, 1980; Zurawski *et al.*, 1981, 1982*a*). The calculated free energy of formation, ΔG, expressed in Kcal is indicated.

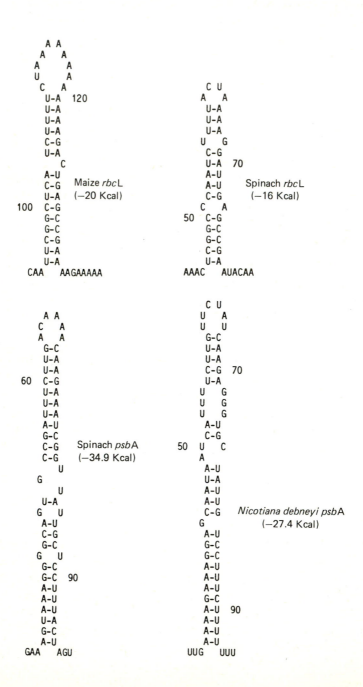

or enzymes are the assembled products of the nuclear and chloroplast genomes. The control mechanisms regulating the interplay between chloroplast and nucleus are poorly understood. Their elucidation and description in molecular terms will require more information on the structure and organization of chloroplast and nuclear genes and on their expression. This article focuses on the chloroplast side of this aspect, reviewing chloroplast genes and gene expression. Special emphasis has been laid on the transcription processes within chloroplasts. Work on other plastid types, such as chromoplasts and etioplasts, has been initiated (Thompson *et al.*, 1981). This research will widen the experimental approach to study different levels of plastid gene expression. The range of organisms from which plastids are studied has been limited to a few species. It is generally assumed that chloroplasts are similar with respect to types of genes present in their genome. However, nuclear genomes of plants vary to a much larger extent than plastid genomes. It has to be expected that nuclear genomes in different species may have developed different mechanisms to control plastid replication, development and basic functions (Stubbe, 1971). This is easily seen in the ways by which plants of different systematic categories respond to light and etiolation, or by the differential expression of genes within different tissues of the same plant.

The study of chloroplast transcription will contribute to this understanding. The clarification of the following points presented in the form of a series of questions would be helpful.

Which chloroplast genes overlap or how closely are they spaced?

How common are introns and what is their function?

How common are large spacer sequences within polycistronic units and what is their function?

How common are polycistronic transcriptional units?

How is a precursor RNA molecule processed into mature RNA(s)?

Are all the chloroplast promoters basically alike or are there families of promoters in order to switch on families of genes by signals from different compartments?

Are there factors resembling bacterial repressors or inducers, or sigma-type RNA polymerase factors in chloroplasts, and where are they encoded?

The importance of resolving these questions clearly requires the establishment of homologous transcriptional systems. Test genes for the screening of factors could be the already available *rbc*L and *psb*A genes which appear to be differently regulated in maize (Link *et al.*, 1978). The study of transcription *in vitro* has to be supplemented with other techniques. The many known strains of mutants from the alga *Chlamydomonas* (Gillham, 1978; Gillham *et al.*, 1979) might be useful to complement the physical approach.

Alternatively, it is possible to work with selected lines of plant cells, as for example interspecific somatic hybrids (Aviv *et al.*, 1980). Among the higher plants there are several species which have been extensively studied genetically, such as tobacco, *Petunia, Arabidopsis, Pelargonium*, maize and *Oenothera* (for a review see Kirk & Tilney-Bassett, 1978). Biochemical studies should be extended to these species.

We wish to thank Drs W. Bottomley, R. Hallick, R. Helling, K. Henningsen, H. Kössel, R. Martin, M. Mubumbila, A. Steinmetz, R. Schantz, P. Sitte, J. Thompson, D. Thurlow, J. H. Weil, P. Whitfeld, and G. Zurawski for helpful discussions and/or the permission to use unpublished results. We are indebted to the technical staffs of EMBL and IBMC, especially to Mrs A. Saunders, Mr C. Christensen, Mrs G. Niedergang and Ms M. Schneider, for their skilled technical assistance. Original results reported here were supported by the Deutsche Forschungsgemeinschaft. H.J.B. was recipient of a Heisenberg grant.

References

Alt, J., Westhoff, P., Sears, B. B., Nelson, N., Hurt, E., Hauska, G. & Herrmann, R. G. (1983*a*). Genes and transcripts for the polypeptides of the cytochrome *b*6/*f* complex from spinach thylakoid membranes. *EMBO J.*, **2**, 979–86.

Alt, J., Winter, P., Sebald, W., Moser, J. G., Schedel, R., Westhoff, P. & Herrmann, R. G. (1983*b*). Localization and nucleotide sequence of the gene for the ATP synthetase proteolipid subunit on the spinach plastid chromosome. *Curr. Genet.*, **7**, 129–38.

Andrews, W. H. & Rawson, J. R. Y. (1982). Expression of cloned chloroplast DNA from *Euglena gracilis* in an *in vitro* DNA-dependent transcription–translation system prepared from *E. coli. Plasmid*, **8**, 148–63.

Apel, K. & Bogorad, L. (1976). Light-induced increase in the activity of maize plastid DNA-dependent RNA polymerase. *Eur. J. Biochem.*, **67**, 615–20.

Aviv, D., Fluhr, R., Edelman, M. & Galun, E. (1980). Progeny analysis of the interspecific somatic hybrids: *Nicotiana tabacum* (CMS) + *Nicotiana sylvestris* with respect to nuclear and chloroplast markers. *Theoret. appl. Genet.*, **56**, 145–50.

Barnett, W. E., Pennington, C. J. & Fairfield, S. A. (1969). Induction of *Euglena* transfer RNAs by light. *Proc. natn. Acad. Sci. USA*, **63**, 1261–8.

Barnett, W. E., Schwartzbach, S. D. & Hecker, L. I. (1978). The transfer RNAs of eukaryotic organelles. *Prog. Nucleic Acid Res. molec. Biol.*, **21**, 143–79.

Bartlett, S. G., Boynton, J. E. & Gillham, N. W. (1981). Genetics of photosynthesis and the chloroplast. In *Genetics as a Tool in Microbiology*. Society for General Microbiology Symposium 31, ed. S. W. Glover & D. A. Hopwood, pp. 379–412. Cambridge University Press, Cambridge.

Becker, W. M. (1979). RNA polymerase in plants. In *Nucleic Acids in Plants*, vol. 1, ed. T. C. Hall & J. W. Davies, pp. 111–41. CRC Press, Boca Raton.

Bedbrook, J. R. & Kolodner, R. (1979). The structure of chloroplast DNA. *A. Rev. Pl. Physiol.*, **30**, 593–620.

Bedbrook, J. R., Link, G., Coen, D. M., Bogorad, L. & Rich, A. (1978). Maize plastid gene expressed during photoregulated development. *Proc. natn. Acad. Sci. USA*, **75**, 3060–4.

Bennett, J. & Ellis, R. J. (1973). Solubilization of the membrane-bound deoxyribonucleic acid-dependent ribonucleic acid polymerase of pea chloroplasts. *Biochem. Soc. Trans.*, **1**, 892–4.

Bennett, J. & Milewska, Y. (1976). Incorporation of ^{32}P-orthophosphate into nucleoside 5′-triphosphate and RNA by isolated pea chloroplasts. In *Genetics and Biogenesis of Chloroplasts and Mitochondria*, ed. Th. Bücher, W. Neupert, W. Sebald & S. Werner, pp. 637–40. North-Holland, Amsterdam and New York.

Blair, G. E. & Ellis, R. J. (1973). Protein synthesis in chloroplasts. I. Light-driven synthesis of the large subunit of fraction I protein by isolated pea chloroplasts. *Biochim. biophys. Acta*, **319**, 223–34.

Blanc, M., Briat, J. F. & Laulhere, J. P. (1981). Influence of the ionic environment on the *in vitro* transcription of the spinach plastid DNA by a selectively bound RNA-polymerase DNA complex. *Biochim. biophys. Acta*, **655**, 374–82.

Bogorad, L., Bedbrook, J. R., Coen, D. M., Kolodner, R. & Link, G. (1978). Genes for chloroplast proteins and rRNAs. In *Chloroplast Development*, ed. G. Akoyunoglou & J. H. Argyroudi-Akoyunoglou, pp. 541–51. Elsevier, Amsterdam.

Bogorad, L., Gubbins, E. J., Krebbers, E. T., Larrinua, I. M., Muskavitch, K. M. T., Rodermel, S. R. & Steinmetz, A. (1983). The organization and expression of maize plastid genes. In *Genetic Engineering*, ed. L. D. Owens, pp. 35–53. Granada, London, Toronto and Sydney.

Bohnert, H. J., Crouse, E. J. & Schmitt, J. M. (1982). Organization and expression of plastid genomes. In *Encyclopedia of Plant Physiology, New Series*, vol. 14B, ed. B. Parthier & D. Boulter, pp. 475–530. Springer Verlag, Berlin, Heidelberg and New York.

Bohnert, H. J., Driesel, A. J., Crouse, E. J., Gordon, K., Herrmann, R. G., Steinmetz, A., Mubumbila, M., Keller, M., Burkard, G. & Weil, J. H. (1979). Presence of a transfer RNA gene in the spacer sequence between the 16S and 23S rRNA genes of spinach chloroplast DNA. *FEBS Letts.*, **103**, 52–6.

Bohnert, H. J., Driesel, A. J. & Herrmann, R. G. (1976). Characterization of the RNA compounds synthesized by isolated chloroplasts. In *Genetics and Biogenesis of Chloroplasts and Mitochondria*, ed. Th. Bücher, W. Neupert, W. Sebald & S. Werner, pp. 629–36. North-Holland, Amsterdam and New York.

Bohnert, H. J., Driesel, A. J. & Herrmann, R. G. (1977). Transcription and processing of transcripts in isolated unbroken chloroplasts. In *Nucleic Acids and Protein Synthesis in Plants*, ed. J. H. Weil & L. Bogorad, pp. 213–18. Centre National de la Recherche Scientifique, Paris.

Bohnert, H. J., Schmitt, J. M. & Herrmann, R. G. (1974). Structural and functional aspects of the plastome. III. DNA and RNA synthesis by isolated chloroplasts. *Port. Acta biol.*, **14**, 71–90.

Bottomley, W. & Bohnert, H. J. (1982). The biosynthesis of chloroplast proteins. In *Encyclopedia of Plant Physiology, New Series*, vol. 14B, ed. B. Parthier & D. Boulter, pp. 531–96. Springer Verlag, Berlin, Heidelberg, New York.

Bottomley, W., Smith, H. J. & Bogorad, L. (1971). RNA polymerases of maize: Partial purification and properties of the chloroplast enzyme. *Proc. natn. Acad. Sci. USA*, **68**, 2412–16.

Bottomley, W., Spencer, D. & Whitfeld, P. R. (1974). Protein synthesis in isolated spinach chloroplasts: comparison of light-driven and ATP-driven synthesis. *Archs. Biochem. Biophys.*, **164**, 106–17.

Bottomley, W. & Whitfeld, P. (1979). Cell-free transcription and translation of total spinach chloroplast DNA. *Eur. J. Biochem.*, **93**, 31–9.

Bowman, C. M. & Dyer, T. A. (1979). 4.5S Ribonucleic acid, a novel ribosome component in the chloroplasts of flowering plants. *Biochem. J.*, **183**, 605–13.

Brantner, J. H. & Dure, L. S. (1975). The developmental biochemistry of cotton seed embryogenesis and germination. VI. Levels of cytosol and chloroplast aminoacyl-tRNA synthetases during cotyledon development. *Biochim. biophys. Acta*, **414**, 99–114.

Briat, J. F., Dron, M., Loiseaux, S. & Mache, R. (1982a). Structure and transcription of the spinach chloroplast rDNA leader region. *Nuc. Acids Res.*, **10**, 6865–78.

Briat, J. F., Laulhere, J. P. & Mache, R. (1979). Transcription activity of a DNA–protein complex isolated from spinach plastids. *Eur. J. Biochem.*, **98**, 285–92.

Briat, J. F., Gigot, C., Laulhere, J. P. & Mache, R. (1982b). Visualization of a spinach plastid transcriptionally active DNA–protein complex in a highly condensed structure. *Pl. Physiol.*, **69**, 1205–11.

Briat, J. F. & Mache, R. (1980). Properties and characterization of a spinach chloroplast RNA polymerase isolated from a transcriptionally active DNA–protein complex. *Eur. J. Biochem.*, **111**, 503–9.

Buetow, D. E. (1982). Molecular biology of chloroplasts. In *Photosynthesis: Development, Carbon Metabolism and Plant Productivity*, vol. II, ed. Govindjee, pp. 43–88. Academic Press, New York.

Bujard, H. (1980). The interaction of *E. coli* RNA polymerase with promoters. *Trends Biochem. Sci.*, **5**, 274–8.

Burkard, G., Steinmetz, A., Keller, M., Mubumbila, M., Crouse, E. J. & Weil, J. H. (1982). Resolution of chloroplast tRNAs by two-dimensional gel electrophoresis. In *Methods in Chloroplast Molecular Biology*, ed. M. Edelman, R. B. Hallick & N. H. Chua, pp. 347–57. Elsevier, Amsterdam.

Burkard, G., Vaultier, J. P. & Weil, J. H. (1972). Differences in the level of plastid-specific tRNAs in chloroplasts and etioplasts of *Phaseolus vulgaris*. *Phytochemistry*, **11**, 1351–3.

Calaghan, J. L., Pirtle, R. M., Pirtle, I. L., Kashdan, M. A., Vreman, H. J.

& Dudock, B. S. (1980). Homology between chloroplast and prokaryotic initiator tRNA, nucleotide sequence of spinach chloroplast methionine initiator tRNA. *J. biol. Chem.*, **255**, 9981–4.

Camerino, G., Carbonera, D., Sanangelantoni, A. M., Riccardi, G. & Ciferri, O. (1982*a*). *In vitro* translation of chloroplast mRNAs. *Pl. Sci. Lett.*, **27**, 191–202.

Camerino, G., Savi, A. & Ciferri, O. (1982*b*). A chloroplast system capable of translating heterologous mRNAs. *FEBS Letts.*, **150**, 94–8.

Canaday, J., Guillemaut, P., Gloeckler, R. & Weil, J. H. (1980*a*). Comparison of the nucleotide sequences of chloroplast tRNAs^{Phe} and tRNAs$_3^{Leu}$ from spinach and bean. *Pl. Sci. Lett.*, **20**, 57–62.

Canaday, J., Guillemaut, P., Gloeckler, R. & Weil, J. H. (1981). The nucleotide sequence of spinach chloroplast tryptophan transfer RNA. *Nucleic Acids Res.*, **9**, 47–53.

Canaday, J., Guillemaut, P. & Weil, J. H. (1980*b*). The nucleotide sequences of initiator transfer RNAs from bean cytoplasm and chloroplasts. *Nucleic Acids Res.*, **8**, 999–1008.

Carritt, B. & Eisenstadt, J. M. (1973). RNA synthesis in isolated chloroplasts: characterization of the newly synthesized RNA. *FEBS Lett.*, **36**, 116–20.

Chambon, P. (1975). Eukaryotic nuclear RNA polymerases. *A. Rev. Biochem.*, **44**, 613–38.

Chang, S. H., Brum, C. K., Silberklang, M., RajBhandary, U. L., Hecker, L. I. & Barnett, W. E. (1976). The first nucleotide sequence of an organelle transfer RNA: chloroplastic tRNA^{Phe}. *Cell*, **9**, 717–23.

Chelm, B. K., Gray, P. W. & Hallick, R. B. (1978). Mapping of transcribed regions of *Euglena gracilis* chloroplast DNA. *Biochemistry* (*Wash.*), **17**, 4239–44.

Chelm, B. K. & Hallick, R. B. (1976). Changes in the expression of the chloroplast genome of *Euglena gracilis* during chloroplast development. *Biochemistry* (*Wash.*), **15**, 593–9.

Chelm, B. K., Hallick, R. B. & Gray, P. W. (1979). Transcription program of the chloroplast genome of *Euglena gracilis* during chloroplast development. *Proc. natn. Acad. Sci. USA*, **76**, 2258–62.

Chu, N. M., Oishi, K. K. & Tewari, K. K. (1981). Physical mapping of the pea chloroplast DNA and localization of the ribosomal RNA genes. *Plasmid*, **6**, 279–92.

Chua, N. H. & Schmidt, G. W. (1978). Post-translational transport into intact chloroplasts of a precursor to the small subunit of ribulose-1,5-bisphosphate carboxylase. *Proc. natn. Acad. Sci. USA*, **72**, 6110–14.

Coen, D. M., Bedbrook, J. R., Bogorad, L. & Rich, A. (1977). Maize chloroplast DNA fragment encoding the large subunit of ribulose bisphosphate carboxylase. *Proc. natn. Acad. Sci. USA*, **74**, 5487–91.

Crouse, E. J., Schmitt, J. M., Bohnert, H. J., Gordon, K., Driesel, A. J. & Herrmann, R. G. (1978). Intramolecular compositional heterogeneity of *Spinacia* and *Euglena* chloroplast DNAs. In *Chloroplast Development*, ed. G. Akoyunoglou & J. H. Argyroudi-Akoyunoglou, pp. 565–72. Elsevier/North-Holland, Amsterdam.

Delihas, N. & Andersen, J. (1982). Generalized structures of the 5S ribosomal RNAs. *Nucleic Acids Res.*, **10**, 7323–44.

Delihas, N., Andersen, J., Sprouse, H. M. & Dudock, B. (1981). The nucleotide sequence of the chloroplast 5S ribosomal RNA from spinach. *Nucleic Acids Res.*, **9**, 2801–5.

Deno, H., Kato, A., Shinozaki, K. & Sugiura, M. (1982). Nucleotide sequences of tobacco chloroplast genes for elongator tRNA^{Met} and tRNA^{Val} (UAC): the tRNA^{Val} (UAC) gene contains a long intron. *Nucleic Acids Res.*, **10**, 7511–20.

Deno, H., Shinozaki, K. & Sugiura, M. (1983). Nucleotide sequence of tobacco chloroplast gene for the α subunit of proton-translocating ATPase. *Nucleic Acids Res.*, **11**, 2185–91.

Detchon, P. & Possingham, J. V. (1973). Chloroplast ribosomal ribonucleic acid synthesis in cultured spinach leaf tissue. *Biochem. J.*, **136**, 829–36.

Dix, K. P. & Rawson, J. R. Y. (1983). *In vivo* transcription products of the chloroplast DNA of *Euglena gracilis*. *Curr. Genet.*, **7**, 265–72.

Driesel, A. J., Crouse, E. J., Gordon, K., Bohnert, H. J., Herrmann, R. G., Steinmetz, A., Mubumbila, M., Keller, M., Burkard, G. & Weil, J. H. (1979). Fractionation and identification of spinach chloroplast transfer tRNAs and mapping of their genes on the restriction map of chloroplast DNA. *Gene*, **6**, 285–306.

Driesel, A. J., Speirs, J. & Bohnert, H. J. (1980). Spinach chloroplast mRNA for a 32 000 dalton polypeptide: size and localization on the physical map of the chloroplast DNA. *Biochim. biophys. Acta*, **610**, 297–310.

Dron, M., Rahire, M. & Rochaix, J. D. (1982*a*). Sequence of the chloroplast DNA region of *Chlamydomonas reinhardtii* containing the gene of the large subunit of ribulose bisphosphate carboxylase and parts of its flanking genes. *J. molec. Biol.*, **162**, 775–93.

Dron, M., Rahire, M. & Rochaix, J. D. (1982*b*). Sequence of the chloroplast 16S rRNA gene and its surrounding regions of *Chlamydomonas reinhardtii*. *Nucleic Acids Res.*, **10**, 7609–20.

Dron, M., Rahire, M., Rochaix, J. D. & Metz, L. (1983). First DNA sequence of a chloroplast mutation: a missense alteration in the ribulose bisphosphate carboxylase large subunit gene. *Plasmid*, **9**, 321–4.

Dyer, T. A. (1982). RNA sequences. In *Encyclopedia of Plant Physiology, New Series*, vol. 14B, ed. B. Parthier & D. Boulter, pp. 171–91. Springer-Verlag, Berlin, Heidelberg and New York.

Dyer, T. A. & Bedbrook, J. R. (1980). The organization in higher plants of the genes coding for chloroplast ribosomal RNA. In *Genome Organization and Expression in Plants*, ed. C. J. Leaver, pp. 305–11. Plenum Press, New York and London.

Dyer, T. A. & Bowman, C. M. (1979). Nucleotide sequences of chloroplast 5S ribosomal ribonucleic acid in flowering plants. *Biochem. J.*, **183**, 595–604.

Dyer, T. A. & Leaver, C. J. (1981). RNA: structure and metabolism. In *The Biochemistry of Plants*, vol. 6, ed. A. Marcus, pp. 111–68. Academic Press, New York and London.

Edelman, M. (1981). Nucleic acids of chloroplasts and mitochondria. In *The Biochemistry of Plants*, vol. 6, ed. A. Marcus, pp. 249–301. Academic Press, New York and London.

Edelman, M., Hallick, R. B. & Chua, N. H. (1982). *Methods in Chloroplast Molecular Biology*. Elsevier Biomedical Press, Amsterdam, New York and Oxford.

Edelman, M. & Reisfeld, A. (1980). Synthesis, processing and functional probing of P-32000, the major membrane protein translated within the chloroplast. In *Genome Organization and Expression in Plants*, ed. C. J. Leaver, pp. 353–62. Plenum Press, New York and London.

Edwards, K., Bedbrook, J., Dyer, T. & Kössel, H. (1981). 4.5S rRNA from *Zea mays* chloroplasts shows structural homology with the 3′ end of procaryotic 23S rRNA. *Biochem. Int.*, **2**, 533–8.

Edwards, K. & Kössel, H. (1981). The rRNA operon from *Zea mays* chloroplasts: nucleotide sequence of 23S rDNA and its homology with *E. coli* 23S rDNA. *Nucleic Acids Res.*, **9**, 2853–69.

El-Gewely, M. R., Helling, R. B. & Dibbits, J. G. T. (1984). Sequences and evolution of the regions between the *rrn* operons in the chloroplast genome of *Euglena gracilis bacillaris*. *Molec. gen. Genet.*, **194**, 432–43.

El-Gewely, M. R., Helling, R. B., Farmerie, W. & Barnett, W. E. (1982). Location of a phenylalanine tRNA gene on the physical map of the *Euglena gracilis* chloroplast genome. *Gene*, **17**, 337–9.

Ellis, R. J. (1976). Protein and nucleic acid synthesis by chloroplasts. In *The Intact Chloroplast*, ed. J. Barber, pp. 335–64. Elsevier, Amsterdam and New York.

Ellis, R. J. (1977). Protein synthesis by isolated chloroplasts. *Biochim. biophys. Acta*, **463**, 185–215.

Ellis, R. J. (1981). Chloroplast proteins: synthesis, transport and assembly. *A. Rev. Pl. Physiol.*, **32**, 111–37.

Ellis, R. J. (1982). Inhibitors for studying chloroplast transcription and translation *in vivo*. In *Methods in Chloroplast Molecular Biology*, ed. M. Edelman, R. Hallick & N.-H. Chua, pp. 559–64. Elsevier, Biomedical Press, Amsterdam.

Ellis, R. J. (1983). Chloroplast protein synthesis: principles and problems. *Subcell. Biochem.*, **9**, 235–61.

Ellis, R. J., Smith, S. M. & Barraclough, R. (1980). Synthesis, transport and assembly of chloroplast proteins. In *Genome Organization and Expression in Plants*, ed. C. J. Leaver, pp. 321–5. Plenum Press, New York and London.

Eneas-Filho, J., Hartley, M. R. & Mache, R. (1981). Pea chloroplast ribosomal proteins: characterization and site of synthesis. *Molec. gen. Genet.*, **184**, 484–8.

Erdmann, V. A., Huysmans, E., Vandenberghe, A. & De Wachter, R. (1983). Collection of published 5S and 5.8S ribosomal RNA sequences. *Nucleic Acids Res.*, **11**, r105–r133.

Francis, M. A. & Dudock, B. S. (1982). Nucleotide sequence of a spinach chloroplast isoleucine tRNA. *J. biol. Chem.*, **257**, 11195–8.

Francis, M., Kashdan, M., Sprouse, H., Otis, L. & Dudock, B. (1982). Nucleotide sequence of a spinach chloroplast proline tRNA. *Nuc. Acids Res.*, **10**, 2755–8.

Galling, G. (1982). Use (and misuse) of inhibitors in gene expression. In *Encyclopedia of Plant Physiology, New Series*, vol. 14B, ed. B. Parthier & D. Boulter, pp. 663–77. Springer-Verlag, Berlin, Heidelberg and New York.

Gauss, D. H. & Sprinzl, M. (1983*a*). Compilation of tRNA sequences. *Nucleic Acids Res.*, **11**, r1–r53.

Gauss, D. H. & Sprinzl, M. (1983*b*). Compilation of sequences of tRNA genes. *Nucleic Acids Res.*, **11**, r55–r103.

Gillham, N. W. (1978). *Organelle Heredity*. Raven Press, New York.

Gillham, N. W. & Boynton, J. E. (1981). Evolution of organelle genomes and protein-synthesizing systems. *Annls NY Acad. Sci.*, **361**, 20–43.

Gillham, N. W., Boynton, J. E., Grant, D. M., Shepherd, H. S. & Wurtz, E. A. (1979). Genetic analysis of chloroplast DNA function in *Chlamydomonas*. In *Extrachromosomal DNA*, ICN–UCLA Symposia on Molecular and Cellular Biology, vol. 15, ed. D. J. Cummings, P. Borst, I. B. Dawid, S. M. Weisman & C. F. Fox, pp. 75–96. Academic Press, New York.

Gillham, N. W., Boynton, J. E. & Harris, E. H. (1984). Evolution of plastid DNA. In *Evolution of Genome Size*, ed. T. Cavalier-Smith. John Wiley and Sons, London, New York, Sydney and Toronto. (In press.)

Gold, L., Pribnow, D., Schneider, T., Shinedling, S. & Storma, G. (1981). Translational initiation in prokaryotes. *A. Rev. Microbiol.*, **35**, 365–403.

Gózdzicka-Józefiak, A. & Augustyniak, J. (1983). Number of tRNA genes in wheat chloroplast DNA. Two different genes for valine tRNAs. *FEBS Lett.*, **156**, 51–4.

Graf, L., Kössel, H. & Stutz, E. (1980). Sequencing of 16S–23S spacer in a ribosomal RNA operon of *Euglena gracilis* chloroplast DNA reveals two tRNA genes. *Nature*, **286**, 908–10.

Graf, L., Roux, E., Stutz, E. & Kössel, H. (1982). Nucleotide sequence of a *Euglena gracilis* chloroplast gene coding for the 16S rRNA: homologies to *E. coli* and *Zea mays* chloroplast 16S rRNA. *Nucleic Acids Res.*, **10**, 6369–81.

Gray, M. W. & Doolittle, W. F. (1982). Has the endosymbiont hypothesis been proven? *Microbiol. Rev.*, **46**, 1–42.

Gray, P. W. & Hallick, R. B. (1979). Isolation of *Euglena gracilis* chloroplast 5S ribosomal RNA and mapping the 5S rRNA gene on chloroplast DNA. *Biochemistry (Wash.)*, **18**, 1820–5.

Grierson, D. (1982). RNA processing and other post-transcriptional modifications. In *Encyclopedia of Plant Physiology, New Series*, vol. 14B, ed. B. Parthier & D. Boulter, pp. 192–223. Springer-Verlag, Berlin, Heidelberg and New York.

Grierson, D. & Loening, U. (1974). Ribosomal RNA precursors and the synthesis of chloroplast and cytoplasmic ribosomal ribonucleic acid in leaves of *Phaseolus vulgaris*. *Eur. J. Biochem.*, **44**, 501–7.

Grossman, A., Bartlett, S. & Chua, N. H. (1980). Energy-dependent uptake of cytoplasmically-synthesized polypeptides by chloroplasts. *Nature*, **285**, 625–8.

Gruissem, W., Prescott, D. M., Greenberg, B. M. & Hallick, R. B. (1982). Transcription of *E. coli* and *Euglena* chloroplast tRNA gene clusters and processing of polycistronic transcripts in a HeLa cell-free system. *Cell*, **30**, 81–92.

Gruol, D. J. & Haselkorn, R. (1976). Counting the genes for stable RNA in the nucleus and chloroplasts of *Euglena. Biochim. biophys. Acta*, **477**, 82–95.

Guderian, R. H., Pulliam, R. L. & Gordon, M. P. (1972). Characterization and fractionation of tobacco leaf transfer RNA. *Biochim. biophys. Acta*, **262**, 50–65.

Guilfoyle, T. J. (1981). DNA and RNA polymerase. In *The Biochemistry of Plants*, vol. 6, ed. A. Marcus, pp. 207–47. Academic Press, New York and London.

Guillemaut, P., Burkard, G., Steinmetz, A. & Weil, J. H. (1973). Comparative studies on the tRNAs[Met] from the cytoplasm, chloroplasts and mito-chondria of *Phaseolus vulgaris. Pl. Sci. Lett.*, **1**, 141–9.

Guillemaut, P. & Keith, G. (1977). Primary structure of bean chloroplastic tRNA[Phe]: comparison with *Euglena* chloroplastic tRNA[Phe]. *FEBS Lett.*, **84**, 351–6.

Guillemaut, P. & Weil, J. H. (1982). The nucleotide sequence of the maize and spinach chloroplast isoleucine transfer RNA encoded in the 16S to 23S rDNA spacer. *Nucleic Acids Res.*, **10**, 1653–9.

Haff, L. A. & Bogorad, L. (1976a). Hybridization of maize chloroplast DNA with transfer ribonucleic acids. *Biochemistry (Wash.)*, **15**, 4105–9.

Haff, L. A. & Bogorad, L. (1976b). Poly(adenylic acid)-containing RNA from plastids of maize. *Biochemistry (Wash.)*, **15**, 4110–15.

Hall, T. C. (1979). Plant messenger RNA. In *Nucleic Acids in Plants*, vol. I, ed. T. C. Hall & J. W. Davies, pp. 217–51. CRC Press, Boca Raton.

Hallick, R. B. (1984). Chloroplast DNA. In *The Biology of Euglena*, vol. IV, ed. D. E. Buetow. Academic Press, New York and London. (In press.)

Hallick, R. B., Gray, P. W., Chelm, B. K., Rushlow, K. E. & Orozco, E. M. (1978). *Euglena gracilis* chloroplast DNA structure, gene mapping, and RNA transcription. In *Chloroplast Development*, ed. G. Akoyunoglou & J. H. Argyroudi-Akoyunoglou, pp. 619–22. Elsevier/North-Holland, Amsterdam.

Hallick, R. B., Lipper, C., Richards, O. C. & Rutter, W. J. (1976). Isolation of a transcriptionally active chromosome from chloroplasts of *Euglena gracilis. Biochemistry (Wash.)*, **15**, 3039–45.

Hartley, M. R. (1979). The synthesis and origin of chloroplast low molecular weight ribosomal ribonucleic acid in spinach. *Eur. J. Biochem.*, **96**, 311–20.

Hartley, M. R. & Ellis, R. J. (1973). Ribonucleic acid synthesis in chloroplasts. *Biochem. J.*, **134**, 249–62.

Hartley, M. R. & Head, C. (1979). The synthesis of chloroplast high molecular weight ribosomal ribonucleic acid in spinach. *Eur. J. Biochem.*, **96**, 301–9.

Hartley, M. R., Head, C. W. & Gardiner, J. (1977). The synthesis of chloroplast RNA. In *Nucleic Acids and Protein Synthesis in Plants*, ed. L. Bogorad & J. H. Weil, pp. 419–23. Centre National de la Recherche Scientifique, Paris.

Hartley, M. R., Wheeler, A. & Ellis, R. J. (1975). Protein synthesis in chloroplasts. V. Translation of messenger RNA for the large subunit of fraction I protein in a heterologous cell-free system. *J. molec. Biol.*, **91**, 67–77.

Hawley, D. K. & McClure, W. R. (1983). Compilation and analysis of *Escherichia coli* promoter DNA sequences. *Nucleic Acids Res.*, **11**, 2237–55.

Heber, U. & Heldt, H. W. (1981). The chloroplast envelope: structure, function and role in leaf metabolism. *A. Rev. Pl. Physiol.*, **32**, 139–68.

Heizmann, P. (1974). Maturation of chloroplast rRNA in *Euglena gracilis*. *Biochem. biophys. Res. Commun.*, **56**, 112–18.

Herrmann, R. G. & Possingham, J. V. (1980). Plastid DNA – the plastome. In *Results and Problems in Cell Differentiation*, vol. 10, ed. J. Reinert, pp. 45–96. Springer-Verlag, Berlin, Heidelberg and New York.

Hiatt, V. S. & Snyder, L. A. (1973). Phenylalanine transfer RNA species in early development of barley. *Biochim. biophys. Acta*, **324**, 57–68.

Highfield, P. E. & Ellis, R. J. (1978). Synthesis and transport of the small subunit of chloroplast ribulose bisphosphate carboxylase. *Nature*, **271**, 420–4.

Hollingsworth, M. J. & Hallick, R. B. (1982). *Euglena gracilis* chloroplast transfer RNA transcription units. Nucleotide sequence analysis of a $tRNA^{Tyr}$–$tRNA^{His}$–$tRNA^{Met}$–$tRNA^{Trp}$–$tRNA^{Glu}$–$tRNA^{Gly}$ gene cluster. *J. biol. Chem.*, **257**, 12795–9.

Howe, C. J., Auffret, A. D., Doherty, A., Bowman, C. M., Dyer, T. A. & Gray, J. C. (1983a). Location and nucleotide sequence of the gene for the proton-translocating subunit of wheat chloroplast ATP synthase. *Proc. natn. Acad. Sci. USA*, **79**, 6903–7.

Howe, C. J., Bowman, C. M., Dyer, T. A. & Gray, J. C. (1982). Localization of wheat chloroplast genes for the beta and epsilon subunits of ATP synthase, *Molec. gen. Genet.*, **186**, 525–30.

Howe, C. J., Bowman, C. M., Dyer, T. A. & Gray, J. C. (1983b). The genes for the alpha and proton-translocating subunits of wheat chloroplast ATP synthase are close together on the same strand of chloroplast DNA. *Molec. gen. Genet.*, **190**, 51–5.

Howell, S. H., Heizmann, P., Gelvin, S. & Walker, L. L. (1977). Identification and properties of the messenger RNA activity in *Chlamydomonas reinhardtii* coding for the large subunit of ribulose-1,5-bisphosphate carboxylase. *Pl. Physiol.*, **59**, 464–70.

Howell, S. H. & Walker, L. L. (1977). Transcription of the nuclear and chloroplast genomes during the vegetative cell cycle in *Chlamydomonas reinhardtii*. *Devl Biol.*, **56**, 11–23.

Ingle, J. (1968). Synthesis and stability of chloroplast ribosomal RNAs. *Pl. Physiol.*, **43**, 1448–54.

Jolly, S. J. & Bogorad, L. (1980). Preferential transcription of cloned maize chloroplast DNA sequences by maize chloroplast RNA polymerase. *Proc. natn. Acad. Sci. USA*, **77**, 822–6.

Jolly, S. O., McIntosh, L., Link, G. & Bogorad, L. (1981). Differential transcription *in vivo* and *in vitro* of two adjacent maize chloroplast genes: the large subunit of ribulose bisphosphate carboxylase and the 2.2-kilobase gene. *Proc. natn. Acad. Sci. USA*, **78**, 6821–5.

Jones, D. S. (1980). The isolation, characterization and structure of chloroplast $tRNA_f^{Met}$ and $tRNA_m^{Met}$ from the green alga *Scenedesmus obliquus*. *EMBO–FEBS tRNA Workshop, Abstract*, Strasbourg, July 16–21.

Joussaume, M. (1973). Mise en évidence de deux formes de RNA polymerase dépendante du DNA dans les chloroplastes isolés de feuilles de poire. *Physiol. Veget.*, **11**, 69–82.

Kahn, A. & von Wettstein, D. (1961). Macromolecular physiology of plastids. *J. Ultrastruct. Res.*, **5**, 557–74.

Karabin, G. D. & Hallick, R. B. (1983). *Euglena gracilis* chloroplast transfer RNA transcription units. Nucleotide sequence analysis of a $tRNA^{Thr}$–$tRNA^{Gly}$–$tRNA^{Met}$–$tRNA^{Ser}$–$tRNA^{Gln}$ gene cluster. *J. biol. Chem.*, **258**, 5512–18.

Karwowska, U., Góździcka-Józefiak, A. & Augustyniak, J. (1979). Chloroplast-specific leucine tRNAs from wheat. *Acta biochim. polon.*, **26**, 319–26.

Kashdan, M. A. & Dudock, B. S. (1982*a*). Structure of a spinach chloroplast threonine tRNA gene. *J. biol. Chem.*, **257**, 1114–16.

Kashdan, M. A. & Dudock, B. S. (1982*b*). The gene for a spinach chloroplast isoleucine tRNA has a methionine anticodon. *J. biol. Chem.*, **257**, 11191–4.

Kashdan, M. A., Pirtle, R. M., Pirtle, I. L., Calagan, J. L., Vreman, H. J. & Dudock, B. S. (1980). Nucleotide sequence of a spinach chloroplast threonine tRNA. *J. biol. Chem.*, **255**, 8831–5.

Kato, A., Shimada, H., Kusada, M. & Sugiura, M. (1981). The nucleotide sequences of two $tRNA^{Asn}$ genes from tobacco chloroplasts. *Nucleic Acids Res.*, **9**, 5601–7.

Keller, M., Burkard, G., Bohnert, H. J., Mubumbila, M., Gordon, K., Steinmetz, A., Heiser, D., Crouse, E. J. & Weil, J. H. (1980). Transfer RNA genes associated with the 16S and 23S rRNA genes of *Euglena* chloroplast DNA. *Biochem. biophys. Res. Commun.*, **95**, 47–54.

Keus, R. J. A., Dekker, A. F., van Roon, M. A. & Groot, G. S. P. (1983*a*). The nucleotide sequences of the regions flanking the genes coding for 23S, 16S and 4.5S ribosomal RNA on chloroplast DNA from *Spirodela oligorhiza*. *Nucleic Acids Res.*, **11**, 6465–74.

Keus, R. J. A., Roovers, D. J., Dekker, A. F. & Groot, G. S. P. (1983*b*). The nucleotide sequence of the 4.5S and 5S RNA genes and flanking regions from *Spirodela oligorhiza* chloroplasts. *Nucleic Acids Res.*, **11**, 3405–10.

Kidd, G. H. & Bogorad, L. (1980). A facile procedure for purifying maize chloroplast RNA polymerase from whole cell homogenates. *Biochim. biophys. Acta*, **609**, 14–30.

Kirk, J. T. O. (1964). DNA-dependent RNA synthesis in chloroplast preparations. *Biochem. biophys. Res. Commun.*, **14**, 393–7.

Kirk, J. T. O. (1971). Will the real chloroplast DNA please stand up? In *Autonomy and Biogenesis of Mitochondria and Chloroplasts*, ed. N. K. Boardman, A. W. Linnane & R. M. Smillie, pp. 267–76. North-Holland, Amsterdam.

Kirk, J. T. O. & Tilney-Bassett, R. A. E. (1978). *The Plastids: their chemistry, structure, growth and inheritance*, 2nd edn. Elsevier/North-Holland Biomedical Press, Amsterdam, New York and Oxford.

Koch, W., Edwards, K. & Kössel, H. (1981). Sequencing of the 16S–23S spacer in a ribosomal RNA operon of *Zea mays* chloroplast DNA reveals two split tRNA genes. *Cell*, **25**, 203–13.

Koller, B., Delius, H. & Dyer, T. A. (1982). The organization of the chloroplast DNA in wheat and maize in the region containing the LS gene. *Eur. J. Biochem.*, **122**, 17–23.

Kozak, M. (1983). Comparison of initiation of protein synthesis in procaryotes, eucaryotes, and organelles. *Microbiol. Rev.*, **47**, 1–45.

Krebbers, E. T., Larrinua, I. M., McIntosh, L. & Bogorad, L. (1982). The maize chloroplast genes for the β and ϵ subunits of the photosynthetic coupling factor CF_1 are fused. *Nucleic Acids Res.*, **10**, 4985–5002.

Kumagai, I., Pieler, T., Subramanian, A. R. & Erdmann, V. A. (1982). Nucleotide sequence and secondary structure analysis of spinach chloroplast 4.5S RNA. *J. biol. Chem.*, **257**, 12924–8.

Kuntz, M., Keller, M., Crouse, E. J., Burkard, G. & Weil, J. H. (1982). Fractionation and identification of *Euglena gracilis* cytoplasmic and chloroplastic tRNAs and mapping of tRNA genes on chloroplast DNA. *Curr. Genet.*, **6**, 63–9.

Langridge, P. (1981). Synthesis of the large subunit of spinach ribulose bisphosphate carboxylase may involve a precursor polypeptide. *FEBS Lett.*, **123**, 85–9.

Leaver, C. J. (1979). Ribosomal RNA in plants. In *Nucleic Acids in Plants*, vol. I, ed. T. C. Hall & J. W. Davies, pp. 193–215. CRC Press, Boca Raton.

Leis, J. P. & Keller, E. B. (1970). Protein chain-initiating methionine tRNAs in chloroplasts and cytoplasm of wheat leaves. *Proc. natn. Acad. Sci. USA*, **67**, 1593–9.

Link, G. (1981*a*). Enhanced expression of a distinct plastid DNA region in mustard seedlings by continuous far-red light. *Planta*, **152**, 379–80.

Link, G. (1981*b*). Cloning and mapping of the chloroplast DNA sequences for two messenger RNAs from mustard (*Sinapis alba* L.). *Nucleic Acids Res.*, **9**, 3681–94.

Link, G. (1982). Phytochrome control of plastid mRNA in mustard (*Sinapis alba* L.). *Planta*, **154**, 81–6.

Link, G., Coen, D. M. & Bogorad, L. (1978). Differential expression of the gene for the large subunit of ribulose bisphosphate carboxylase in maize cell types. *Cell*, **15**, 725–31.

Loiseaux, S., Rozier, C. & Dalmon, J. (1980). Plastidal origin of a large ribosomal precursor molecule in the brown alga *Pylaiella littoralis* (L.) Kjellm. *Pl. Sci. Lett.*, **18**, 381–8.

Losick, R. & Chamberlin, M. (1976). *RNA Polymerase*. Cold Spring Harbor Laboratory, Cold Spring Harbor.

Machatt, M. A., Ebel, J. P. & Branlant, C. (1981). The 3'-terminal region of bacterial 23S ribosomal RNA: structure and homology with the 3'-terminal region of eukaryotic 28S rRNA and with chloroplast 4.5S rRNA. *Nucleic Acids Res.*, **9**, 1533–49.

Mache, R., Jalliffier-Verne, M., Rozier, C. & Loiseaux, S. (1978). Molecular weight determination of precursor, mature and post-mature plastid ribosomal RNA from spinach using fully denaturing conditions. *Biochim. biophys. Acta*, **517**, 390–9.

MacKay, R. M. (1981). The origin of plant chloroplast 4.5S ribosomal RNA. *FEBS Lett.*, **123**, 17–18.

Malnoë, P. & Rochaix, J. D. (1978). Localization of 4S RNA genes on the chloroplast genome of *Chlamydomonas reinhardtii*. *Molec. gen. Genet.*, **166**, 269–75.

Malnoë, P., Rochaix, J. D., Chua, N. H. & Spahr, P. F. (1979). Characterization of the gene and messenger RNA of the large subunit of ribulose-1,5-diphosphate carboxylase in *Chlamydomonas reinhardtii*. *J. molec. Biol.*, **133**, 417–34.

Matsuda, Y. & Surzycki, S. J. (1980). Chloroplast gene expression in *Chlamydomonas reinhardtii*. *Molec. gen. Genet.*, **180**, 463–74.

McCoy, J. M. & Jones, D. S. (1980). The nucleotide sequence of *Scenedesmus obliquus* chloroplast tRNA^{Met}. *Nucleic Acids Res.*, **8**, 5089–93.

McCrea, J. M. & Hershberger, C. L. (1976). Chloroplast DNA codes for transfer RNA. *Nucleic Acids Res.*, **3**, 2005–18.

McIntosh, L., Poulsen, C. & Bogorad, L. (1980). Chloroplast gene sequence for the large subunit of ribulose bisphosphate carboxylase of maize. *Nature*, **288**, 556–60.

Meeker, R. & Tewari, K. K. (1980). Transfer ribonucleic acid genes in the chloroplast deoxyribonucleic acid of pea leaves. *Biochemistry (Wash.)*, **19**, 5973–81.

Meeker, R. & Tewari, K. K. (1982). Divergence of tRNA genes in chloroplast DNA of higher plants. *Biochim. biophys. Acta*, **696**, 66–75.

Merrick, W. C. & Dure, L. S. (1972). The developmental biochemistry of cotton seed embryogenesis and germination. IV. Levels of cytoplasmic and chloroplastic transfer ribonucleic acid species. *J. biol. Chem.*, **247**, 7988–99.

Miller, M. J. & McMahon, D. (1974). Synthesis and maturation of chloroplast and cytoplasmic ribosomal RNA in *Chlamydomonas reinhardtii*. *Biochim. biophys. Acta*, **366**, 35–44.

Milner, J. J., Hershberger, C. L. & Buetow, D. E. (1979). *Euglena gracilis* chloroplast DNA codes for polyadenylated RNA. *Pl. Physiol.*, **64**, 818–21.

Miyata, T., Kikuno, R. & Ohshima, Y. (1982). A pseudogene cluster in the leader region of the *Euglena* chloroplast 16S–23S rRNA genes. *Nucleic Acids Res.*, **10**, 1771–80.

Montandon, P.-E. & Stutz, E. (1983). Nucleotide sequence of a *Euglena gracilis* chloroplast genome region coding for the elongation factor Tu; evidence for a spliced mRNA. *Nucleic Acids Res.*, **17**, 5877–92.

Mubumbila, M., Burkard, G., Keller, M., Steinmetz, A., Crouse, E. J. & Weil, J. H. (1980). Hybridization of bean, spinach, maize and *Euglena* chloroplast tRNAs with homologous and heterologous chloroplast DNAs. An approach to the study of homology between chloroplast tRNAs from various species. *Biochim. biophys. Acta*, **609**, 31–9.

Mubumbila, M., Gordon, K. H. J., Crouse, E. J., Burkard, G. & Weil, J. H. (1983). Construction of the physical map of the chloroplast DNA of *Phaseolus vulgaris* and localization of ribosomal and transfer RNA genes. *Gene*, **21**, 257–66.

Munsche, D. & Wollgiehn, R. (1973). Die Synthese von ribosomaler RNA in Chloroplasten von *Nicotiana rustica*. *Biochim. biophys. Acta*, **249**, 106–17.

Ness, P. J. & Woolhouse, H. W. (1980). RNA synthesis in *Phaseolus* chloroplasts. I. Ribonucleic acid synthesis in chloroplast preparations from *Phaseolus vulgaris* L. leaves and solubilization of the RNA polymerase. *J. exp. Bot.*, **31**, 223–33.

Oishi, K. K. & Tewari, K. K. (1983). Characterization of the gene and mRNA of the large subunit of ribulose-1,5-bisphosphate carboxylase in pea plants. *Molec. cell. Biol.*, **3**, 587–95.

Oishi, K. K., Sumnicht, T. & Tewari, K. K. (1981). Messenger ribonucleic acid transcripts of pea chloroplast deoxyribonucleic acid. *Biochemistry (Wash.)*, **20**, 5710–17.

Orozco, Jr, E. M. & Hallick, R. B. (1982a). *Euglena gracilis* chloroplast transfer RNA transcription units. I. Physical map of transfer RNA gene loci. *J. biol. Chem.*, **257**, 3258–64.

Orozco, Jr, E. M. & Hallick, R. B. (1982b). *Euglena gracilis* chloroplast transfer RNA transcription units. II. Nucleotide sequence analysis of a tRNA^{Val}–tRNA^{Asn}–tRNA^{Arg}–tRNA^{Leu} gene cluster. *J. biol. Chem.*, **257**, 3265–75.

Orozco, Jr, E. M., Gray, P. W. & Hallick, R. B. (1980a). *Euglena gracilis* chloroplast ribosomal RNA transcription units. I. The location of transfer RNA, 5S, 16S, and 23S ribosomal RNA genes. *J. biol. Chem.*, **255**, 10991–6.

Orozco, Jr, E. M., Rushlow, K. E., Dodd, J. B. & Hallick, R. B. (1980b). *Euglena gracilis* ribosomal RNA transcription units. II. Nucleotide sequence homology between the 16S–23S ribosomal RNA spacer and the 16S ribosomal RNA leader regions. *J. biol. Chem.*, **255**, 10997–1003.

Osorio-Almeida, M. L., Guillemaut, P., Keith, G., Canaday, J. & Weil, J. H. (1980). Primary structure of three leucine transfer RNAs from bean chloroplast. *Biochem. biophys. Res. Commun.*, **92**, 102–8.

Palmer, J. D. & Thompson, W. F. (1981). Rearrangements in the chloroplast genomes of mung bean and pea. *Proc. natn. Acad. Sci. USA*, **78**, 5533–7.

Palmer, J. D., Edwards, H., Jorgensen, R. A. & Thompson, W. F. (1982). Novel evolutionary variation in transcription and location of two chloroplast genes. *Nucleic Acids. Res.*, **10**, 6819–32.

Parthier, B. & Krauspe, R. (1974). Chloroplast and cytoplasmic transfer RNA of *Euglena gracilis*. *Biochem. Physiol. Pfl.*, **165**, 1–17.

Pieler, T., Erdmann, V. A., Digweed, M. & Delihas, N. (1982). Size heterogeneity in *Spinacia oleracea* (spinach) chloroplast 5S ribosomal RNA. *Nucleic Acids Res.*, **10**, 6579–80.

Pirtle, R., Calagan, J., Pirtle, I., Kashdan, M., Vreman, H. & Dudock, B. (1981). The nucleotide sequence of spinach chloroplast methionine elongator tRNA. *Nucleic Acids Res.*, **9**, 183–8.

Posner, H. B. & Rosner, A. (1975). Effect of chloramphenicol on RNA synthesis in *Spirodela* chloroplasts. *Pl. Cell Physiol.*, **16**, 361–5.

Poulsen, C. (1981). Comments on the structure and function of the large subunit of the enzyme ribulose bisphosphate carboxylase-oxygenase. *Carlsberg Res. Commun.*, **46**, 259–78.

Poulsen, C. (1983). The barley chloroplast genome: physical structure and transcriptional activity *in vivo*. *Carlsberg Res. Commun.*, **48**, 57–80.

Preddie, D. L., Preddie, E. C., Guerrini, A. M. & Cremona, T. (1973). Two isoaccepting species of tryptophan-tRNA from *Chlamydomonas reinhardtii*. *Can. J. Bot.*, **51**, 951–4.

Rawson, J. R. Y. (1975). A measurement of the fraction of chloroplast DNA transcribed in *Euglena*. *Biochem. biophys. Res. Commun.*, **62**, 539–45.

Rawson, J. R. Y. & Boerma, C. L. (1976). A measurement of the fraction of chloroplast DNA transcribed during chloroplast development in *Euglena gracilis*. *Biochemistry (Wash.)*, **15**, 588–92.

Rawson, J. R. Y., Boerma, C. L., Andrews, W. H. & Wilkerson, C. F. (1981). Complexity and abundance of ribonucleic acid transcribed from restriction endonuclease fragments of *Euglena* chloroplast deoxyribonucleic acid during chloroplast development. *Biochemistry (Wash.)*, **20**, 2639–44.

Reger, B. J., Fairfield, S. A., Epler, J. L. & Barnett, W. E. (1970). Identification and origin of some chloroplast aminoacyl-tRNA synthetases and tRNA. *Proc. natn. Acad. Sci. USA*, **67**, 1207–13.

Rochaix, J. D. (1981). Organization, function and expression of the chloroplast DNA of *Chlamydomonas reinhardtii*. *Experientia*, **37**, 323–32.

Rochaix, J. D. (1984). Genome evolution in prokaryotes and eukaryotes. In *International Review of Cytology, Special Volume*, ed. D. M. Reammey & P. Chambon. Academic Press, New York, London, Toronto, Sydney and San Francisco. (In press.)

Rochaix, J. D. & Darlix, J. L. (1982). Composite structure of the chloroplast 23S ribosomal RNA genes of *Chlamydomonas reinhardtii*. Evolutionary and functional implications. *J. molec. Biol.*, **159**, 383–95.

Rochaix, J. D. & Malnoë, P. (1978). Anatomy of the chloroplast ribosomal DNA of *Chlamydomonas reinhardtii*. *Cell*, **15**, 661–70.

Rosenberg, M. & Court, D. (1979). Regulatory sequences involved in the promotion and termination of RNA transcription. *A. Rev. Genet.*, **13**, 319–53.

Rosner, A., Jakob, K. M., Gressel, J. & Sagher, D. (1975). The early synthesis and possible function of a 0.5×10^6 M_r RNA after transfer of dark-grown *Spirodela* plants to light. *Biochem. biophys. Res. Commun.*, **67**, 383–91.

Rushlow, K. E., Orozco, Jr, E. M., Lipper, C. & Hallick, R. B. (1980). Selective *in vitro* transcription of *Euglena* chloroplast ribosomal RNA genes by a transcriptionally active chromosome. *J. biol. Chem.*, **255**, 3786–92.

Sagher, D., Grosfeld, H. & Edelman, M. (1976). Large subunit ribulose bisphosphate carboxylase messenger RNA from *Euglena* chloroplasts. *Proc. natn. Acad. Sci. USA*, **73**, 722–6.

Sano, H., Spaeth, E. & Burton, W. G. (1979). Messenger RNA of the large subunit of ribulose-1,5-bisphosphate carboxylase from *Chlamydomonas reinhardtii*. *Eur. J. Biochem.*, **93**, 173–80.

Schiemann, J., Wollgiehn, R. & Parthier, B. (1977). Isolation of a transcription-active RNA polymerase–DNA complex from *Euglena* chloroplasts. *Biochem. Physiol. Pfl.*, **171**, 474–8.

Schiemann, J., Wollgiehn, R. & Parthier, B. (1978). DNA-dependent RNA polymerase in *Euglena gracilis* broken chloroplast. *Biochem. Physiol. Pfl.*, **172**, 507–19.

Schmitt, J. M., Bohnert, H. J., Gordon, K. H. J., Herrmann, R. G., Bernardi, G. & Crouse, E. J. (1981). Compositional heterogeneity of the chloroplast DNAs from *Euglena gracilis* and *Spinacia oleracea*. *Eur. J. Biochem.*, **117**, 375–82.

Schmitt, J. M. & Herrmann, R. G. (1977). Fractionation of cell organelles in silica sol gradients. *Meth. Cell Biol.*, **15**, 177–200.

Schwartzbach, S. D., Hecker, L. I. & Barnett, W. E. (1976). Transcriptional origin of *Euglena* chloroplast tRNA. *Proc. natn. Acad. Sci. USA*, **73**, 1984–8.

Schwarz, Zs. & Kössel, H. (1979). Sequencing of the 3′-terminal region of a 16S rRNA gene from *Zea mays* chloroplast reveals homology with *E. coli* 16S rRNA. *Nature*, **279**, 520–2.

Schwarz, Zs. & Kössel, H. (1980). The primary structure of 16S rDNA from *Zea mays* chloroplast is homologous to *E. coli* 16S rRNA. *Nature*, **283**, 739–42.

Schwarz, Zs., Jolly, S. O., Steinmetz, A. & Bogorad, L. (1981a). Overlapping divergent genes in the maize chloroplast chromosome and *in vitro* transcription of the gene for tRNA[His]. *Proc. natn. Acad. Sci. USA*, **78**, 3423–7.

Schwarz, Zs., Kössel, H., Schwarz, E. & Bogorad, L. (1981b). A gene coding for tRNA[Val] is located near 5′-terminus of 16S rRNA gene in *Zea mays* chloroplast genome. *Proc. natn. Acad. Sci. USA*, **78**, 4748–52.

Scott, N. S. (1976). Precursors of chloroplast ribosomal RNAs in *Euglena gracilis*. *Phytochem.*, **15**, 1207–13.

Selden, R. F., Steinmetz, A., McIntosh, L., Bogorad, L., Burkard, G., Mubumbila, M., Kuntz, M., Crouse, E. J. & Weil, J. H. (1983). Transfer RNA genes of *Zea mays* chloroplast DNA. *Pl. molec. Biol.*, **2**, 141–53.

Shine, J. & Dalgarno, L. (1974). The 3′-terminal sequence of *Escherichia coli* 16S ribosomal RNA: complementarity of nonsense triplets and ribosome binding sites. *Proc. natn. Acad. Sci. USA*, **71**, 1342–6.

Shinozaki, K. & Sugiura, M. (1982a). The nucleotide sequence of the tobacco chloroplast gene for the large subunit of ribulose-1,5-bisphosphate carboxylase/oxygenase. *Gene*, **20**, 91–102.

Shinozaki, K. & Sugiura, M. (1982b). Sequence of the intercistronic region between the ribulose-1,5-bisphosphate carboxylase/oxygenase large subunit and the coupling factor β subunit gene. *Nucleic Acids Res.*, **10** 4923–34.

Shinozaki, K., Deno, H., Kato, A. & Sugiura, M. (1983). Overlap and cotranscription of the genes for the beta and epsilon subunits of tobacco chloroplast ATPase. *Gene*, **24**, 147–55.

Siebenlist, U., Simpson, R. B. & Gilbert, W. (1980). *E. coli* RNA polymerase interacts homologously with two different promoters. *Cell*, **20**, 269–81.

Smith, H. J. & Bogorad, L. (1974). The polypeptide subunit structure of the DNA-dependent RNA polymerase of *Zea mays* chloroplasts. *Proc. natn. Acad. Sci. USA*, **71**, 4839–42.

Speirs, J. & Grierson, D. (1978). Isolation and characterization of 14S RNA from spinach chloroplasts. *Biochim. biophys. Acta*, **521**, 619–33.

Spielmann, A. & Stutz, E. (1983). Nucleotide sequence of soybean chloroplast DNA regions which contain the *psb*A and *trn*H genes and cover the ends of the large single copy region and one end of the inverted repeats. *Nucleic Acids Res.*, **11**, 7157–67.

Sprouse, H. M., Kashdan, M., Oris, L. & Dudock, B. (1981). Nucleotide sequence of a spinach chloroplast valine tRNA. *Nucleic Acids Res.*, **9**, 2543–7.

Steinback, K. E. (1981). Proteins of the chloroplast. In *The Biochemistry of Plants*, vol. 6, ed. A. Marcus, pp. 303–19. Academic Press, New York and London.

Steinmetz, A., Gubbins, E. J. & Bogorad, L. (1982). The anticodon of the maize chloroplast gene for tRNA$_{UAA}^{Leu}$ is split by a large intron. *Nucleic Acids Res.*, **10**, 3027–37.

Steinmetz, A. A., Krebbers, E. T., Schwarz, Zs., Gubbins, E. J. & Bogorad, L. (1983). Nucleotide sequences of five maize chloroplast transfer RNA genes and their flanking regions. *J. biol. Chem.*, **258**, 5503–11.

Stiegler, G. L., Matthews, H. M., Bingham, S. E. & Hallick, R. B. (1982). The gene for the large subunit of ribulose-1,5-bisphosphate carboxylase in *Euglena gracilis* chloroplast DNA: location, polarity, cloning, and evidence for an intervening sequence. *Nucleic Acids Res.*, **10**, 3427–44.

Stubbe, W. (1971). Origin and continuity of plastids. In *Origin and Continuity of Cell Organelles*, vol. 3, ed. J. Reinert & H. Ursprung, pp. 65–81. Springer-Verlag, Berlin, Heidelberg and New York.

Subramanian, A. R., Steinmetz, A. & Bogorad, L. (1983). Maize chloroplast DNA encodes a protein sequence homologous to the bacterial ribosome assembly protein S4. *Nucleic Acids Res.*, **11**, 5277–86.

Sugita, M. & Sugiura, M. (1983). A putative gene of tobacco chloroplast coding for ribosomal protein similar to *E. coli* ribosomal protein S19. *Nucleic Acids Res.*, **11**, 1913–18.

Sugiura, M., Shinozaki, K., Deno, H., Kamogashira, T. & Sugita, M. (1983). Structure of tobacco chloroplast tRNA genes. *Int. tRNA Workshop, Abstract*. Hakone, Japan.

Surzycki, S. J. (1969). Genetic functions of the chloroplast of *Chlamydomonas reinhardtii*: effect of rifampicin on chloroplast DNA-dependent RNA polymerase. *Proc. natn. Acad. Sci. USA*, **63**, 1327–34.

Surzycki, S. J. & Shellenbarger, D. L. (1976). Purification and characterization of a putative sigma factor from *Chlamydomonas reinhardtii*. *Proc. natn. Acad. Sci. USA*, **73**, 3961–5.

Swamy, G. S. & Pillay, D. T. N. (1982). Characterization of *Glycine max* cytoplasmic, chloroplastic and mitochondrial tRNAs and synthetases for phenylalanine, tryptophan and tyrosine. *Pl. Sci. Lett.*, **25**, 73–84.

Takaiwa, F. & Sugiura, M. (1980*a*). Nucleotide sequences of the 4.5S and 5S ribosomal RNA genes from tobacco chloroplasts. *Molec. gen. Genet.*, **180**, 1–4.

Takaiwa, F. & Sugiura, M. (1980*b*). Cloning and characterization of 4.5S and 5S RNA genes in tobacco chloroplasts. *Gene*, **10**, 95–103.

Takaiwa, F. & Sugiura, M. (1980*c*). The nucleotide sequence of 4.5S ribosomal RNA from tobacco chloroplasts. *Nucleic Acids Res.*, **8**, 4125–9.

Takaiwa, F. & Sugiura, M. (1981). Heterogeneity of 5S RNA species in tobacco chloroplasts. *Molec. gen. Genet.*, **182**, 385–9.

Takaiwa, F. & Sugiura, M. (1982*a*). The complete nucleotide sequence of a 23S rRNA gene from tobacco chloroplasts. *Eur. J. Biochem.*, **124**, 13–19.

Takaiwa, F. & Sugiura, M. (1982*b*). The nucleotide sequence of chloroplast 5S ribosomal RNA from a fern, *Dryopteris acuminata. Nucleic Acids Res.*, **10**, 5369–73.

Takaiwa, F. & Sugiura, M. (1982*c*). Nucleotide sequence of the 16S–23S spacer region in an rRNA gene cluster from tobacco chloroplast DNA. *Nucleic Acids Res.*, **10**, 2665–76.

Takaiwa, F., Kusuda, M. & Sugiura, M. (1982). The nucleotide sequence of chloroplast 4.5S rRNA from a fern, *Dryopteris acuminata. Nucleic Acids Res.*, **10**, 2257–60.

Tewari, K. K. & Goel, A. (1983). Solubilization and partial purification of RNA polymerase from pea chloroplasts. *Biochemistry (Wash.)*, **22**, 2142–8.

Thompson, J. A., Hansmann, P., Knoth, R., Link, G. & Falk, H. (1981). Electron microscopical localization of the 23S and 16S rRNA genes within an inverted repeat for two chromoplast DNAs. *Curr. Genet.*, **4**, 25–8.

Tiboni, O., Panzeri, L., Di Pasquale, G., Sora, S. & Ciferri, O. (1981). Expression of a spinach chloroplast gene in *Escherichia coli. EMBO Workshop on Chloroplast DNA*, Abstract, Arolla, Switzerland.

Tohdoh, N., Shinozaki, K. & Sugiura, M. (1981). Sequence of a putative promoter region for the rRNA genes of tobacco chloroplast DNA. *Nucleic Acids Res.*, **9**, 5399–406.

Tohdoh, N. & Sugiura, M. (1982). The complete nucleotide sequence of a 16S ribosomal RNA gene from tobacco chloroplasts. *Gene*, **17**, 213–18.

van Ee, J. H., Vos, Y. J., Bohnert, H. J. & Planta, R. J. (1982). Mapping of genes on the chloroplast DNA of *Spirodela oligorhiza. Pl. molec. Biol.*, **1**, 117–31.

Wallace, D. C. (1982). Structure and evolution of organelle genomes. *Microbiol. rev.*, **46**, 208–40.

Watanabe, A. & Price, C. A. (1982). Translation of mRNAs for subunits of chloroplast coupling factor I in spinach. *Proc. natn. Acad. Sci. USA*, **79**, 6304–8.

Weil, J. H. (1979). Cytoplasmic and organellar tRNAs in plants. In *Nucleic Acids in Plants*, ed. T. C. Hall & J. W. Davies, pp. 143–92. CRC Press, Boca Raton.

Weil, J. H. & Parthier, B. (1982). Transfer RNA and aminoacyl-tRNA synthetases in plants. In *Encyclopedia of Plant Physiology, New Series*, vol. 14A, ed. D. Boulter & B. Parthier, pp. 65–112. Springer-Verlag, Berlin, Heidelberg and New York.

Westhoff, P., Alt, J., Nelson, N., Bottomley, W., Bünemann, H. & Herrmann, R. G. (1983). Genes and transcripts for the P_{700} chlorophyll a apoprotein and subunit 2 of the photosystem I reaction center complex from spinach thylakoid membranes. *Pl. molec. Biol.*, **2**, 95–107.

Westhoff, P., Nelson, N., Bünemann, H. & Herrmann, R. G. (1981). Localization of genes for coupling factor subunits on the spinach plastid chromosome. *Curr. Genet.*, **4**, 109–20.

Wheeler, A. M. & Hartley, M. R. (1975). Major mRNA species from spinach chloroplasts do not contain poly(A). *Nature*, **257**, 66–7.

Whitfeld, P. R. (1977). Chloroplast RNA. In *The Ribonucleic Acids*, ed. P. R. Stewart & D. S. Letham, pp. 297–332. Springer, New York.

Whitfeld, P. R. & Bottomley, W. (1980). Mapping of the gene for the large subunit of ribulose bisphosphate carboxylase on spinach chloroplast DNA. *Biochem. Int.*, **1**, 172–8.

Whitfeld, P. R. & Bottomley, W. (1983). Organization and structure of chloroplast genes. *A. Rev. Pl. Physiol.*, **34**, 279–310.

Whitfeld, P. R., Leaver, C. J., Bottomley, W. & Atchison, B. A. (1978). Low-molecular-weight (4.5S) ribonucleic acid in higher plant chloroplast ribosomes. *Biochem. J.*, **175**, 1103–12.

Wildeman, A. G. & Nazar, R. W. (1980). Nucleotide sequence of wheat chloroplastid 4.5S ribonucleic acid. Sequence homology in 4.5S RNA species. *J. biol. Chem.*, **255**, 11896–900.

Wollgiehn, R. (1982). RNA polymerase and regulation of transcription. In *Encyclopedia of Plant Physiology, New Series*, vol. 14B, ed. B. Parthier & D. Boulter, pp. 125–70. Springer-Verlag, Berlin, Heidelberg and New York.

Wollgiehn, R. & Parthier, B. (1979). RNA synthesis in isolated chloroplasts of *Euglena gracilis*. *Pl. Sci. Lett.*, **16**, 203–10.

Wollgiehn, R. & Parthier, B. (1980). RNA and protein synthesis in plastid differentiation. In *Results and Problems in Cell Differentiation*, vol. 10, ed. J. Reinert, pp. 97–145. Springer-Verlag, Berlin, Heidelberg and New York.

Zech, M., Hartley, M. R. & Bohnert, H. J. (1981). Binding sites of *E. coli* DNA-dependent RNA polymerase on spinach chloroplast DNA. *Curr. Genet.*, **4**, 37–46.

Zenke, G., Edwards, K., Langridge, P. & Kössel, H. (1982). The rRNA operon from maize chloroplasts: analysis of *in vivo* transcription products in relation to its structure. In *Cell Function and Differentiation*, Part B, ed. G. Akoyunoglou *et al.*, pp. 309–19. Alan R. Liss, New York.

Zurawski, G., Bohnert, H. J., Whitfeld, P. R. & Bottomley, W. (1982a). Nucleotide sequence of the gene for the M_r 32000 thylakoid membrane protein from *Spinacia oleracea* and *Nicotiana debneyi* predicts a totally conserved primary translation product of M_r 38950. *Proc. natn. Acad. Sci. USA*, **79**, 7699–703.

Zurawski, G., Bottomley, W. & Whitfeld, P. R. (1982b). Structure of the genes for the β and ϵ subunits of spinach chloroplast ATPase indicates a dicistronic mRNA and an overlapping translation stop/start signal. *Proc. natn. Acad. Sci. USA*, **79**, 6260–4.

Zurawski, G., Perrot, B., Bottomley, W. & Whitfeld, P. R. (1981). The structure of the gene for the large subunit of ribulose-1,5-bisphosphate carboxylase from spinach chloroplast DNA. *Nucleic Acids Res.*, **9**, 3251–70.

J. C. GRAY, A. L. PHILLIPS AND A. G. SMITH

Protein synthesis by chloroplasts

Chloroplasts contain ribosomes and all the components necessary for the polymerisation of amino acids into discrete products. Studies on the characteristics of the chloroplast protein synthetic system, including tRNAs, aminoacyl-tRNA synthetases, ribosomes, initiation factors and elongation factors, have shown that it is distinct from the cytoplasmic system. In general, the components of the chloroplast system are much more similar to those from prokaryotic organisms than they are to the cytoplasmic system. A large amount of effort has been expended on the identification of the proteins synthesised on chloroplast ribosomes, using a variety of approaches, and this has been successful to the extent that the identity of about 30 polypeptides is now known. This work has also shown that a large number of chloroplast components are synthesised not on chloroplast ribosomes but on cytoplasmic ribosomes and subsequently transferred to the chloroplast. A fundamental problem is to understand why particular proteins are synthesised on chloroplast ribosomes and if there is an underlying pattern to explain the sites of synthesis of chloroplast proteins. It is also important to establish if there are any differences between species in the identity of proteins synthesised on chloroplast ribosomes.

Although only a relatively small proportion of chloroplast polypeptides is synthesised on chloroplast ribosomes, it appears likely that the assembly of multi-subunit proteins and complexes takes place exclusively in the chloroplast. Knowledge of the assembly processes, particularly where they involve products of protein synthesis on both chloroplast and cytoplasmic ribosomes, is a vital part of our understanding of chloroplast biogenesis.

Identification of proteins synthesised on chloroplast ribosomes

A number of experimental approaches have been used to establish the identity of the polypeptides synthesised on chloroplast ribosomes. All of these approaches have advantages and limitations.

Selective protein synthesis inhibitors in vivo

A range of antibiotics such as chloramphenicol, lincomycin, strepto-mycin and spectinomycin, inhibit protein synthesis on chloroplast ribosomes but have no effect on cytoplasmic protein synthesis, whereas other compounds, such as cycloheximide and anisomycin, inhibit protein synthesis on cyto-plasmic, but not chloroplast, ribosomes. By examining the effects of such inhibitors on the synthesis of chloroplast proteins *in vivo*, it is possible to determine the sites of synthesis of the proteins. Early experiments investigated the effects of selective protein synthesis inhibitors on the appearance of specific enzyme activities, particularly during the greening of etiolated tissues. Many of these experiments gave erroneous or inconclusive results because of several factors which were not appreciated at the time. First, in many cases the increase in enzyme activity was not shown to result entirely from synthesis of new enzyme protein. Secondly, many of the inhibitors were later shown not to be specific for protein synthesis. Chloramphenicol, for example, may inhibit a variety of processes other than protein synthesis, such as ion uptake, oxidative phosphorylation and photophosphorylation (Ellis, 1969). Thirdly, reciprocal effects of inhibitors of protein synthesis on chloroplast and cytoplasmic ribosomes were frequently not reported. Only if the appearance of proteins is prevented by protein synthesis inhibitors at one site but not by inhibitors at the other, is it possible to establish the site of synthesis by this approach. Fourthly, in many experiments the inhibitors were present for extended periods leading to general metabolic distortion of cell function.

These objections may be overcome in carefully controlled short-term experiments in which the effects of a range of inhibitors on the incorporation of radioactive precursors into specific chloroplast polypeptides are examined. Such experiments have been carried out with whole higher plants, e.g. *Spirodela* (Nechushtai *et al.*, 1981), cut shoots (Criddle *et al.*, 1970; Ellis, 1975) and isolated cells of higher plants (Nishimura & Akazawa, 1978; Barraclough & Ellis, 1980) and with whole algal cells (Chua & Gillham, 1977; Nechushtai & Nelson, 1981). Measuring the incorporation of radioactive precursors into specific polypeptides ensures that only the synthesis of the polypeptide chain is being examined. Using a range of inhibitors ensures that the expected reciprocal effects of chloroplast and cytoplasmic protein synthesis inhibitors may be examined, and may help with the identification of non-specific effects, which are not likely to be the same for different inhibitors. Non-specific effects of chloramphenicol may be examined by the use of different stereo-isomers. D-*threo*-Chloramphenicol is the only isomer which inhibits protein synthesis whereas all four stereo-isomers give non-specific effects (Ellis, 1969). The usual experimental approach is to compare the effects of D-*threo*- and

L-*threo*-chloramphenicol; direct inhibition of protein synthesis will be shown only by the D-*threo*-isomer.

The advantage of inhibitor experiments is that they are carried out *in vivo* so that controls operating in whole cells should still be functioning. This may not be the case in cell-free systems *in vitro*. One feature of chloroplast protein synthesis is that synthesis or assembly in one compartment may depend on protein synthesis in the other compartment. This should be recognisable in inhibitor experiments because the synthesis or assembly of a specific component would be inhibited by inhibitors of protein synthesis on both chloroplast and cytoplasmic ribosomes. This has been shown to be the case with the subunits of ribulose-1,5-bisphosphate (RuBP) carboxylase (Iwanij, Chua & Siekevitz, 1975; Nishimura & Akazawa, 1978); ribosomal proteins in *Euglena* (Freyssinet, 1978) and the photosystem I reaction centre polypeptide (Ellis, 1975).

Protein synthesis by isolated chloroplasts in vitro

Isolated intact chloroplasts from a number of plants, including pea (Blair & Ellis, 1973), spinach (Bottomley, Spencer & Whitfeld, 1974; Morgenthaler & Mendiola-Morgenthaler, 1976), maize (Grebanier, Steinback & Bogorad, 1979), *Euglena* (Vasconcelos, 1976) and cucumber (Walden & Leaver, 1980) are able to use light as an energy source for the incorporation of radioactively labelled amino acids into discrete polypeptides. One advantage of using light as an energy source is that it is possible to use fairly crude preparations of chloroplasts, containing broken chloroplasts with some contamination by nuclei and other organelles. Only intact chloroplasts are able to generate ATP in the light and to use this for protein synthesis. Further purification of intact chloroplasts, for example by centrifugation through silica sol gradients (Morgenthaler & Mendiola-Morgenthaler, 1976; Vasconcelos, 1976; Ortiz, Reardon & Price, 1980) gives chloroplast preparations showing the same overall pattern of incorporation of labelled amino acids into discrete polypeptides. Isolated chloroplasts can also use added ATP (as the Mg–ATP complex) in the light or the dark as an energy source for protein synthesis (Ellis, 1977). Isolated chloroplasts appear to have a requirement for K^+ in the incubation medium, which cannot be substituted by Na^+ but can be replaced by Rb^+ or by NH_4^+ in the dark (J. Edwards, A. Davidson & J.C. Gray, unpublished). In general, divalent cations and particularly Ca^{2+}, are inhibitory to protein synthesis by isolated chloroplasts.

The rate of incorporation of labelled amino acids by isolated chloroplasts decreases during incubation in the light at 20 °C, such that there is little protein synthesis occurring after 20–30 min. Lucchini & Bianchetti (1980) have suggested that in *Euglena* chloroplasts this is because of a decrease in the rate of initiation of new rounds of protein synthesis and have shown that the addition of N^{10}-formyltetrahydrofolate stimulates, and maintains at high levels, the rate of incorporation of [^{14}C]leucine. However, the addition of N^{10}-formyltetrahydrofolate does not stimulate protein synthesis by isolated pea chloroplasts (B. Spreckley & J. C. Gray, unpublished), where there is direct evidence that reinitiation of protein synthesis is taking place (Highfield & Ellis, 1976). Inhibition of protein synthesis in isolated pea (*Pisum sativum*) chloroplasts to the same extent (70% inhibition) by a range of prokaryotic initiation inhibitors, including kanamycin, neomycin and lincomycin, also suggests that a large part of the incorporation of labelled amino acids is the result of reinitiation (N. Dearnaley & J. C. Gray, unpublished).

Fractionation of chloroplasts after incorporation of labelled amino acids *in vitro* demonstrates that polypeptides in the stroma, thylakoid and envelope fractions are synthesised by isolated chloroplasts (Blair & Ellis, 1973; Bottomley *et al.*, 1974; Eaglesham & Ellis, 1974; Joy & Ellis, 1975; Morgenthaler & Mendiola-Morgenthaler, 1976). Estimates of the numbers of polypeptides synthesised by isolated chloroplasts by electrophoretic analysis of labelled products suggest that about 100 polypeptides are made. This estimate comprises about 80 stromal polypeptides (Ellis, Highfield & Silverthorne, 1977), about 20 thylakoid membrane polypeptides (A. Doherty & J. C. Gray, unpublished) and a few envelope components (Joy & Ellis, 1975; Vasconcelos, 1976). This total number of polypeptides agrees well with the estimated coding capacity of chloroplast DNA (Bedbrook & Kolodner, 1979). However, the estimate of the numbers of polypeptides synthesised in pea chloroplasts is based on the assumption that each labelled band or spot separated by electrophoresis represents a separate polypeptide. There is, as yet, no evidence that this is the case and some of the labelled polypeptides may represent discrete premature termination products. It is clear that the nature of the chloroplast incubation medium affects the assembly of RuBP carboxylase *in vitro* (Barraclough & Ellis, 1980) and the addition of ATP to the medium is needed for assembly of the cytochrome *b/f* complex in pea (A. L. Phillips & J. C. Gray, unpublished; see Fig. 2). This suggests that isolated pea chloroplasts in the light do not produce enough ATP for the assembly of chloroplast proteins. ATP deficiency may result in changes in the fidelity of chloroplast protein synthesis.

One advantage of the use of isolated chloroplasts for the identification of products of chloroplast protein synthesis is the very high levels of incorporation of radioactive amino acids that can be achieved. With isolated pea chloroplasts, 15% incorporation of label can be obtained on incubation of 1.2 mg chlorophyll with 120 μCi [^{35}S]methionine in the light at 20 °C for 60 min (A. L. Phillips, unpublished). This means it is possible to detect the synthesis of components which account for only 0.02% of the total incorporation

(Doherty & Gray, 1979). Identification of products of chloroplast protein synthesis should be made on the basis of as large a number of criteria as possible. Co-migration of radioactive material with authentic protein in a one-dimensional polyacrylamide gel is rarely sufficient. Two-dimensional gel electrophoresis, immunoprecipitation with monospecific antibodies or analysis of proteolytic fragments of the labelled material are needed to confirm the identity of the products. The nature and identity of the products of chloroplast protein synthesis are discussed below.

Chloroplast ribosomes are located free in the stroma and bound to the thylakoid membrane. Tao & Jagendorf (1973) have estimated that about 80% of pea chloroplast ribosomes are free and 20% are membrane-bound. Free ribosomes, supplemented with a high-speed post-ribosomal supernatant from isolated chloroplasts, have been shown to synthesise a number of discrete products including the LSU of RuBP carboxylase (Ellis, 1977). A proportion of the ribosomes attached to thylakoid membranes may be detached only by treatments, such as puromycin, which release the nascent polypeptide chain (Chua *et al.*, 1973; Margulies & Michaels, 1974), indicating that the ribosomes are attached to the membrane by the polypeptide chain. Alscher, Patterson & Jugendorf (1978) have found that the products remain associated with the membrane suggesting that the membrane-bound ribosomes are synthesising membrane proteins, while Ellis (1977) has shown that the thylakoid-bound ribosomes synthesise a set of proteins different from those synthesised by free ribosomes. Further, the unidentified membrane protein D-2 from *Chlamydomonas* thylakoids is synthesised on membrane-bound ribosomes (Herrin, Michaels & Hickey, 1981). Recent studies have also shown that cytochrome *f* is synthesised on membrane-bound ribosomes from pea chloroplasts (J. C. Gray, unpublished; see Fig. 1).

One limitation of the use of isolated chloroplasts *in vitro* is that it is difficult to supply them with components synthesised on cytoplasmic ribosomes. If cytoplasmic products control the translation or assembly of chloroplast products then it is clear that these processes may not be detected unless there are suitable concentrations of the cytoplasmic products in the isolated chloroplasts.

Translation of chloroplast RNA in vitro

Three cell-free systems have been used for the translation of chloroplast RNA. The first to be used was a cell-free system from *Escherichia coli* which translated added spinach chloroplast RNA to produce the large subunit (LSU) of RuBP carboxylase (Hartley, Wheeler & Ellis, 1975). The *E. coli* system has subsequently been used for the translation of chloroplast RNA from *Chlamydomonas* (Gelvin, Heizmann & Howell, 1977; Howell *et al.*, 1977) and cucumber (Walden & Leaver, 1980), and in each case the LSU of RuBP carboxylase was identified as a product by immunoprecipitation. The wheat-germ translation system has also been used for the translation of mRNA for LSU of RuBP carboxylase from *Euglena* chloroplasts (Sagher, Grosfeld & Edelman, 1976). However, the wheat-germ system is not particularly efficient at translating chloroplast RNA (Bottomley, Higgins & Whitfeld, 1976), and the rabbit reticulocyte lysate is much more useful for this. It has been used with RNA from spinach (Driesel, Speirs & Bohnert, 1980; Silverthorne & Ellis, 1980; Westhoff *et al.*, 1981), maize (Bedbrook *et al.*, 1978), mustard (Link, 1981) and *Chlorogonium* (Westhoff & Zetsche, 1981). Products identified include the LSU of RuBP carboxylase, the 32 kD herbicide-binding protein and the α, β and ϵ subunits of ATP synthase. Positive identification of the products of translation *in vitro* requires the use of immunoprecipitation with monospecific antibodies, and comparison of proteolytic degradation products with those of authentic material.

One advantage of translation *in vitro* is that it is free from translational controls operating *in vivo*, and hence if mRNA is present in chloroplasts, albeit in a non-translatable form, its extraction and translation *in vitro* should enable the identity of its products to be determined. The major limitation as a method for the identification of products of chloroplast protein synthesis is the difficulty in obtaining chloroplast RNA free from cytoplasmic RNA contamination. The levels of contamination are such that minor products of translation *in vitro* by the wheat-germ or rabbit reticulocyte lysate systems cannot be stated unequivocally to be products of chloroplast RNA. One possible solution is the use of the *E. coli* translation system because of the very poor translation of cytoplasmic RNA in this system (Bottomley *et al.*, 1976). One further problem in the translation of chloroplast RNA is the inhibition of the reticulocyte lysate system by low molecular weight RNA. Bedbrook *et al.* (1978) found it necessary to remove, by sucrose density gradient centrifugation, all RNA with a sedimentation coefficient less than 6S, to achieve reasonable levels of translation.

Transcription and translation of chloroplast DNA

This cannot yet be regarded as a primary method for the identification of products of chloroplast protein synthesis, since it is possible, although improbable, that transcripts from chloroplast DNA leave the chloroplast and are translated on cytoplasmic ribosomes. However, there is no experimental evidence for the movement of RNA across the chloroplast envelope and all proteins whose genes have been located in chloroplast DNA have been shown to be synthesised on chloroplast ribosomes.

The first system to be used was a linked two-stage assay: in the first stage chloroplast DNA, or cloned restriction fragments of chloroplast DNA, was incubated with *E. coli* RNA polymerase and in the second stage the products of the first reaction were incubated with the rabbit reticulocyte lysate and [^{35}S]methionine to label the translation products (Coen *et al.*, 1977; Bedbrook *et al.*, 1979). This system was used to identify the LSU of RuBP carboxylase (Coen *et al.*, 1977), and the 32 kD herbicide binding protein (Bedbrook *et al.*, 1979), as products of maize chloroplast DNA. This linked system has now been replaced by a much simpler, and more efficient, coupled transcription–translation system derived from *E. coli*. This system, which requires only a single incubation, was first used with chloroplast DNA by Bottomley & Whitfeld (1979). They showed that spinach, tobacco and *Oenothera* chloroplast DNA directed the synthesis of a large number of discrete polypeptides on incubation with the coupled system. The LSU of RuBP carboxylase was identified as the major product of spinach chloroplast DNA (Bottomley & Whitfeld, 1979). This system has also been used to identify products from cloned restriction fragments and linear fragments of chloroplast DNA. The products identified include the LSU of RuBP carboxylase (Whitfeld & Bottomley, 1980; Erion *et al.*, 1981), the α, β, ϵ and DCCD-binding subunits of ATP synthase (Howe *et al.*, 1982*a*, 1982*b*; C. J. Howe, unpublished), cytochrome *f* (Willey *et al.*, 1982) and a 15.5 kD polypeptide of the cytochrome complex (A. L. Phillips & J. C. Gray, unpublished).

Products of chloroplast protein synthesis

The products of chloroplast protein synthesis include components located in the stroma and in the envelope and thylakoid membrane fractions. Experiments with isolated chloroplasts suggest that as many as 100 different polypeptides may be synthesised on chloroplast ribosomes. The identity of about 30 of these polypeptides is now known. They are listed in Table 1. None of the envelope membrane proteins synthesised by chloroplasts has been identified, and a large number of products that remain in the stromal fraction have yet to be identified.

Stromal proteins

For the purposes of this article, stromal proteins are regarded as proteins remaining in solution after removal of thylakoid membranes by centrifugation. Protein components of ribosomes are therefore included in this section. Analysis by two-dimensional gel electrophoresis of the stromal fraction from isolated pea chloroplasts incubated with [^{35}S]methionine in the light demonstrated the labelling of about 80 discrete polypeptides (Ellis *et al.*, 1977). Whether these all represent different gene products or whether some are premature termination products or breakdown products is not yet clear.

However, by far the major component is the LSU of RuBP carboxylase (Blair & Ellis, 1973). The only other components identified are elongation factors (Ciferri, Di Pasquale & Tiboni, 1979) and ribosomal proteins (Eneas-Filho, Hartley & Mache, 1981). In addition the increase in some stromal enzyme activities during greening has been reported to be inhibited by chloramphenicol. These products are discussed in more detail below.

RuBP carboxylase. RuBP carboxylase is the major stromal protein and is the most abundant protein in the leaves of most plants. It is composed of two different subunits; large subunits (LSU; 52 kD) which contain the catalytic site and small subunits (SSU; 12–15 kD) whose function is unknown. Separate sites of synthesis of the two subunits were first shown by inhibitor experiments with greening barley shoots (Criddle *et al.*, 1970) and it has become clear subsequently that the LSU is the major soluble product of protein synthesis by chloroplast ribosomes during chloroplast development (Blair & Ellis, 1973), whereas the SSU is synthesised on cytoplasmic ribosomes in a precursor form (Dobberstein, Blobel & Chua, 1977; Highfield & Ellis, 1978). The precursor form of the SSU is imported into the chloroplasts, processed and assembled with LSUs into the complete enzyme (Chua & Schmidt, 1978; Smith & Ellis, 1979).

A very large amount of research has been concerned with the synthesis of RuBP carboxylase, mainly because it is the major product of protein synthesis

Table 1. *Identity of the products of chloroplast protein synthesis*

Stromal proteins
Ribulose bisphosphate carboxylase: large subunit
Elongation factor G
Elongation factor Tu
Ribosomal proteins: S6, S9, S13, S19, S22, S23/24
                     L4, L22, L24, L25, L29

Thylakoid proteins
Photosystem I: P700–chlorophyll *a* protein (CPI)
                  17 kD and 11 kD polypeptides
Photosystem II: Chlorophyll–protein CPa polypeptides 49 kD and 45 kD
                  32 kD herbicide-binding protein
                  Cytochrome *b*-559
Cytochrome complex: cytochrome *f*
                     cytochrome *b*-563
                     15.5 kD polypeptide
ATP synthase: CF_1 subunits α, β and ϵ
                CF_0 subunits I and III (DCCD-binding proteolipid)

DCCD = dicyclohexylcarbodiimide. For other abbreviations see text.

in developing leaves. The involvement of chloroplast ribosomes in the synthesis of RuBP carboxylase was suggested by the inhibition by chloramphenicol of the appearance of RuBP carboxylase activity during the greening of maize leaves (Graham *et al.*, 1970), *Chlamydomonas* (Armstrong *et al.*, 1971; Margulies, 1971) and *Euglena* (Lord, Armitage & Merrett, 1975). Criddle *et al.* (1970) showed that chloramphenicol, but not cycloheximide, inhibited the incorporation of [³H]arginine into the LSU in greening barley shoots, whereas cycloheximide, but not chloramphenicol, inhibited incorporation into the SSU. This differential inhibition of incorporation of labelled amino acids into the LSU and SSU by protein synthesis inhibitors has now been shown in a number of higher plant systems including pea shoots (Cashmore, 1976), soybean cells (Barraclough & Ellis, 1979) and tobacco protoplasts (Hirai & Wildman, 1977). In all cases incorporation into the LSU was inhibited by chloroplast protein synthesis inhibitors such as chloramphenicol, lincomycin and spectinomycin. However, in spinach protoplasts (Nishimura & Akazawa, 1978) and in green algae, such as *Chlamydomonas* (Iwanij *et al.*, 1975) the syntheses of the LSU and of the SSU are very tightly coupled, such that inhibitors of protein synthesis on either chloroplast or cytoplasmic ribosomes prevent synthesis of both subunits.

The first direct demonstration of the synthesis of the LSU on chloroplast ribosomes was by Blair & Ellis (1973), who showed that the LSU was the major product of light-driven protein synthesis by isolated pea chloroplasts. Synthesis of the LSU has also been demonstrated in chloroplasts isolated from spinach (Bottomley *et al.*, 1974; Morgenthaler & Mendiola-Morgenthaler, 1976), *Euglena* (Vasconcelos, 1976) and cucumber (Walden & Leaver, 1980). The LSU has been shown to be synthesised on free ribosomes from pea chloroplasts (Ellis, 1977).

The LSU of RuBP carboxylase was the first product of the translation of isolated chloroplast RNA to be identified. Hartley *et al.* (1975) demonstrated the synthesis of a 52 kD polypeptide, identified as the LSU by peptide mapping, after translation of spinach chloroplast RNA in a cell-free system from *E. coli*. The LSU has subsequently been identified as a product of translation of chloroplast RNA from spinach (Silverthorne & Ellis, 1980), *Euglena* (Sagher *et al.*, 1976), maize (Coen *et al.*, 1977), mustard (Link, 1981) and cucumber (Walden & Leaver, 1980) in a variety of cell-free systems. The mRNA for *Chlamydomonas* LSU has been purified from immunoprecipitated polysomes (Gelvin *et al.*, 1977; Sano, Spaeth & Burton, 1979) and has been shown to be about 1500 nucleotides long (Malnoe *et al.*, 1979). A similar size for the mRNA for the LSUs from maize (Link & Bogorad, 1980), pea (Smith & Ellis, 1981), mustard (Link, 1981) and wheat (Koller, Delius & Dyer, 1982) has been determined by 'Northern' hybridisation (Alwine, Kemp & Stark,

1977) of chloroplast RNA with labelled cloned gene sequences. The mRNA for spinach LSU does not contain large tracts of poly(A) (Wheeler & Hartley, 1975).

LSU as a product of transcription and translation of chloroplast DNA *in vitro* has been shown for several plants. Linked transcription–translation of maize chloroplast DNA by *E. coli* RNA polymerase and a rabbit reticulocyte lysate originally showed that the LSU gene was located in chloroplast DNA (Coen *et al.*, 1977; Bedbrook *et al.*, 1979). This location has been confirmed by the use of a coupled transcription–translation system from *E. coli* with chloroplast DNA from spinach, tobacco and *Oenothera* (Bottomley & Whitfeld, 1979). The use of these systems has led to the localisation of the gene for LSU (Bedbrook *et al.*, 1979; Whitfeld & Bottomley, 1980; Erion *et al.*, 1981) and the determination of the complete nucleotide sequences of the genes for LSU from maize (McIntosh, Poulsen & Bogorad, 1980) and spinach (Zurawski *et al.*, 1981).

Comparison of the amino acid sequences predicted from the nucleotide sequences of the LSU genes with the known amino acid sequence of parts of the barley protein indicates that there is a 13 amino acid sequence following the initiation codons of the maize and spinach genes which is not represented at the amino-terminus of the barley LSU (McIntosh *et al.*, 1980). Langridge (1981) has shown that the translation product of spinach chloroplast RNA in the *E. coli* system has a slightly higher molecular weight than the labelled product accumulating in isolated chloroplasts incubated with [^{35}S]methionine. The higher molecular weight product synthesised *in vitro* has been shown to be processed to the smaller product by the stromal fraction of chloroplasts. It is not clear if this early processing takes place *in vivo* or if it has any role in the assembly of the whole enzyme.

Assembly of RuBP carboxylase appears to take place solely in the chloroplasts and assembly in isolated chloroplasts has been demonstrated (Barraclough & Ellis, 1980). This suggests that there must be a pool of free small subunits in isolated chloroplasts. Although free small subunits have been demonstrated in leaf extracts (Roy, Costa & Adari, 1978), they have not yet been shown in isolated chloroplasts. Assembly of RuBP carboxylase appears to depend to a great degree on the composition of the chloroplast suspension medium. Barraclough & Ellis (1980) have shown that assembly takes place in isolated pea chloroplasts in sorbitol media but not in KCl media. In both media, newly synthesised LSU is bound to a high molecular weight complex (> 600 kD) containing, in addition to LSU, another polypeptide of subunit molecular weight 58 kD. This complex may act to keep the large subunit in soluble form prior to assembly, but no direct evidence to support this hypothesis is yet available (Ellis, 1981).

Protein synthesis components. A large number of proteins are involved in chloroplast protein synthesis. To date, only elongation factors G and T and several ribosomal proteins have been shown to be synthesised on chloroplast ribosomes. Ciferri & Tiboni (1976) first showed that the appearance on greening and the labelling of elongation factor G with [^{35}S]methionine were inhibited by chloramphenicol but not by cycloheximide in *Chlorella*. Elongation factor G (EF-G) was later shown to be synthesised in isolated spinach chloroplasts incubated in the light with [^{35}S]methionine, although the incorporation into EF-G accounted for only 0.04% of the total radioactivity incorporated by the isolated chloroplasts (Ciferri *et al.*, 1979). In contrast, EF-G appears to be synthesised on cytoplasmic ribosomes in *Euglena gracilis*; the light-induced appearance of EF-G is prevented by cycloheximide but, not by spectinomycin or streptomycin (Breitenberger, Graves & Spremulli, 1979).

Elongation factor T (EF-T) is composed of two polypeptides, EF-Tu which is unstable at high temperatures, and EF-Ts which is stable. EF-Tu is synthesised in isolated spinach chloroplasts incubated in the light with [^{35}S]methionine (Ciferri *et al.*, 1979). The labelled polypeptide was identified by two-dimensional gel electrophoresis and immunodiffusion, and accounted for 0.2% of the total radioactivity incorporated by the isolated chloroplasts. EF-Ts has also been reported briefly to be synthesised in isolated chloroplasts (Ciferri, 1978), although once again there is contradictory evidence from *Euglena gracilis*. In *Euglena*, the appearance of EF-Ts during greening is inhibited about 80% by cycloheximide at 10 μg ml^{-1} but is unaffected by chloramphenicol at 100 μg ml^{-1}, suggesting that EF-Ts is synthesised on cytoplasmic ribosomes (Fox *et al.*, 1980). This apparent difference in the sites of synthesis of elongation factors G and Ts in different organisms is of great interest and should be investigated in more detail.

Chloroplast ribosomes are composed of about 58 ribosomal proteins, of which 34 proteins are found in the 50S subunit and 24 proteins in the 30S subunit. Inhibitor studies indicate that both chloroplast and cytoplasmic ribosomes are involved in the synthesis of these ribosomal proteins. With *Euglena*, Freyssinet (1978) has shown that lincomycin inhibits the incorporation of $^{35}SO_4^{2-}$ into 9 ribosomal proteins, whereas cycloheximide inhibits incorporation into a further 12 ribosomal proteins. Incorporation into 6 ribosomal proteins was inhibited by both lincomycin and cycloheximide, suggesting that assembly of ribosomal subunits was also being inhibited. The synthesis of ribosomal proteins has been investigated using light-driven protein synthesis by isolated pea chloroplasts (Eneas-Filho *et al.*, 1981). Incorporation of [^{35}S]methionine into 15 ribosomal proteins, separated by two-dimensional gel electrophoresis, has been demonstrated. Six ribosomal proteins, S6, S9, S13, S19, S22 and S23 or S24, associated with the 30S subunit were labelled in two independent experiments, and in addition there was some evidence for the labelling of proteins S3, S11 and S17. Five ribosomal proteins, L4, L22, L24, L25 and L29, associated with the 50S subunit were labelled in both experiments and protein L32 was labelled in only one experiment. These data indicate that more than one-quarter of the chloroplast ribosomal proteins are synthesised within the chloroplast.

Other stromal proteins. The syntheses of a number of stromal proteins, including enzymes of the Calvin cycle, have been examined by studying the effects of protein synthesis inhibitors on the increases in enzyme activities during greening. The results suggest that the majority of stromal enzymes examined to date are synthesised on cytoplasmic ribosomes. However, equivocal results have been obtained with nitrite reductase. Although it has been suggested that these may be taken to indicate a chloroplast site of synthesis (Schrader, Beevers & Hageman, 1967; Sawhney & Naik, 1972), recent work in this laboratory has shown that wheat nitrite reductase (61.5 kD) is synthesised as a higher molecular weight form (63 kD) in the rabbit reticulocyte lysate programmed with wheat poly(A)-RNA (I. S. Small & J. C. Gray, unpublished). This is an indication that nitrite reductase is synthesised on cytoplasmic ribosomes *in vivo*.

Inhibitor studies with *Chlamydomonas* and *Euglena* have suggested that components of the fatty acid synthetase system may be synthesised on chloroplast ribosomes. They both show that the increase in fatty acid synthetase activity on greening can be prevented by chloramphenicol (Ernst-Fonberg & Bloch, 1971; Sirevag & Levine, 1972). The synthesis of components of fatty acid synthetase requires further study.

Thylakoid membrane proteins

The thylakoid membrane is composed essentially of five functionally distinct complexes. These are photosystem I, photosystem II, the light-harvesting chlorophyll complex (which normally serves as an antenna complex for photosystem II but which may transfer light energy to photosystem I), the cytochrome complex, and the ATP synthase complex. With the exception of the light-harvesting chlorophyll complex, components of each of these complexes have been shown to be synthesised on chloroplast ribosomes. The polypeptides of the light-harvesting complex are synthesised on cytoplasmic ribosomes (Ellis, 1975).

Analysis by gel electrophoresis of thylakoid membranes obtained from isolated pea chloroplasts after incubation with [^{35}S]methionine resolves more than 20 discrete labelled polypeptides (A. Doherty & J. C. Gray, unpublished).

However, as with the stromal products it is not clear if these are all separate gene products, or if some represent premature termination or degradation products. A very careful inhibitor study of the synthesis of *Chlamydomonas* thylakoid membrane proteins suggests that only nine polypeptides resolved by gel electrophoresis are synthesised on chloroplast ribosomes (Chua & Gillham, 1977).

The identities of the components of the thylakoid membrane complexes synthesised on chloroplast ribosomes are discussed below.

Photosystem I. The photosystem I complex contains the P700–chlorophyll *a* protein (CPI) with a subunit polypeptide of 60–70 kD, and a number of smaller polypeptides, whose molecular weights vary somewhat between species.

Initial inhibitor experiments on the site of synthesis of CPI were inconclusive. Machold & Aurich (1972) showed that the incorporation of [^{3}H]leucine into CPI in *Vicia* leaves was severely inhibited by chloramphenicol, but also reported a slight inhibition of labelling by cycloheximide. Ellis (1975) similarly showed that labelling of membrane-located CPI with [^{35}S]methionine in greening pea shoots was inhibited by chloramphenicol and cycloheximide. However, in further studies with pea shoots Cashmore (1976) showed that although cycloheximide and anisomycin prevented the appearance of labelled CPI in the thylakoid membrane, a labelled polypeptide of similar electrophoretic mobility appeared in the stromal fraction, suggesting that a continuous supply of some cytoplasmic product was needed for the incorporation of CPI into the thylakoid membrane. This suggestion is supported by experiments with *Chlamydomonas* described by Chua & Gillham (1977). Incorporation of radioactivity from [^{14}C]acetate into the membrane-located polypeptide of CPI is inhibited by chloramphenicol and spectinomycin, and to a variable extent by anisomycin. The level of inhibition by anisomycin could be decreased markedly if cells were preincubated with chloramphenicol prior to incubation with [^{14}C]acetate and anisomycin. The preincubation with chloramphenicol would permit the pool size of the putative cytoplasmic product to increase relative to that of the CPI polypeptide, so that, during the incubation with [^{14}C]acetate and anisomycin, CPI would be labelled and there would be sufficient cytoplasmic product for assembly into the thylakoid membrane.

The synthesis of CPI by isolated chloroplasts has been reported for *Vicia* (Hachtel, 1975), spinach (Zielinski & Price, 1980), *Acetabularia* (Green, 1980) and *Euglena* (Ortiz & Stutz, 1980). In most cases identification was made on the basis of a characteristic change in electrophoretic mobility on heating CPI in the presence of sodium dodecylsulphate (SDS). Recent studies with isolated pea chloroplasts have used peptide mapping techniques to identify CPI as a product of chloroplast protein synthesis (A. G. Smith & J. C. Gray, unpublished). However, in rye seedlings which have ribosome-deficient chloroplasts, grown at 32 °C, the apoprotein of CPI has been reported to be located in chloroplast membranes, suggesting that the polypeptide is synthesised on cytoplasmic ribosomes (Feierabend, Meschede & Vogel, 1980). The site of synthesis of CPI in rye needs to be investigated directly.

Recent studies on the synthesis of the photosystem I complex by isolated pea chloroplasts have indicated that two polypeptides, of 17 kD and 11 kD, in addition to CPI, are labelled with [^{35}S]methionine in the light (A. G. Smith & J.C. Gray, unpublished). These polypeptides appear to be similar to photosystem I polypeptides labelled in the presence of cycloheximide, but not chloramphenicol, in *Spirodela* (Nechushtai *et al.*, 1981) and *Chlamydomonas* (Nechushtai & Nelson, 1981). However, because of the variable molecular weights of the individual polypeptides, and the confusing numbering system for these polypeptides, direct comparisons are not possible.

Photosystem II. Photosystem II as a membrane complex is not particularly well-defined but the following polypeptides appear to be part of the complex: two polypeptides of 49 kD and 45 kD, which make up the reaction centre and migrate as the green band CPa on electrophoresis in the presence of lithium dodecylsulphate at 4 °C in the dark; a polypeptide of 32 kD, which is the apoprotein of the electron transfer component B, and which binds 3-(3,4-dichlorophenyl-1,1-dimethyl)urea (DCMU) and other herbicides; a polypeptide of 10 kD which is probably the apoprotein of the high potential form of cytochrome *b*-559 and two polypeptides of 34 kD and 16 kD of unknown function, but which may function in the water photolysis reaction (Metz & Miles, 1982). Several of these polypeptides have been shown to be synthesised on chloroplast ribosomes.

The reaction centre complex CPa has been shown to be labelled when isolated *Euglena* chloroplasts are incubated in the light with [^{35}S]methionine (Ortiz & Stutz, 1980). Further electrophoresis after boiling the complex in SDS revealed two labelled polypeptides of 56 kD and 48 kD (Ortiz & Stutz, 1980); these probably correspond to the two polypeptides associated with the reaction centre. Inhibitor experiments with *Chlamydomonas* indicate that two polypeptides of 49 kD and 46 kD, designated polypeptides 5 and 6, are synthesised by chloroplast ribosomes (Chua & Gillham, 1977). Incorporation of radioactivity from [^{14}C]acetate into these polypeptides was inhibited by chloramphenicol and spectinomycin, but not by anisomycin (Chua & Gillham, 1977). The sites of synthesis of the reaction-centre polypeptides have not been investigated in higher plants, although Zielinski & Price (1980) have reported that isolated spinach chloroplasts incorporate [^{35}S]methionine into two polypeptides of 48 kD and 42 kD which may be related to the reaction-centre polypeptides.

The polypeptide of 32 kD associated with photosystem II is the apoprotein of an electron transfer component B, a plastoquinone molecule involved in electron transfer between Q, the primary acceptor of photosystem II, and the plastoquinone pool. This polypeptide is exposed at the stromal surface of the thylakoid membrane and is the site of action of a number of herbicides, including DCMU and atrazine (Steinback *et al.*, 1981; Mattoo *et al.*, 1982). This polypeptide has a number of names, most of which were coined before its function was known. These include peak D (Eaglesham & Ellis, 1974), the photogene product (Steinback *et al.*, 1981) and D-1 in *Chlamydomonas* (Chua & Gillham, 1977). This polypeptide is the major membrane protein synthesised by chloroplast ribosomes in terms of the amount of incorporated [^{35}S]methionine.

The synthesis of this polypeptide was first shown with isolated pea chloroplasts incubated in the light with [^{35}S]methionine (Eaglesham & Ellis, 1974). A heavily labelled polypeptide at 32 kD which did not correspond to a staining band was ascribed to a polypeptide with rapid turnover. This polypeptide is labelled with [^{35}S]methionine in greening pea shoots in the presence of cycloheximide but not in the presence of lincomycin, which indicates that it is synthesised on chloroplast ribosomes *in vivo* (Ellis, 1975). The polypeptide is first synthesised as a precursor of 34–35 kD, which is processed *in vivo* (Grebanier *et al.*, 1978; Reisfeld, Mattoo & Edelman, 1982) and in isolated pea chloroplasts (Ellis & Barraclough, 1978) but not in maize chloroplasts *in vitro* (Grebanier *et al.*, 1978).

Translation of chloroplast RNA from spinach (Hartley *et al.*, 1975; Driesel *et al.*, 1980; Silverthorne & Ellis, 1980), maize (Bedbrook *et al.*, 1978), mustard (Link, 1981) and *Spirodela* (Reisfeld *et al.*, 1982) in a variety of cell-free systems gives a product of about 35 kD. The mRNA for this polypeptide is about 1200 nucleotides long in maize and mustard (Link, 1981), and similar-sized RNA molecules have been purified from *Spirodela* plastids (Rosner *et al.*, 1975) and spinach chloroplasts (Speirs & Grierson, 1978).

Cytochrome *b*-559 has been reported to be synthesised in isolated spinach chloroplasts (Zielinski & Price, 1980) and details are included here, although it is not clear that the polypeptide synthesised by isolated chloroplasts is a component of photosystem II. Thylakoid membranes contain two cytochrome *b*-559 components which differ in their mid-point redox potential and their distribution in the membrane. The high potential cytochrome *b*-559 is associated with photosystem II where it may act as an alternative electron donor to the reaction centre. The low potential cytochrome *b*-559 is extracted from the thylakoid membrane by detergents such as digitonin which leave photosystem II in a particulate state; its function is unknown. The identity of the cytochrome b-559 synthesised by isolated spinach chloroplasts is unknown. Zielinski & Price (1980) reported the incorporation of [^{35}S]methionine and [^{3}H]leucine by isolated spinach chloroplasts into a polypeptide of 6 kD which is characteristic of cytochrome b-559 prepared by extraction with Triton X-100 and 4 mol l^{-1} urea. The majority of the radioactivity in the cytochrome b-559 preparation, however, migrated with an apparent molecular weight of 7 kD, slightly slower than the staining polypeptide. The identity of this band is unknown but Zielinski & Price (1980) have suggested it may be a precursor form of cytochrome b-559.

Cytochrome complex. A membrane complex which catalyses the transfer of electrons from plastoquinol to plastocyanin may be extracted by mild detergent treatment of thylakoid membranes. The complex contains cytochrome f, cytochrome b-563, the Rieske iron-sulphur centre and two polypeptides of unknown function (Hurt & Hauska, 1981). Three of the five polypeptides have been shown to be synthesised on chloroplast ribosomes.

Inhibitor studies with *Chlamydomonas* found that the appearance of cytochrome b-563 during greening was prevented by chloramphenicol and spectinomycin, but not by cycloheximide (Armstrong *et al.*, 1971). Similar results were obtained with cytochrome b-563 and cytochrome f in greening bean leaves; the appearance of the two cytochromes was inhibited by D-*threo* chloramphenicol but not by L-*threo* chloramphenicol (Gregory & Bradbeer, 1973). Confirmation of the site of synthesis of cytochrome f was obtained by experiments with isolated pea chloroplasts. Cytochrome f was identified by immunoprecipitation and peptide mapping as a minor product of protein synthesis by isolated pea chloroplasts (Doherty & Gray, 1979). Cytochrome f has subsequently been shown to be synthesised by membrane-bound ribosomes (J. C. Gray, unpublished; Fig. 1) and coupled transcription and translation of pea chloroplast DNA gives a higher molecular weight product (Willey *et al.*, 1982). This, and the absence of N-formylmethionine at the amino-terminus of cytochrome f (Ho & Krogmann, 1980), suggests that cytochrome f may be synthesised initially with an amino-terminal extension to provide a recognition signal for binding to the thylakoid membrane.

Experiments on the synthesis of the cytochrome complex by isolated pea chloroplasts have shown that three polypeptides, cytochrome f (38 kD), cytochrome b-563 (19 kD) and a 15.5 kD polypeptide of unknown function, are labelled by [^{35}S]methionine and assembled into the complex in the thylakoid membrane (A. L. Phillips & J. C. Gray, unpublished). Assembly of labelled cytochrome b-563 into the complex depends on the addition of

Fig. 1. Synthesis of cytochrome f by thylakoid-bound ribosomes. Washed pea thylakoid membranes (2 mg chlorophyll) were incubated with 40 μCi [^{35}S]methionine, 2 mmol l^{-1} ATP, 0.2 mmol l^{-1} GTP, 2.5 mmol l^{-1} MgCl$_2$, 100 mmol l^{-1} KCl, 25 mmol l^{-1} tricine-KOH pH 8.0 for 60 min at 20 °C. Cytochrome f was extracted with ethylacetate:ethanol:ammonia (2:1:0.015, v/v) and immunoprecipitated with monospecific antibodies to charlock cytochrome f. Polypeptides were separated by SDS–polyacrylamide gel electrophoresis.

(a) Fluorograph of total products of the incubation; (b) fluorograph of the cytochrome f immunoprecipitate; (c) stained polypeptides of the immunoprecipitate. Cytochrome f migrates as a polypeptide of 38 kD, which does not appear to correspond to the major labelled bands in the total products.

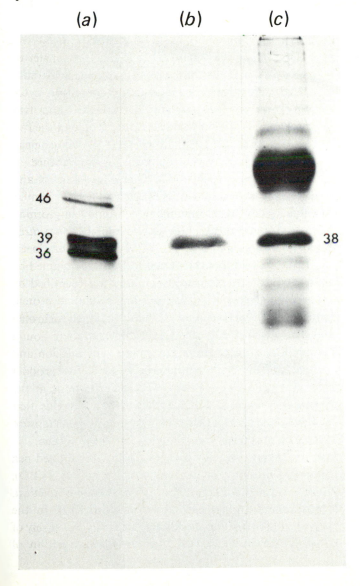

5 mmol l^{-1} Mg-ATP after 15 min incubation in the light (Fig. 2). It appears that assembly is energy-dependent and that isolated pea chloroplasts are deficient in an energy supply suitable for assembly.

ATP synthase. The chloroplast ATP synthase is similar in structure to the proton-translocating ATP synthase complexes of mitochondria and bacteria. Each is composed of a peripheral coupling factor possessing latent ATPase activity and an intrinsic membrane complex catalysing proton-translocation through the membrane. The chloroplast coupling factor (CF$_1$) is composed of five different sorts of polypeptide chains: α (58 kD), β (56 kD), γ (37 kD), δ (20 kD) and ϵ (14 kD). There is some variation in subunit molecular weight between plants. The membrane component (CF$_0$) is composed of three types of polypeptide chains I (18 kD), II (16 kD) and III (8 kD) (Nelson, Nelson & Schatz, 1980). By analogy with F$_0$ components of bacteria, additional polypeptides assigned to CF$_0$ are probably impurities. Subunit III is an exceedingly hydrophobic polypeptide which may be extracted from chloroplast membranes into organic solvents and is thus known as the proteolipid. This polypeptide is involved in proton translocation and reacts covalently with dicyclohexylcarbodiimide, which inhibits proton translocation and ATP synthesis in chloroplast membranes.

Early studies showed that the appearance of ATPase activity in greening bean leaves was prevented by both chloramphenicol and cycloheximide (Horak & Hill, 1972). Later experiments, with pea shoots (Bouthyette & Jagendorf, 1978), *Spirodela* (Nechushtai *et al.*, 1981) and *Chlamydomonas* (Nechushtai & Nelson, 1981) showed that the incorporation of radioactivity into the α, β and ϵ subunits of CF$_1$ was inhibited by chloramphenicol, but not by cycloheximide, whereas the reverse was true for γ and δ subunits. These experiments also demonstrated that labelling of subunit III of CF$_0$ was inhibited by chloramphenicol, and of subunit II by cycloheximide.

The results of these inhibitor studies are in agreement with earlier work on the synthesis of subunits of the ATP synthase in isolated chloroplasts. Mendiola-Morgenthaler, Morgenthaler & Price (1976) showed that α, β and ϵ subunits were synthesised by isolated spinach chloroplasts and this was confirmed with isolated pea chloroplasts (Ellis, 1977). No synthesis of the γ or δ subunits by isolated chloroplasts was observed. In these studies the labelled subunits were assembled into the CF$_1$ complex which could be extracted from the thylakoid membrane by low ionic strength buffers. Nelson *et al.* (1980) also showed that the α, β and ϵ subunits synthesised in isolated spinach chloroplasts were assembled into CF$_1$, and in addition showed that the CF$_1$ complex was associated with the membrane-located CF$_0$ complex. They also showed that CF$_0$ subunit I, and possibly subunit III, were synthesised and assembled into the ATP synthase complex in these chloroplasts. Subunit III, the dicyclohexylcarbodiimide-binding proteolipid, had previously been shown to be synthesised and inserted into chloroplast membranes by isolated pea chloroplasts (Doherty & Gray, 1980). It thus appears that isolated spinach and pea chloroplasts are able to synthesise

Fig. 2. Effect of Mg-ATP on the assembly of cytochrome *b*-563 into the cytochrome complex in isolated chloroplasts. Isolated pea chloroplasts (0.9 mg chlorophyll) were incubated in the light with 120 μCi [^{35}S]methionine for (*a*) 75 min at 20 °C, or (*b*) 75 min at 20 °C but with Mg-ATP added to a final concentration of 5 mmol l^{-1} after 15 min incubation. At the end of the incubations, the cytochrome complex was isolated and analysed by SDS–polyacrylamide gel electrophoresis. Each preparation contained four staining bands at 38 kD (cytochrome *f*), 34 kD, 19 kD (cytochrome *b*-563) and 15 kD. Tracks a and b are fluorographs of these gels and show that the 19 kD polypeptide of the complex is labelled only in the presence of added Mg-ATP.

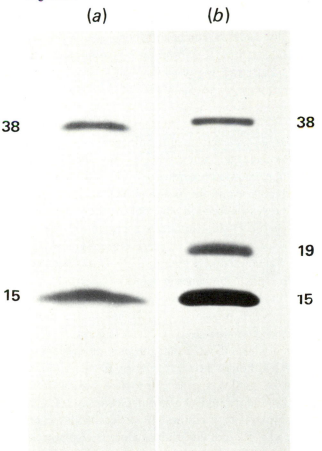

subunits α, β, ϵ, I and III of the ATP synthase and assemble them into the ATP synthase complex. However, this property is not shared by isolated maize chloroplasts, where newly synthesised α and β subunits are not assembled correctly and cannot be extracted from chloroplast membranes by treatments which remove the CF_1 complex (Grebanier *et al.*, 1978).

Translation of spinach chloroplast RNA in the nuclease-treated rabbit reticulocyte lysate results in the synthesis of the α, β and ϵ subunits of CF_1 (Westhoff *et al.*, 1981). The subunit polypeptides were identified by immunoprecipitation and peptide mapping. The α, β and ϵ subunits of CF_1 and subunit III of CF_0 have been identified as products of coupled transcription and translation of cloned fragments of wheat chloroplast DNA (Howe *et al.*, 1982*a, b*, 1983).

The site of synthesis of the γ subunit of CF_1 requires clarification. Although inhibitor studies suggested that the γ subunit is synthesised on cytoplasmic ribosomes, and synthesis of γ could not be detected in isolated spinach or pea chloroplasts, Nelson *et al.* (1980) reported that the γ subunit was labelled with [^{35}S]methionine by isolated spinach chloroplasts. However, it was subsequently reported that the γ subunit was synthesised as a higher molecular form on translation of spinach poly(A)-RNA in the rabbit reticulocyte lysate, supporting a cytoplasmic site of synthesis (Westhoff *et al.*, 1981).

Differences between species

One feature to emerge from studies on the sites of synthesis of chloroplast proteins is the remarkable similarity between species. For example, the LSU of RuBP carboxylase and the 32 kD herbicide-binding protein have each been shown to be synthesised on chloroplast ribosomes from a large number of species. Because of this similarity, reported differences between species in the site of synthesis of chloroplast proteins are of great interest. Three examples of possible differences have emerged. EF-G has been reported to be synthesised on chloroplast ribosomes in spinach (Ciferri *et al.*, 1979) and *Chlorella* (Ciferri & Tiboni, 1976) but on cytoplasmic ribosomes in *Euglena* (Breitenberger *et al.*, 1979). Similarly, EF-Ts has been reported to be synthesised on cytoplasmic ribosomes in *Euglena* (Fox *et al.*, 1980) but, in a brief review, Ciferri (1978) indicated that this subunit was synthesised on chloroplast ribosomes, although no details were given. Another possible difference in the site of synthesis of a chloroplast protein concerns the apoprotein of the P700-chlorophyll *a* protein. Studies with a number of species indicate a chloroplast site of synthesis but Feierabend *et al.* (1980) have reported the presence of a polypeptide with similar electrophoretic mobility in membranes of chloroplast-ribosome deficient plastids from rye leaves grown at high temperatures. The polypeptide has not been identified unequivocally and further investigation is necessary.

It is most important that these reported differences in the site of synthesis of these proteins are investigated fully. If differences in the site of synthesis can be confirmed, it may lead to important insights into the evolution of chloroplasts.

We would like to thank A. Doherty, I. S. Small, C. J. Howe, J. Edwards, A. Davidson, B. Spreckley and N. Dearnaley for permission to cite their unpublished work and for stimulating discussions. We thank the Science and Engineering Research Council, the Agricultural Research Council and the Rank Prize Funds for financial support.

References

Alscher, R., Patterson, R. & Jagendorf, A. T. (1978). Activity of thylakoid-bound ribosomes in pea chloroplasts. *Pl. Physiol.*, **62**, 88–93.

Alwine, J. C., Kemp, P. J. & Stark, G. R. (1977). Method for detection of specific RNAs in agarose gels by transfer to diazobenzyloxymethyl paper and hybridization with DNA probes. *Proc. natn. Acad. Sci. USA*, **74**, 5350–4.

Armstrong, J. J., Surzycki, S. J., Moll, B. & Levine, R. P. (1971). Genetic transcription and translation specifying chloroplast components in *Chlamydomonas reinhardtii. Biochemistry*, **10**, 692–701.

Barraclough, R. & Ellis, R. J. (1979). The biosynthesis of ribulose bisphosphate carboxylase. Uncoupling of the synthesis of the large and small subunits in isolated soybean leaf cells. *Eur. J. Biochem.*, **94**, 165–77.

Barraclough, R. & Ellis, R. J. (1980). Protein synthesis in chloroplasts. IX. Assembly of newly-synthesised large subunits into ribulose bisphosphate carboxylase in isolated intact pea chloroplasts. *Biochim. biophys. Acta*, **608**, 19–31.

Bedbrook, J. R., Coen, D. M., Beaton, A. R., Bogorad, L. & Rich, A. (1979). Location of the single gene for the large subunit of ribulose bisphosphate carboxylase on the maize chloroplast chromosome. *J. biol. Chem.*, **254**, 905–10.

Bedbrook, J. R. & Kolodner, R. (1979). The structure of chloroplast DNA. *A. Rev. Pl. Physiol.*, **30**, 593–620.

Bedbrook, J. R., Link, G., Coen, D. M., Bogorad, L. & Rich, A. (1978). Maize plastid gene expressed during photoregulated development. *Proc. natn. Acad. Sci. USA*, **75**, 3060–4.

Blair, G. E. & Ellis, R. J. (1973). Protein synthesis in chloroplasts. I. Light-driven synthesis of the large subunit of Fraction I protein by isolated pea chloroplasts. *Biochim. biophys. Acta*, **319**, 223–34.

Bottomley, W., Higgins, T. J. V. & Whitfeld, P. R. (1976). Differential recognition of chloroplast and cytoplasmic messenger RNA by 70S and 80S ribosomal systems. *FEBS Lett.*, **63**, 120–4.

Bottomley, W., Spencer, D. & Whitfeld, P. R. (1974). Protein synthesis in isolated spinach chloroplasts: comparison of light-driven and ATP-driven synthesis. *Archs. Biochem. Biophys.*, **164**, 106–17.

Bottomley, W. & Whitfeld, P. R. (1979). Cell-free transcription and translation of total spinach chloroplast DNA. *Eur. J. Biochem.*, **93**, 31–9.

Bouthyette, P.-Y. & Jagendorf, A. T. (1978). The site of synthesis of pea chloroplast coupling factor. *Pl. Cell Physiol.*, **19**, 1169–74.

Breitenberger, C. A., Graves, M. C. & Spremulli, L. L. (1979). Evidence for the nuclear location of the gene for chloroplast elongation factor G. *Archs. Biochem. Biophys.*, **194**, 265–70.

Cashmore, A. (1976). Protein synthesis in plant leaf tissue. The sites of synthesis of the major proteins. *J. biol. Chem.*, **251**, 2848–53.

Chua, N.-H., Blobel, G., Siekevitz, P. & Palade, G. (1973). Attachment of chloroplast polysomes to thylakoid membranes in *Chlamydomonas reinhardtii*. *Proc. natn. Acad. Sci. USA*, **70**, 1554–8.

Chua, N.-H. & Gillham, N. W. (1977). The sites of synthesis of the principal thylakoid membrane polypeptides in *Chlamydomonas reinhardtii*. *J. Cell Biol.*, **74**, 441–52.

Chua, N.-H. & Schmidt, G. W. (1978). Post-translational transport into intact chloroplasts of a precursor to the small subunit of ribulose-1,5-bisphosphate carboxylase. *Proc. natn. Acad. Sci. USA*, **75**, 6110–14.

Ciferri, O. (1978). The chloroplast DNA mystery. *Trends Biochem. Sci.*, **3**, 256–8.

Ciferri, O., Di Pasquale, G. & Tiboni, O. (1979). Chloroplast elongation factors are synthesised in the chloroplast. *Eur. J. Biochem.*, **102**, 331–5.

Ciferri, O. & Tiboni, O. (1976). Evidence for the synthesis in the chloroplast of elongation factor G. *Pl. Sci. Lett.*, **7**, 455–66.

Coen, D. M., Bedbrook, J. R., Bogorad, L. & Rich, A. (1977). Maize chloroplast DNA fragment encoding the large subunit of ribulose bisphosphate carboxylase. *Proc. natn. Acad. Sci. USA*, **74**, 5487–91.

Criddle, R. S., Dau, B., Kleinkopf, G. E. & Huffaker, R. C. (1970). Differential synthesis of ribulose diphosphate carboxylase subunits. *Biochem. biophys. Res. Commun.*, **41**, 621–7.

Dobberstein, B., Blobel, G. & Chua, N.-H. (1977). *In vitro* synthesis and processing of a putative precursor for the small subunit of ribulose-1,5-bisphosphate carboxylase of *Chlamydomonas reinhardtii*. *Proc. natn. Acad. Sci. USA*, **74**, 1082–5.

Doherty, A. & Gray, J. C. (1979). Synthesis of cytochrome *f* by isolated pea chloroplasts. *Eur. J. Biochem.*, **98**, 87–92.

Doherty, A. & Gray, J. C. (1980). Synthesis of a dicyclohexylcarbodiimide-binding proteolipid by isolated pea chloroplasts. *Eur. J. Biochem.*, **108**, 131–6.

Driesel, A. J., Speirs, J. & Bohnert, H.-J. (1980). Spinach chloroplast mRNA for a 32 000 dalton polypeptide. Size and localization on the physical map of the chloroplast DNA. *Biochim. biophys. Acta*, **610**, 297–310.

Eaglesham, A. R. J. & Ellis, R. J. (1974). Protein synthesis in chloroplasts. II. Light-driven synthesis of membrane proteins by isolated pea chloroplasts. *Biochim. biophys. Acta*, **335**, 396–407.

Ellis, R. J. (1969). Chloroplast ribosomes: stereo-specificity of inhibition by chloramphenicol. *Science*, **163**, 477–8.

Ellis, R. J. (1975). Inhibition of chloroplast protein synthesis by lincomycin and 2-(4-methyl-2,6-dinitroanilino)-*N*-methylpropionamide. *Phytochemistry*, **14**, 89–93.

Ellis, R. J. (1977). Protein synthesis by isolated chloroplasts. *Biochim. biophys. Acta*, **463**, 185–215.

Ellis, R. J. (1981). Chloroplast proteins: synthesis, transport, and assembly. *A. Rev. Pl. Physiol.*, **32**, 111–37.

Ellis, R. J. & Barraclough, R. (1978). Synthesis and transport of chloroplast proteins inside and outside the cell. In *Chloroplast Development*, ed. G. Akoyunoglou & J. H. Argyroudi-Akoyunoglou, pp. 185–94. Elsevier/North-Holland, Amsterdam.

Ellis, R. J., Highfield, P. E. & Silverthorne, J. (1977). The synthesis of chloroplast proteins by subcellular systems. In *Proc. 4th Int. Congr. Photosynthesis*, ed. D. O. Hall, J. Coombs & T. W. Goodwin, pp. 497–506. Biochemical Society Press, London.

Eneas-Filho, J., Hartley, M. R. & Mache, R. (1981). Pea chloroplast ribosomal proteins: characterization and site of synthesis. *Molec. gen. Genet.*, **184**, 484–8.

Erion, J. L., Tarnowski, J., Weissbach, H. & Brot, N. (1981). Cloning, mapping and *in vitro* transcription–translation of the gene for the large subunit of ribulose-1,5-bisphosphate carboxylase from spinach chloroplasts. *Proc. natn. Acad. Sci. USA*, **78**, 3459–63.

Ernst-Fonberg, M. L. & Bloch, K. (1971). A chloroplast-associated fatty acid synthetase system in *Euglena*. *Archs. Biochem. Biophys.*, **143**, 392–400.

Feierabend, J., Meschede, D. & Vogel, K.-D. (1980). Comparison of the polypeptide compositions of the internal membrane of chloroplasts, etioplasts and ribosome-deficient heat-bleached plastids from rye leaves. *Z. Pfl-Physiol.*, **98**, 61–78.

Fox, L., Erion, J., Tarnowski, J., Spremulli, L., Brot, N. & Weissbach, H. (1980). *Euglena gracilis* chloroplast EF-Ts. Evidence that it is a nuclear-coded gene product. *J. biol. Chem.*, **255**, 6018–19.

Freyssinet, G. (1978). Determination of the site of synthesis of some *Euglena* cytoplasmic and chloroplast ribosomal proteins. *Expl. Cell Res.*, **115**, 207–19.

Gelvin, S., Heizmann, P. & Howell, S. H. (1977). Identification and cloning of the chloroplast gene coding for the large subunit of ribulose-1,5-bisphosphate carboxylase from *Chlamydomonas reinhardtii*. *Proc. natn. Acad. Sci. USA*, **74**, 3193–7.

Graham, D., Hatch, M. D., Slack, G. R. & Smillie, R. M. (1970). Light-induced formation of enzymes of the C_4-dicarboxylic acid pathway of photosynthesis in detached leaves. *Phytochemistry*, **9**, 531–2.

Grebanier, A. E., Coen, D. M., Rich, A. & Bogorad, L. (1978). Membrane proteins synthesised but not processed by isolated maize chloroplasts. *J. Cell Biol.*, **78**, 734–46.

Grebanier, A. E., Steinback, K. E. & Bogorad, L. (1979). Comparison of the molecular weights of proteins synthesised by isolated chloroplasts with those which appear during greening in *Zea mays*. *Pl. Physiol.*, **63**, 436–9.

Green, B. R. (1980). Protein synthesis by isolated *Acetabularia* chloroplasts. *Biochim. biophys. Acta*, **609**, 107–20.

Gregory, P. & Bradbeer, J. W. (1973). Plastid development in primary leaves of *Phaseolus vulgaris*: the light-induced development of chloroplast cytochromes. *Planta*, **109**, 317–26.

Hachtel, W. (1975). *In vitro* synthesis of membrane proteins by isolated chloroplasts of *Vicia faba. Ber. dt. bot. Ges.*, **89**, 185–92.

Hartley, M. R., Wheeler, A. M. & Ellis, R. J. (1975). Protein synthesis in chloroplasts. V. Translation of messenger RNA for the large subunit of Fraction I protein in a heterologous cell-free system. *J. molec. Biol.*, **91**, 67–77.

Herrin, D., Michaels, A. & Hickey, E. (1981). Synthesis of a chloroplast membrane polypeptide on thylakoid-bound ribosomes during the cell-cycle of *Chlamydomonas reinhardtii* 137$^+$. *Biochim. biophys. Acta*, **655**, 136–45.

Highfield, P. E. & Ellis, R. J. (1976). Protein synthesis in chloroplasts. VII. Initiation of protein synthesis in isolated intact pea chloroplasts. *Biochim. biophys. Acta*, **447**, 20–7.

Highfield, P. E. & Ellis, R. J. (1978). Synthesis and transport of the small subunit of ribulose bisphosphate carboxylase. *Nature*, **271**, 420–4.

Hirai, A. & Wildman, S. G. (1977). Kinetic analysis of Fraction 1 protein biosynthesis in young protoplasts of tobacco leaves. *Biochim. biophys. Acta*, **479**, 39–52.

Ho, K. K. & Krogmann, D. W. (1980). Cytochrome *f* from spinach and cyanobacteria. *J. biol. Chem.*, **255**, 3855–61.

Horak, A. & Hill, R. D. (1972). ATPase of bean plastids. Its properties and site of formation. *Pl. Physiol.*, **49**, 365–70.

Howe, C. J., Auffret, A., Doherty, A., Bowman, C. M., Dyer, T. A. & Gray, J. C. (1982a). Location and nucleotide sequence of the gene for the proton-translocating subunit of wheat chloroplast ATP synthase. *Proc. natn. Acad. Sci. USA*, **79**, 6903–7.

Howe, C. J., Bowman, C. M., Dyer, T. A. & Gray, J. C. (1982b). Localization of wheat chloroplast genes for the beta and epsilon subunits of ATP synthase. *Molec. gen. Genet.*, **186**, 525–30.

Howe, C. J., Bowman, C. M., Dyer, T. A. & Gray, J. C. (1983). The genes for the alpha and proton-translocating subunits of wheat chloroplast ATP synthase are close together on the same strand of chloroplast DNA. *Molec. gen. Genet.*, **190**, 51–5.

Howell, S. H., Heizmann, P., Gelvin, S. & Walker, L. L. (1977). Identification and properties of messenger RNA activity in *Chlamydomonas reinhardtii* coding for the large subunit of ribulose 1,5-bisphosphate carboxylase. *Pl. Physiol.*, **59**, 464–70.

Hurt, G. & Hauska, G. (1981). A cytochrome f/b_6 complex of five polypeptides with plastoquinol-plastocyanin-oxidoreductase activity from spinach chloroplasts. *Eur. J. Biochem.*, **117**, 591–9.

Iwanij, V., Chua, N.-H. & Siekevitz, P. (1975). Synthesis and turnover of ribulose bisphosphate carboxylase and of its subunits during the cell cycle of *Chlamydomonas reinhardtii*. *J. Cell Biol.*, **64**, 572–85.

Joy, K. W. & Ellis, R. J. (1975). Protein synthesis in chloroplasts. IV.

Polypeptides of the chloroplast envelope. *Biochim. biophys. Acta*, **378**, 143–51.

Koller, B., Delius, H. & Dyer, T. A. (1982). The organisation of the chloroplast DNA in wheat and maize in the region containing the LS gene. *Eur. J. Biochem.*, **122**, 17–23.

Langridge, P. (1981). Synthesis of the large subunit of spinach ribulose bisphosphate carboxylase may involve a precursor polypeptide. *FEBS Lett.*, **123**, 85–9.

Link, G. (1981). Cloning and mapping of the chloroplast DNA sequences for two messenger RNAs from mustard (*Sinapis alba* L.). *Nucleic Acids Res.*, **9**, 3681–94.

Link, G. & Bogorad, L. (1980). Sizes, locations and directions of transcription of two genes on a cloned maize chloroplast DNA sequence. *Proc. natn. Acad. Sci. USA*, **77**, 1832–6.

Lord, J. M., Armitage, T. L. & Merrett, M. J. (1975). Ribulose 1,5-diphosphate carboxylase synthesis in *Euglena*. *Pl. Physiol.*, **56**, 600–4.

Lucchini, G. & Bianchetti, R. (1980). Initiation of protein synthesis in isolated mitochondria and chloroplasts. *Biochim. biophys. Acta*, **608**, 54–61.

Machold, O. & Aurich, O. (1972). Sites of synthesis of chloroplast lamellar proteins in *Vicia faba*. *Biochim. biophys. Acta*, **281**, 103–12.

McIntosh, L., Poulsen, C. & Bogorad, L. (1980). Chloroplast gene sequence for the large subunit of ribulose bisphosphate carboxylase of maize. *Nature*, **288**, 556–60.

Malnoe, P., Rochaix, J.-D., Chua, N.-H. & Spahr, P.-F. (1979). Characterisation of the gene and messenger RNA of the large subunit of ribulose 1,5-bisphosphate carboxylase in *Chlamydomonas reinhardtii*. *J. molec. Biol.*, **133**, 417–34.

Margulies, M. M. (1971). Concerning the sites of synthesis of proteins of chloroplast ribosomes and of Fraction I protein (ribulose-1,5-diphosphate carboxylase). *Biochem. biophys. Res. Commun.*, **44**, 539–45.

Margulies, M. M. & Michaels, A. (1974). Ribosomes bound to chloroplast membranes in *Chlamydomonas reinhardtii*. *J. Cell Biol.*, **60**, 65–77.

Mattoo, A. K., Marder, J. B., Gressel, J. & Edelman, M. (1982). Presence of the rapidly-labelled 32000 dalton chloroplast membrane protein in triazine resistant biotypes. *FEBS Lett.*, **140**, 36–40.

Mendiola-Morgenthaler, L. R., Morgenthaler, J.-J. & Price, C. A. (1976). Synthesis of coupling factor CF_1 protein by isolated spinach chloroplasts. *FEBS Lett.*, **62**, 96–100.

Metz, J. & Miles, D. (1982). Use of a nuclear mutant of maize to identify components of photosystem II. *Biochim. biophys. Acta*, **681**, 95–102.

Morgenthaler, J. J. & Mendiola-Morgenthaler, L. R. (1976). Synthesis of soluble, thylakoid and envelope membrane proteins by spinach chloroplasts purified from gradients. *Archs. Biochem. Biophys.*, **172**, 51–8.

Nechushtai, R. & Nelson, N. (1981). Purification properties and biogenesis of *Chlamydomonas reinhardtii* photosystem I reaction centre. *J. biol. Chem.*, **256**, 11624–8.

Nechushtai, R., Nelson, N., Mattoo, K. & Edelman, M. (1981). Site of synthesis of subunits of photosystem I reaction centre and the proton-translocating ATPase in *Spirodela. FEBS Lett.*, **125**, 115–19.

Nelson, N., Nelson, H. & Schatz, G. (1980). Biosynthesis and assembly of the proton-translocating adenosine triphosphatase complex from chloroplasts. *Proc. natn. Acad. Sci. USA*, **77**, 1361–4.

Nishimura, M. & Akazawa, T. (1978). Biosynthesis of ribulose 1,5-bisphosphate carboxylase in spinach leaf protoplasts. *Pl. Physiol.*, **62**, 97–100.

Ortiz, W., Reardon, E. M. & Price, C. A. (1980). Preparation of chloroplasts from *Euglena* highly active in protein synthesis. *Pl. Physiol.*, **66**, 291–4.

Ortiz, W. & Stutz, E. (1980). Synthesis of polypeptides of the chlorophyll-protein complexes in isolated chloroplasts of *Euglena gracilis. FEBS Lett.*, **116**, 298–302.

Reisfeld, A., Mattoo, A. K. & Edelman, M. (1982). Processing of a chloroplast-translated membrane protein *in vivo. Eur. J. Biochem.*, **124**, 125–9.

Rosner, A., Jakob, K. M., Gressel, J. & Sagher, D. (1975). The early synthesis and possible function of a 0.5×10^6 M_r RNA after transfer of dark-grown *Spirodela* plants to light. *Biochem. biophys. Res. Commun.*, **67**, 383–91.

Roy, H., Costa, K. A. & Adari, H. (1978). Free subunits of ribulose 1,5-bisphosphate carboxylase in pea leaves. *Pl. Sci. Lett.*, **11**, 159–68.

Sagher, D., Grosfeld, H. & Edelman, M. (1976). Large subunit ribulose bisphosphate carboxylase mRNA from *Euglena* chloroplasts. *Proc. natn. Acad. Sci. USA*, **73**, 722–6.

Sano, H., Spaeth, E. & Burton, W. G. (1979). Messenger RNA of the large subunit of ribulose 1,5-bisphosphate carboxylase from *Chlamydomonas reinhardtii. Eur. J. Biochem.*, **93**, 173–80.

Sawhney, S. K. & Naik, M. S. (1972). Role of light in the synthesis of nitrate reductase and nitrite reductase in rice seedlings. *Biochem. J.*, **130**, 475–85.

Schrader, L. E., Beevers, L. & Hageman, R. H. (1967). Differential effects of chloramphenicol on the induction of nitrate and nitrite reductase in green leaf tissue. *Biochem. biophys. Res. Commun.*, **26**, 14–17.

Silverthorne, J. & Ellis, R. J. (1980). Protein synthesis in chloroplasts. VIII. Differential synthesis of chloroplast proteins during spinach leaf development. *Biochim. biophys. Acta*, **607**, 319–30.

Sirevag, R. & Levine, R. P. (1972). Fatty acid synthetase from *Chlamydomonas reinhardtii*. Sites of transcription and translation. *J. biol. Chem.*, **247**, 2586–91.

Smith, S. M. & Ellis, R. J. (1979). Processing of small subunit precursor of ribulose bisphosphate carboxylase and its assembly into whole enzyme are stromal events. *Nature*, **278**, 662–4.

Smith, S. M. & Ellis, R. J. (1981). Light-stimulated accumulation of transcripts of nuclear and chloroplast genes for ribulose bisphosphate carboxylase. *J. molec. appl. Genet.*, **1**, 127–37.

Speirs, J. & Grierson, D. (1978). Isolation and characterisation of 14S RNA from spinach chloroplasts. *Biochim. biophys. Acta*, **521**, 619–33.

Steinback, K. E., McIntosh, L., Bogorad, L. & Arntzen, C. J. (1981). Identification of the triazine receptor protein as a chloroplast gene product. *Proc. natn. Acad. Sci. USA*, **78**, 7463–7.

Tao, K.-L. & Jagendorf, A. T. (1973). The ratio of free to membrane-bound chloroplast ribosomes. *Biochim. biophys. Acta*, **324**, 518–32.

Vasconcelos, A. C. (1976). Synthesis of proteins by isolated *Euglena gracilis* chloroplasts. *Pl. Physiol.*, **58**, 719–21.

Walden, R. & Leaver, C. J. (1980). Synthesis of chloroplast proteins during germination and early development of cucumber. *Pl. Physiol.*, **67**, 1090–6.

Westhoff, P., Nelson, N., Bunemann, H. & Herrmann, R. G. (1981). Localization of genes for coupling factor subunits on the spinach plastid chromosome. *Curr. Genet.*, **4**, 109–20.

Westhoff, P. & Zetsche, K. (1981). Regulation of the synthesis of ribulose 1,5-bisphosphate carboxylase and its subunits in the flagellate *Chlorogonium elongatum*. *Eur. J. Biochem.*, **116**, 261–7.

Wheeler, A. M. & Hartley, M. R. (1975). Major mRNA species from spinach chloroplasts do not contain poly(A). *Nature*, **257**, 66–7.

Whitfeld, P. R. & Bottomley, W. (1980). Mapping of the gene for the large subunit of ribulose bisphosphate carboxylase on spinach chloroplast DNA. *Biochem. Int.*, **1**, 172–8.

Willey, D. L., Huttly, A. K., Phillips, A. L. & Gray, J. C. (1982). Localization of the gene for cytochrome *f* in pea chloroplast DNA. *Molec. gen. Genet.*, **189**, 85–9.

Zielinski, R. E. & Price, C. A. (1980). Synthesis of thylakoid membrane proteins by chloroplasts isolated from spinach. Cytochrome b-559 and P700–chlorophyll a protein. *J. Cell Biol.*, **85**, 435–45.

Zurawski, G., Perrot, B., Bottomley, W. & Whitfeld, P. R. (1981). The structure of the gene for the large subunit of ribulose bisphosphate carboxylase from spinach chloroplast DNA. *Nucleic Acids Res.*, **9**, 3251–70.

PART III

The formation of thylakoids

J. BENNETT, G. I. JENKINS, A. C. CUMING,
R. S. WILLIAMS AND M. R. HARTLEY

Photoregulation of thylakoid biogenesis: the case of the light-harvesting chlorophyll *a*/*b* complex

Chloroplasts belong to the family of plant organelles known as the plastids (Kirk & Tilney-Bassett, 1978). In the algae and the lower land-plants the chloroplast is the predominant plastid type, but in the higher plants, and in the angiosperms especially, plastid diversity is very marked and many cell types contain non-green plastids (Thomson & Whatley, 1980). Plastids contain their own distinctive genome, which specifies plastid ribosomal RNAs, plastid transfer RNAs and a minority of plastid polypeptides (Whitfeld & Bottomley, 1983). The majority of plastid polypeptides are encoded in nuclear DNA and are synthesized on cytoplasmic ribosomes prior to entry into the plastid (Ellis, 1981; Grossman *et al.*, 1982). Plastid differentiation is therefore presumed to depend on the co-ordinated regulation of gene expression in both the plastid and the nucleus. Very little is known about the fundamental endogenous regulators of plastid differentiation in angiosperms, but considerable progress is being made in understanding how light, an exogenous regulatory factor, modulates one example of plastid differentiation, the conversion of proplastids of the shoot meristem into the chloroplasts of leaf tissue. In dark-grown seedlings of most angiosperms, this conversion fails to go to completion and a distinct plastid type, the etioplast, is observed. Although the extent to which specific chloroplast components accumulate within etioplasts is rather variable between species, only in rare instances does chlorophyll become detectable (Adamson & Hiller, 1981). Since it is now accepted that most, if not all, of the chlorophyll of the photosynthetic membranes of angiosperms is associated with specific chlorophyll-binding proteins (Thornber, Markwell & Reinman, 1979), we became interested in the co-ordination between the synthesis of chlorophyll and the synthesis of chlorophyll-binding proteins, and in the role of light in regulating the two processes. In this article, we describe our studies on one family of chlorophyll-binding proteins, the light-harvesting chl *a*/*b* proteins (LHCPs). Although there are several different classes of chlorophyll-binding

Table 1. *Chlorophyll-protein complexes of representative green plants*[a]

Complex	Synonyms (including oligomers)	% of total chlorophyll[b]	Chl a/chl b	Apoproteins (kD)	(References)
Photosystem I					
P700–chl a complex	CPI	20	>15	65, 68 (P)	Haworth, Watson & Arntzen (1983) Westhoff et al. (1983)
Peripheral complex	LHC-I	15	3.7	66, 67 (S) 20, 21, 22 (H) 21, 23, 23.5 (P)	Bellemare et al. (1982) Haworth et al. (1983)
Photosystem II					
CPa-1	CP47	3	8	47 (S) 41 (A)	Camm & Green (1983) Green, Camm & van Houten (1982)
CPa-2	CP43	3	10	43 (S) 37 (A)	Camm & Green (1983) Green et al. (1982)
CP29	D-1, 2, 3	5–7	3–4	29 (S, A)	Camm & Green (1980) Green et al. (1982)
Light-harvesting chl a/b complex					
LHC	LHC-II; LHCP-1, 2, 3; AB-1, 2, 3	50–55	0.8–1.1	25, 26, 27, 29 (P) 24, 25, 27 (H)	Mullet et al. (1981) Bellemare et al. (1982)

[a] A, *Acetabularia*; H, *Hordeum*; P, *Pisum*; S, *Spinacia*.
[b] Figures vary with growth conditions (Melis & Harvey, 1981; Leong & Anderson, 1983).

proteins in angiosperms (Table 1), the LHCPs were attractive to study for four reasons. First, the LHCPs are the most abundant chlorophyll-binding proteins in green plants. Secondly, their synthesis is especially tightly controlled by light. Thirdly, the LHCPs are encoded and synthesized outside the chloroplast, while chlorophyll is synthesized inside the organelle; this situation offers an opportunity to study the wider problem of the co-ordination of synthetic events in different subcellular compartments. Finally, the study of LHCP synthesis complements our other studies on the regulation of photosynthesis by the reversible phosphorylation of LHCP (Bennett, 1983). The article is divided into three main sections. In the first, we discuss the range of chlorophyll-binding proteins found in angiosperms and describe the LHCPs in some detail. In the second section, the biosyntheses of LHCPs and chlorophyll are summarized. In the third section, we discuss the photo-regulation of LHCP and chlorophyll synthesis and compare it with other aspects of chloroplast development.

What is the light-harvesting chlorophyll *a/b* complex?

Chlorophyll–protein complexes were discovered when photosynthetic membranes (thylakoids) of green plants were subjected to sodium dodecyl-sulphate (SDS)–polyacrylamide gel electrophoresis without thermal denaturation (Ogawa, Obata & Shibata, 1966; Thornber *et al.*, 1967). In early studies, only three major green bands were seen on unstained gels: two chlorophyll–protein (CP) complexes (CPI and CPII) and a rapidly-migrating band of free chlorophyll. CPI contained only chl *a* and migrated with an apparent molecular weight of 110 kD. CPII contained approximately equal amounts of chl *a* and chl *b* and migrated with an apparent molecular weight of about 35 kD. Similar complexes were also isolated from green algae. The fact that a mutant of *Scenedesmus* lacked both CPI and photosystem I activity led to the conclusion that CPI is a part of photosystem I. However, the notion that CPII is a component of photosystem II, originally proposed on the basis of chlorophyll fluorescence studies, was revised following two important discoveries. First, when thylakoids are treated with either Triton X-100 or digitonin, a chl *b*-deficient photosystem II complex is readily separated by centrifugation from a photochemically inactive complex containing approximately equal amounts of chl *a* and chl *b* (Vernon *et al.*, 1972; Wessels, van Alphen-van Waveren & Voorn, 1973). Secondly, Thornber & Highkin (1974) showed that the chl *b*-less *chlorina f*2 mutant of barley contains CPI but lacks CPII, even though it is known to possess both photosystem I and photosystem II activities. These results suggested that CPII is not identical with photosystem II but rather serves photosystem II in a light-harvesting capacity. It was proposed that CPII should be renamed the light-harvesting chl *a/b* protein.

In this article we use the acronym LHC to denote the light-harvesting chl *a/b* complex and LHCP to denote the polypeptide components of the complex.

LHC is believed to occur in all organisms containing chl *b*. This includes not only all land plants, green algae (Chlorophyceae) and *Euglena* species, but also the only known chl *b*-containing prokaryote, *Prochloron* (Withers *et al.*, 1978; Giddings, Withers & Staehelin, 1980).

At one time it was thought that LHC was the only chl *b*-containing chlorophyll–protein complex. However, new procedures for the solubilization and electrophoretic analysis of chlorophyll–protein complexes have led to the discovery of many additional complexes, including several containing both chl *a* and chl *b* (Table 1). One of the consequences of the existence of several chl *a*- and chl *b*-containing complexes is that there is now no absolute definition of LHC. As pointed out by Green & Camm (1982), LHC can at present be defined only operationally, in terms of the procedures that are currently used to purify the complex. Probably the best of these procedures is that devised by Burke, Ditto & Arntzen (1978). Thylakoids are first washed

Fig. 1. Fractionation of thylakoid components. Pea thylakoids in 6.2 mmol l^{-1} Tris–48 mmol l^{-1} glycine (0.5 mg chlorophyll ml^{-1}) were solubilized with 0.5% Triton X-100 and fractionated by sucrose density gradient centrifugation (0.1–0.7 mol l^{-1} sucrose in Tris–glycine buffer containing 0.02% Triton X-100) at 105000 × g_{max} for 16 h. Fractions were analysed by SDS–polyacrylamide gel electrophoresis. The LHC (bracketed) sediments in a polydisperse manner, indicating the existence of oligomeric LHC. The values on the right indicate the molecular weights of marker proteins.

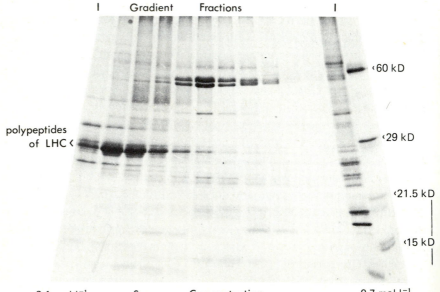

thoroughly in a medium of low ionic strength to convert the intricate system of stacked and unstacked lamellae into large vesicles. Secondly, the vesicles are treated with Triton X-100 under defined conditions to solubilize protein components with minimal disruption of pigment–protein interactions. Thirdly, the solubilized material is subjected to overnight centrifugation through a sucrose density gradient to separate components according to sedimentation coefficient (Fig. 1). Finally, $MgCl_2$ is added to the most highly fluorescent gradient fractions to induce specific aggregation of the LHC, and the aggregates are recovered by low-speed centrifugation. Prepared in this way, LHC contains chl *a*, chl *b*, violaxanthin and lutein (Ryrie, Anderson & Goodchild, 1980). It is also heterogeneous with respect to polypeptide composition. This can be seen at the level of chlorophyll–protein complexes, which may be separated by isoelectric focussing, or at the level of SDS–polyacrylamide gel electrophoresis (Mullet, Baldwin & Arntzen, 1981).

It seems frequently to be the case that green plants contain two or three readily detectable LHCPs, which are structurally related as judged by amino-acid analysis, partial proteolytic cleavage, partial cyanogen bromide cleavage and immunological cross-reaction (Apel, 1977; Chua & Blomberg, 1979; Hoober, Millington & D'Angelo, 1980; Bennett *et al.*, 1981; Schmidt *et al.*, 1981). Bellemare, Bartlett & Chua (1982) showed that at least three LHCP variants may be distinguished in the case of barley, with apparent molecular weights of 24, 25 and 27 kD. In contrast, the putative peripheral chl *a*- and chl *b*-binding polypeptides of photosystem I (Table 1) were shown to have apparent molecular weights of 20, 21 and 22 kD. In the case of pea, we usually restrict our attention to two LHCPs: a major form and a readily detectable minor form, with apparent molecular weights of about 26 and 24 kD, respectively. Both are phosphorylated but *in vitro* the minor LHCP is often labelled to a higher specific activity than the major form (Fig. 2).

For many years, only one form of LHC was discernible on SDS–polyacrylamide gels (that is, CPII), but improvements in technique have resulted in the preservation of three 'oligomers' of LHC (Anderson, Waldron & Thorne, 1978; Markwell, Reinman & Thornber, 1978; Machold, Simpson & Møller, 1979). Markwell *et al.* (1978) found three forms of LHC (AB-1, AB-2 and AB-3) with apparent molecular weights of 80, 60 and 45 kD, and Bennett *et al.* (1981) showed that all three forms contained phosphorylated 26 and 24 kD apoproteins. The existence of LHC oligomers is consistent with the sedimentation behaviour of the LHC released from thylakoids by Triton X-100 (Fig. 1).

Fig. 2. Phosphorylated LHCPs. Pea thylakoids were labelled with [^{32}P]orthophosphate (Bennett *et al.*, 1981), solubilized with Triton X-100 and fractionated by sucrose density gradient centrifugation (see legend to Fig. 1). LHC was precipitated specifically by addition of MgCl$_2$ to appropriate gradient fractions (Burke *et al.*, 1978). LHCPs were analysed by SDS–polyacrylamide gel electrophoresis. A, B: autoradiogram. C, D: Coomassie stain. A, C: 5 µg chl *b* per track. B, D: 10 µg chl *b* per track. The minor 24 kD LHCP is indicated by the arrow.

A   B   C   D

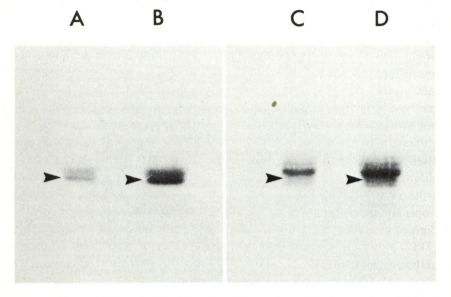

Fig. 3. Biosynthesis of the light-harvesting chl *a/b* complex. ALA, δ-aminolaevulinic acid; proto, protoporphyrin IX; pchlide, protochlorophyllide; chlide, chlorophyllide; IPP, isopentenyl pyrophosphate; GGPP, geranylgeranyldiphosphate; PPP, phytyldiphosphate; *hv*, quantum of light. + indicates stimulation and − indicates inhibition.

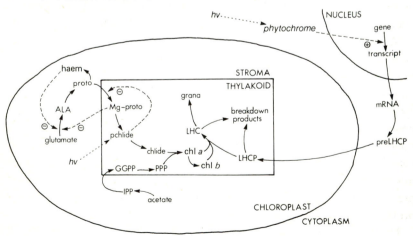

Biosynthesis of the light-harvesting chlorophyll *a*/*b* complex

Biosynthesis of LHC involves the nucleus, the cytoplasm and the plastid (Fig. 3). Like the majority of chloroplast proteins, the LHCPs are encoded in nuclear DNA and synthesized in precursor form on cytoplasmic ribosomes (for references, see Bennett, 1983). The pre-LHCPs are then taken up by chloroplasts and become associated with the thylakoid membrane (Schmidt *et al.*, 1981). At some stage in this process, the pre-LHCPs are cleaved to their mature molecular weights and become ligated with either chl *a* or chl *b*, and presumably also with carotenoids. The pigments, in contrast, are synthesized entirely within the plastid. Since chlorophyll is an ester of the tetrapyrrole chlorophyllide and the terpenoid alcohol phytol, chlorophyll synthesis should be regarded as the culmination of two complex pathways, not just one pathway. Chlorophyllide synthesis involves the conversion of glutamate to δ-aminolaevulinic acid (ALA) and thence via a series of intermediates (including protoporphyrin IX, Mg-protoporphyrin and proto-chlorophyllide) to chlorophyllide (Castelfranco & Beale, 1983). Phytol synthesis involves the conversion of acetate into the C_5 terpenoid precursor, isopentenylpyrophosphate, and thence to the C_{20} molecule, geranylgeranyl-diphosphate (GGPP). Hydrogenation of GGPP yields phytyldiphosphate (PPP). However, it is not clear whether hydrogenation of the unsaturated carbon–carbon double bonds in GGPP takes place before or after esterification of the C_{20} diphosphate with chlorophyllide, a reaction catalysed by chlorophyll synthetase (Rüdiger, Benz & Guthoff, 1980). The final step in chlorophyll synthesis is the oxidation of chl *a* to form chl *b*: a methyl group on ring II of the chlorin head group is converted to a formyl group. Little is known about this reaction, but it probably involves a specific subfraction of the chl *a* population and it is stimulated in the dark by calcium ions (Oelze-Karow, Kasemir & Mohr, 1978; Tanaka & Tsuji, 1982).

Many of the steps in the synthesis of LHC have been reproduced *in vitro*. Gallagher & Ellis (1982) have shown that nuclei isolated from light-grown pea seedlings transcribe LHCP genes *in vitro*. Several studies, beginning with that of Apel & Kloppstech (1978), have shown that poly(A)-mRNA from leaves, when translated *in vitro* in a wheat-germ cell-free system, specifies the pre-LHCPs. Figure 4 shows the total translation products specified by the poly(A)-mRNA from four species of angiosperms, together with the products recognized by antibodies raised in rabbits against the LHCPs of pea. The mRNA of pea specifies an abundant 32 kD precursor and a minor 30 kD precursor, while the mRNAs of broad bean, barley and wheat specify several pre-LHCPs of similar but not identical apparent molecular weights.

The pre-LHCPs are taken up by isolated intact chloroplasts by a post-translational mechanism (Schmidt *et al.*, 1981). They are cleaved to their mature molecular weights and incorporated into thylakoids. At the same time, the 20 kD molecular weight precursor of the small (14 kD) subunit of ribulose-1,5-bisphosphate carboxylase/oxygenase is also taken up by intact chloroplasts, and cleaved to its mature size, but it remains in the stroma where it associates with the large (54 kD) subunit of the enzyme (Chua & Schmidt, 1978; Smith & Ellis, 1979).

Most of the steps in chlorophyll biosynthesis have also been reproduced *in vitro*, including the conversion of glutamate to ALA (Kannangara, Gough & von Wettstein, 1978), the ATP-dependent insertion of Mg^{2+} into protoporphyrin IX (Fuesler, Wright & Castelfranco, 1981), the reduction of protochlorophyllide to chlorophyllide by NADPH-linked protochlorophyllide reductase (Griffiths, 1978; Griffiths & Oliver, this volume), and the conversion

Fig. 4. Detection of LHCP precursors by translation *in vitro* and immunoprecipitation. Poly(A)-mRNA was extracted from leaves of pea (*Pisum sativum*) (A, B), wheat (*Triticum aestivum*) (C, D), barley (*Hordeum vulgare*) (E, F), and broad bean (*Vicia faba*) (G, H), and translated in the wheat-germ cell-free system as described by Cuming & Bennett (1981). Translation products corresponding to LHCP precursors were collected by reaction with anti-LHCP antibodies and protein A–Sepharose (Bennett, 1981). Total translation products (tracks A, C, E, G) and immunoprecipitates (tracks B, D, F, H) were analysed by SDS–polyacrylamide gel electrophoresis and autoradiography. The 30 (lower) and 32 (upper) kD pre-LHCPs of pea are indicated by arrows.

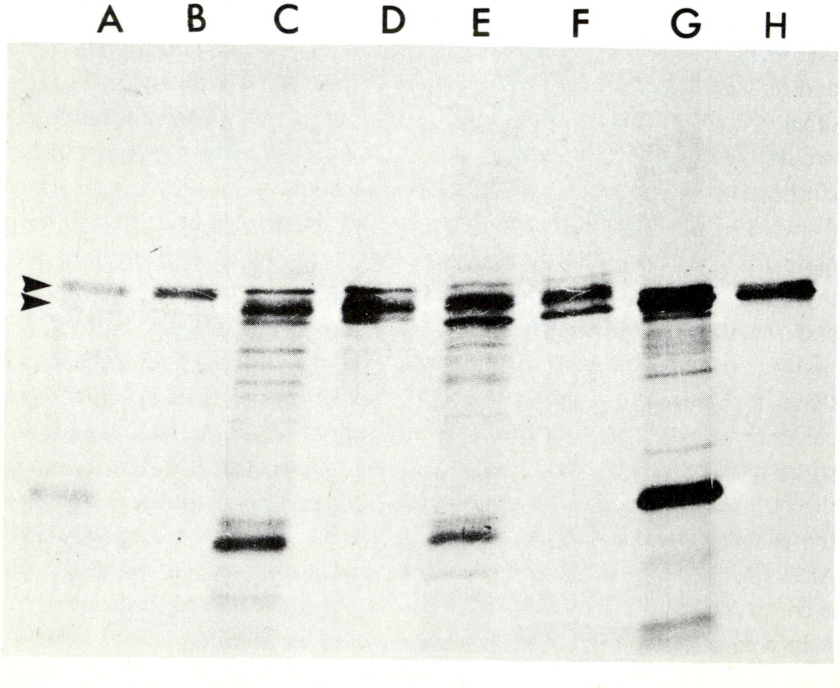

of isopentenyldiphosphate into chlorophyll (Block, Joyard & Douce, 1980).

Several steps in the biosynthesis of LHC have yet to be reproduced *in vitro*. Although Schmidt *et al.* (1981) have presented evidence that pre-LHCPs are cleaved and ligated with chlorophyll in isolated intact chloroplasts, these same reactions have not yet been achieved in simpler systems such as broken chloroplasts or washed thylakoids. It would be interesting to know whether the pre-LHCPs are cleaved by the same protease which cleaves the precursor to the small subunit of ribulose bisphosphate carboxylase/oxygenase. Another important advance would be the isolation of the enzyme responsible for the oxidation of chl *a* to chl *b*.

Photoregulation of the accumulation of the light-harvesting chlorophyll *a/b* complex

As mentioned earlier, most angiosperms when germinated in darkness fail to synthesize normal levels of many chloroplast components (Kirk & Tilney-Bassett, 1978). The transformation of proplastids into chloroplasts is certainly initiated, but the process begins to show signs of derangement at about the stage when the thylakoid membranes would normally begin to appear (Thomson & Whatley, 1980). The etioplasts which form under these conditions are intermediate in size between proplastids and chloroplasts, and contain distinctive paracrystalline arrays of tubules known as prolamellar bodies. Thylakoids as such are absent, but attached to each prolamellar body are short lamellar structures known as prothylakoids, which are enriched for certain thylakoid polypeptides including the CF_1 component of ATP synthase and also protochlorophyllide reductase (Lütz *et al.*, 1981). However, many other thylakoid polypeptides appear to be absent from etioplasts, or are present in greatly reduced amounts. On exposure of etiolated seedlings to continuous white light or normal day/night cycles, the prolamellar bodies disperse, while the prothylakoids lengthen and coalesce until they stretch almost from one end of the organelle to the other. These so-called primary thylakoids subsequently develop localized regions of membrane appression and stacking which eventually give rise to the lateral differentiation of thylakoids into granal lamellae and stromal lamellae. Concomitant with these ultrastructural changes, the thylakoids accumulate chlorophyll, and chlorophyll-binding proteins, and become capable of electron transport from water to NADP, and of ATP synthesis (Wellburn, 1982). At the same time, the stroma of the organelle becomes enriched in the numerous enzymes of the reductive carbon assimilation pathway (Kirk & Tilney-Bassett, 1978). After four to six days, the transformation of the etioplast into a chloroplast is complete.

In contrast, if etiolated plants are exposed to light–dark cycles (such as

2 min of white light every 90–120 min), thylakoid development is limited to the formation of primary thylakoids; the membranes show very little appression and no stacking (Argyroudi-Akoyunoglou, Kondylaki & Akoyunoglou, 1976; Armond *et al.*, 1976). The thylakoids have photosystem I and photosystem II activity and are photosynthetically competent (Akoyunoglou & Argyroudi-Akoyunoglou, 1972), but they contain only about 7% of the normal amount of chl *a* and virtually no chl *b*. They also lack the LHCPs and the peripheral apoproteins of photosystem I (Mullet, Burke & Arntzen, 1980). When such plants are transferred to continuous white light, the missing pigment and protein components appear (unless, as we see later, the period of exposure to light–dark cycles is too prolonged), and the thylakoids differentiate into granal and stromal lamellae (Argyroudi-Akoyunoglou *et al.*, 1976; Armond *et al.*, 1976; Mullet *et al.*, 1980). Thus, the LHCPs belong to a small group of chloroplast polypeptides whose accumulation requires continuous illumination. We have been studying the molecular basis of this stringent light requirement.

Light-dependence of protochlorophyllide reductase

Part of the reason for the stringent light-dependence of LHC accumulation lies in the very unusual light-dependence of protochlorophyllide reductase, the enzyme which catalyses the synthesis of chlorophyllide. Unlike the many chloroplast enzymes which are activated only indirectly by light (Buchanan, 1980), protochlorophyllide reductase requires light as part of its reaction mechanism (Griffiths, 1978). The ternary complex of NADPH–enzyme–protochlorophyllide is converted to NADP–enzyme–chlorophyllide only after excitation of the bound protochlorophyllide molecule (by blue or orange-red light). Thus, continuous chlorophyllide synthesis requires essentially continuous illumination.

It might be expected that when a plant containing a light-dependent protochlorophyllide reductase is transferred from light to darkness, protochlorophyllide would accumulate. However, very little protochlorophyllide accumulates in darkness unless exogenous ALA is fed to the plants (Beale, 1978). This result indicates that the ALA-synthesizing enzyme is subject to photoregulation. Gough, Girnth & Kannangara (1981) and Chereskin & Castelfranco (1982) have shown that haem and Mg-protoporphyrin are capable of inhibiting ALA synthesis to a significant extent at quite low concentrations in angiosperms, while Wang, Boynton & Gillham (1977) have suggested that in the green alga *Chlamydomonas* protochlorophyllide may act as a feedback inhibitor of the first enzyme of the chlorophyll branch of the porphyrin biosynthetic pathway, Mg-chelatase (Fig. 3). If these regulatory elements are combined, it is possible to devise a scheme in which transfer of plants from light to darkness results firstly in the build-up of a small pool of protochlorophyllide, then in the inhibition of Mg-chelatase and the diversion of protoporphyrin IX away from chlorophyll synthesis to haem synthesis, and finally in the inhibition of ALA synthesis as haem accumulates in excess of the amount required for haemoprotein synthesis. When plants are returned to the light, the small pool of protochlorophyllide is rapidly phototransformed to chlorophyllide, the feedback inhibition of Mg-chelatase and ALA synthesis is gradually relieved, and the flux through the chlorophyll biosynthetic pathway resumes. It is not clear at present whether the rate-limiting step for chlorophyll synthesis, under light intensities that are saturating for protochlorophyllide photoreduction, is ALA synthesis, as proposed by Gassman & Bogorad (1967), or the availability of chlorophyll-binding proteins, as proposed by Kirk (1974).

The limited quantity of chlorophyll found in plants exposed to light–dark cycles undoubtedly reflects the fact that chlorophyllide can be synthesized only during the light periods. During the dark periods, this small amount of chlorophyllide is esterified to yield chl *a*, but it is not clear why only an extremely small proportion of that chlorophyll should accumulate as chl *b*. Four possibilities suggest themselves. First, chl *b* synthesis from chl *a* might be light-dependent; this seems unlikely because Tanaka & Tsuji (1982) reported chl *b* synthesis in darkness. Secondly, as proposed by Oelze-Karow *et al.* (1978), the synthesis of chl *b* from chl *a* might occur only after a threshold level of chl *a* has been reached; that threshold might not be reached under light–dark cycles. Thirdly, LHCPs and the other chl *a/b*-binding proteins of the thylakoids might not be synthesized under light–dark cycles, preventing chl *b* accumulation. Fourthly, LHC and other chl *a/b*-binding complexes might be formed under light–dark cycles but might not be stable. To study some of these possibilities, we have employed several specific and sensitive assays for LHCP and LHCP mRNA. In the following sections these assays are briefly described and results obtained with them are discussed.

Assaying LHCP and LHCP mRNA

We have assayed LHCP in pea plants by a radioimmune assay procedure based on that of Vaessen, Kreike & Groot (1981). In our procedure, total leaf proteins are extracted with hot SDS, fractionated by SDS–polyacrylamide gel electrophoresis and transferred electrophoretically to a sheet of nitrocellulose, on which LHCPs are assayed by successive exposure to anti-LHCP antibody and ^{125}I-labelled protein A. Antigen–antibody–protein A complexes are detected by autoradiography. This procedure is described in detail by Bennett, Jenkins & Hartley (1984).

Our initial procedure for assaying LHCP mRNA involved translation *in vitro* of the poly(A)-RNA fraction in the presence of ^{35}S-labelled methionine. Anti-LHCP antibody was used to precipitate pre-LHCPs from total translation products (Fig. 4), and immunoprecipitated radioactivity was expressed as a percentage of total acid-precipitable radioactivity. This procedure is described in detail by Cuming & Bennett (1981). More recently (Jenkins, Hartley &

Fig. 5. Hybridization–release–translation of LHCP mRNA. pFab31 DNA was digested with endonuclease EcoRl, denatured and bound to a DBM paper disc. Pea leaf poly(A)-RNA was hybridized to the filter-bound DNA and the non-hybridized RNA was washed away. Bound RNA was eluted with 90% formamide, recovered by ethanol precipitation and translated in the wheat-germ cell-free system. The radioactively labelled products were analysed by SDS–polyacrylamide gel electrophoresis and autoradiography. Track A: products of translation of total poly(A)-RNA. Track B: products of translation of hybrid-selected RNA. Track C: products of translation of hybrid-selected RNA were mixed with the products of translation of *Xenopus laevis* ovary mRNA, and subsequently immunoprecipitated with anti-LHCP antibodies. The 30 and 32 kD precursor polypeptides to LHCP and the 20 kD precursor to the small subunit of RuBP carboxylase are indicated.

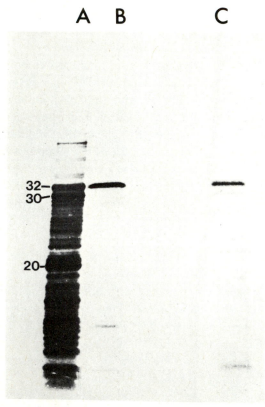

Bennett, 1983; Bennett *et al.*, 1984; Jenkins *et al.*, 1984), we have assayed LHCP mRNA by DNA–RNA hybridization using a ^{32}P-labelled DNA clone (cDNA) prepared against pea leaf LHCP mRNA. The clone (pFab 31) is complementary to the mRNA encoding the 32 kD pre-LHCP but not to the mRNA for the 30 kD pre-LHCP (Fig. 5). The lack of cross-hybridization between these two mRNAs from pea has also been noted by Coruzzi *et al.* (1983), who showed further that the 32 kD pre-LHCP is the precursor of the 26 kD mature LHCP. Presumably, the 30 kD pre-LHCP is the precursor of the 24 kD mature LHCP. We have used the cloned probe in two ways. First, total leaf RNA is fractionated by agarose gel electrophoresis, blotted on to a nitrocellulose sheet and then challenged with ^{32}P-labelled cDNA clone. Autoradiography reveals a single band of hybridization. Secondly, encouraged by the high degree of specificity of the probe, we have developed a dot hybridization procedure that is well-suited to the quantitative assay of a large number of samples. In this approach, ^{32}P-labelled cDNA probe is hybridized to a nitrocellulose sheet bearing replicate dots of total leaf RNA. Each dot is then cut out and counted by liquid scintillation spectrometry. Both assays are described in detail by Bennett *et al.* (1984).

Photoregulation of LHCP and LHCP mRNA accumulation

Using the above assays, we have investigated the photoregulation of the accumulation of LHCP and LHCP mRNA in pea seedlings. Some of our results are summarized in Table 2. LHCP was found in pea seedlings that had been grown in darkness for 8 days and then transferred to continuous white light (100 μmol quanta m^{-2} s^{-1}), but the protein was below the limit of detection both in etiolated seedlings and in etiolated seedlings exposed to 2 min of white light every 2 h for 2 days. This result confirms the conclusion

Table 2. *Photoregulation of the accumulation of LHCP and LHCP mRNA in apical buds of* Pisum sativum

Light regime	LHCP content[a] (%)	LHCP mRNA content[b] (%)
Dark (10 days)	<0.5	0.16
Dark (8 days), light–dark cycles (2 days)[c]	<0.5	1.03
Dark (8 days), continuous light (2 days)	100	2.01

[a] LHCP content per apical bud, expressed as percentage of continuous light control.
[b] LHCP mRNA content expressed as percentage of total translatable poly(A)-mRNA; data from Cuming & Bennett (1981).
[c] 2 min white light per 118 min dark.

that the accumulation of LHCP occurs in continuous illumination but not in darkness or under light–dark cycles (Armond *et al.*, 1976; Argyroudi-Akoyunoglou & Akoyunoglou, 1979; Mullet *et al.*, 1980). In contrast, translatable LHCP mRNA was detectable in all three sets of plants (Cuming & Bennett, 1981). A readily detectable level of LHCP mRNA was found in etiolated pea seedlings, and increases in the level (expressed as a percentage of total translatable poly(A)-mRNA) were recorded on exposure to light–dark cycles (6-fold increase) and continuous illumination (12-fold increase).

The results in Table 2 raise a number of important points. First, the presence of LHCP mRNA in etiolated peas contrasts with the absence of LHCP mRNA from etiolated barley (Apel, 1979). In both cases, LHCP mRNA was assayed *in vitro* by translation and immunoprecipitation. We have reinvestigated this question in the case of peas by means of DNA–RNA hybridization using a ^{32}P-labelled cDNA clone. As shown in Table 3, we confirm the result of Cuming & Bennett (1981): LHCP mRNA is present in dark-grown peas and increases approximately 35-fold in amount per apical bud during 48 h of continuous illumination. The total RNA content of the bud increases 3.5-fold over the same period (Bennett *et al.*, 1984; Jenkins *et al.*, 1984).

Secondly, how does illumination, whether continuous or intermittent, lead to the increases in LHCP mRNA levels recorded in Table 2? It has been reported that the levels of LHCP mRNA in both barley (Apel, 1979; Gollmer & Apel, 1983) and duckweed (Tobin, 1981; Stiekema *et al.*, 1983) are under phytochrome control. Phytochrome is the photoreceptor for many light-dependent responses of plants, including aspects of chloroplast formation (Mohr, 1977). It exists in two interconvertible forms (P_r and P_{fr}), which

Table 3. *Phytochrome control of LHCP mRNA levels in apical buds of* Pisum sativum

Light regime	LHCP content (%)	LHCP mRNA content (%)
48 h white light	100	100
48 h dark	<0.5	2.9
15 min red, 48 h dark	<0.5	11.8
15 min far-red, 48 h dark	<0.5	6.2
15 min red, 15 min far-red, 48 h dark	<0.5	5.5

Plants were grown in darkness for 8 days and then illuminated as shown (red light, 662 nm interference filter; far-red light, 735 nm interference filter). LHCP content per apical bud and hybridizable LHCP mRNA content per bud are both expressed as percentages of control values from continuously illuminated plants.

preferentially absorb red light (650–670 nm) and far-red light (730–760 nm), respectively, according to the equation:

Phytochrome is synthesized as P_r and it is this form which accumulates in etiolated plants. Since many phytochrome-mediated effects of light are elicited in response to a single pulse of red light that is adequate to convert some P_r to P_{fr}, the latter is conventionally referred to as the 'active form' of phytochrome. If P_{fr} is converted back to P_r by a pulse of far-red light before it has been able fully to initiate the response under study, then the response will be greatly reduced in magnitude. The reversibility by far-red light of the effects of red light is characteristic of many phytochrome-mediated responses in plants, and is the basis of the red/far-red photoreversibility test for the involvement of phytochrome in a light-dependent response.

We have investigated the effects of red and far-red light treatments on the level of LHCP and its mRNA in etiolated pea seedlings (Table 3). In this experiment, 8-day-old etiolated seedlings were exposed to 15 min of red or far-red light, or red followed immediately by far-red, and then returned to darkness for 48 h. LHCP accumulation was measured by radioimmunoassay of total leaf protein, and LHCP mRNA was assayed by dot hybridization. All results were calculated per apical bud, and are presented as percentages of the LHCP and LHCP mRNA contents of buds exposed to 48 h of continuous white light. LHCP was below the limit of detection in all cases except the continuously illuminated control. In contrast, LHCP mRNA was detectable in all samples. Red light exhibited a much more powerful inductive effect than far-red light, and the additional effect of red light was abolished by the subsequent far-red treatment. Thus, the results satisfy substantially the demands of the red/far-red photoreversibility test. However, a single red light treatment is only 12% as effective as continuous white light in inducing LHCP mRNA accumulation per apical bud (or 40% as effective if the results are expressed on an RNA basis). It is not yet clear whether this indicates that the red light pulse failed to saturate the phytochrome response, or that phytochrome must be stimulated repeatedly to achieve the maximum level of mRNA, or even that a second photoreceptor is involved (e.g. a general contribution from photosynthesis to mRNA synthesis).

Our assays measure the steady-state level of LHCP mRNA, and do not permit us to decide whether phytochrome controls mRNA synthesis or degradation. In an attempt to tackle this question, several groups are studying transcription in nuclei isolated from plants exposed to different light regimes. Preliminary indications are that light stimulates LHCP gene transcription (Gallagher & Ellis, 1982; see Gallagher *et al.* and Tobin *et al.*, this volume).

Accumulation of LHCP and chlorophyll

The third point raised by the data in Table 2, and reinforced by the data in Table 3, concerns the absence of LHCP from pea seedlings which contain readily detectable levels of LHCP mRNA. There are in fact several circumstances in which plant tissue may exhibit this remarkable phenomenon (Table 4). It is found, as we have already seen, in dark-grown peas, and in peas exposed to light–dark cycles or a single pulse of red light (Cuming & Bennett, 1981; Jenkins *et al.*, 1983; Bennett *et al.*, 1984). It has also been reported in barley and duckweed exposed to one or several red light pulses (Apel, 1979; Apel & Kloppstech, 1980; Tobin, 1981; Slovin & Tobin, 1982; Viro & Kloppstech, 1982) and in mutants of barley (Bellemare *et al.*, 1982) and pea (Schwarz & Kloppstech, 1982) which are devoid of chl *b*. Furthermore, LHCP is absent from the dimly illuminated leaves that are found deep inside lettuce heads (Henriques & Park, 1976). Note also from Table 4 the close correlation between the accumulation of LHCP and the accumulation of chl *b*. The correlation with the accumulation of chl *a* is not as close.

The data in Tables 2–4 strongly suggest that in addition to phytochrome-mediated control over LHCP mRNA levels there is also a translational or post-translational control on the accumulation of LHCP itself. At the present time, only the concept of post-translational control has direct experimental support. Bennett (1981) showed that when etiolated pea seedlings are illuminated for about 16 h to initiate the process of LHC formation, and then returned to the dark, the majority of the LHC accumulated in the light is

Table 4. *Differential accumulation of chl* a, *chl* b, *LHCP and LHCP mRNA in green plants*

Conditions	Chl *a*	Chl *b*	LHCP	LHCP mRNA
Etiolated barley	−	−	−	−
Etiolated pea	−	−	−	+
Dark-grown *Chlamydomonas* at 25 °C	−	−	−	−
Dark-grown *Chlamydomonas* at 38 °C	−	−	−	+
Continuously illuminated green plants	+	+	+	+
Green plants under day/night cycles	+	+	+	+
Pea and bean under light/dark cycles[a]	+	−	−	+
Pea, barley, duckweed in dark after one or several red light pulses	−	−	−	+
Dimly lit leaves in lettuce head	+	−	−	?
Chl *b*-less mutants of barley and pea	+	−	−	+

[a] 2 min of white light every 90 or 120 min for 2 days.
−, not detectable; +, readily detectable.

rapidly broken down in darkness. This applies to both the chlorophyll and the protein components of LHC. However, as judged by labelling *in vivo* with [³⁵S]methionine, LHCP synthesis continues for at least 48 h. Furthermore, the LHCP synthesized in darkness is also subject to turnover. Thus, in immature chloroplasts, LHCP and any associated chlorophylls are unstable in the dark. In mature chloroplasts, in contrast, LHC is remarkably stable, a fact that has led Bennett (1981) to suggest that the stabilization of LHC might depend on some aspect of chloroplast maturation, such as the incorporation of LHC units into granal stacks where they might be physically protected from any proteolytic activity. In immature chloroplasts, such as those present in etiolated plants after 16 h of illumination or in plants exposed to light–dark cycles, the primary thylakoids are not differentiated into granal and stromal lamellae, and might permit ready access of proteases to LHCP. LHC instability has also been detected in immature radish cotyledons (Lichtenthaler *et al.*, 1981) and in immature bean leaves (Argyroudi-Akoyunoglou *et al.*, 1982).

What are the implications of these ideas for plants grown under light–dark cycles? A characteristic feature of such plants is that they exhibit a very high chl *a/b* ratio combined with a low total chlorophyll content. As mentioned earlier, the low total chlorophyll content is because chlorophyllide is synthesized only during the short light periods. There appear to be at least two explanations for the high chl *a/b* ratio. First, the absence of chl *b* could be the result of the inability of plants exposed to light–dark cycles to accumulate the threshold level of chl *a* that Oelze-Karow *et al.* (1978) have postulated must be reached before chl *b* synthesis can occur. The absence of LHCP would then result from turnover. Alternatively, plants grown under light–dark cycles might actually synthesize both chl *a* and chl *b* and ligate them to LHCP, but the instability of the LHC during the dark periods would prevent LHCP and chl *b* from accumulating. Only chl *a* bound to stable proteins, such as the reaction centre polypeptides, would accumulate.

Both Cuming & Bennett (1981) and Tobin & Slovin (1982) have provided evidence for LHCP turnover in plants exposed to light–dark cycles. Cuming & Bennett (1981) were unable to detect LHCP by staining SDS–polyacrylamide gels, but they were able to detect LHCP mRNA in polysomes from plants exposed to light–dark cycles and to show that this mRNA could be translated *in vitro* in a 'run-off' system based on a wheat-germ $100\,000 \times g$ supernatant fraction. Furthermore, when such plants were labelled in darkness with [³⁵S]methionine, ³⁵S-labelled mature LHCP was detectable in small amounts in isolated thylakoids upon addition of anti-LHCP antibodies. These results strongly support the notion that LHCP is synthesized in seedlings exposed to light–dark cycles, and is cleaved to its mature size and inserted into thylakoids, but is thereafter broken down in the absence of two necessary stabilizing factors – chl *b* and continuous illumination.

Slovin & Tobin (1982) have exploited the aquatic mode of growth of duckweed (*Lemna gibba*) to perform pulse-chase experiments. They established that LHCP is subject to turnover under intermittent illumination (a brief pulse of red light every 8 h). However, they concluded that the rate of turnover of LHCP was too slow to account for the very low level of LHCP labelling with [^{35}S]methionine relative to other proteins of comparable abundance. They suggested that there is a special control on the translation of LHCP mRNA that reduces the labelling of the protein under intermittent illumination. Since, in intermittently illuminated leaves, LHCP mRNA is associated with polysomes and may be translated in run-off experiments (Cuming & Bennett, 1981; Slovin & Tobin, 1982; unpublished work cited by Viro & Kloppstech, 1982), any such translational control would presumably have to be exerted after recruitment of LHCP mRNA into polysomes. However, a diligent search for a translational control factor in polysomes and messenger ribonucleoprotein particles of pea leaves exposed to light–dark cycles provided no support for this notion (A. C. Cuming, unpublished results).

Possible evidence for translational control of LHCP synthesis may be provided by a study of bean leaves by Giles, Grierson & Smith (1977). They showed that a mRNA encoding a major 32 kD translation product (possibly a pre-LHCP) was present in etiolated leaves, but became associated with the polysome fraction only after illumination. A single pulse of red light led to the recruitment of this particular mRNA into polysomes but continuous far-red light and continuous white light were more effective. This led Giles *et al.* (1977) to suggest that phytochrome is probably mediating the recruitment, but by way of the high-irradiance reaction rather than the low-irradiance reaction on which the red/far-red photoreversibility test depends (Mohr, 1977). It is not clear whether the data of Giles *et al.* (1977) provide an explanation for the poor LHCP labelling observed by Slovin & Tobin (1982) under intermittent red light, or whether bean and duckweed exhibit two distinct forms of translational control on the synthesis of LHCP.

Photoregulation of LHCP accumulation has also been observed in the green alga *Chlamydomonas reinhardtii*. In both the y-1 strain (Hoober & Stegeman, 1976; Hoober *et al.*, 1982) and the CW15 strain (Sheperd, Ledoigt & Howell, 1983) accumulation of LHCP mRNA is light-dependent at about 25 °C. However, at 38 °C, accumulation of LHCP mRNA in the y-1 strain occurs in the dark. Since accumulation of LHCP is light-dependent at both temperatures, there must exist an additional, post-transcriptional control involving light at least in cells grown at 38 °C. Hoober *et al.* (1982) present evidence that in the y-1 strain, which resembles most angiosperms in having a light-dependent protochlorophyllide reductase, the additional photocontrol is post-translational, and involves LHCP turnover within thylakoids in the absence of chlorophyll synthesis.

The duration of LHC biosynthesis in pea leaves

When 6-day-old etiolated pea seedlings are exposed to a light regime providing 12 h white light and 12 h dark per day, the third leaf pair is the first to develop normally. It synthesizes chlorophyll for about 6 days and then ceases to accumulate further chlorophyll. The period from the second to the fourth day of illumination corresponds to the period of most rapid chlorophyll accumulation. What determines the duration of chlorophyll accumulation in these leaves?

Two general ideas have been advanced to explain this sort of observation (Akoyunoglou & Argyroudi-Akoyunoglou, 1978). First, chlorophyll synthesis may continue until a genetically programmed chlorophyll content is attained. Secondly, certain enzymes involved in chlorophyll biosynthesis may accumulate and remain active for only a limited period after the initiation of leaf expansion; chlorophyll accumulation will cease when the enzyme with the briefest time-span of activity has disappeared. A third possibility, of course, is that there is only a limited time-span for the synthesis of cholorophyll-binding proteins, and once production of these proteins ceases in a leaf, chlorophyll accumulation must also cease.

One way of distinguishing between these three alternatives is to study the ability of pea leaves grown under light–dark cycles (2 min of white light every

Table 5. *Prolonged exposure to light–dark cycles inhibits the ability of pea leaves* (Pisum sativum) *to form chlorophyll and LHCP under continuous illumination*

Light regime after 8 days etiolation but prior to transfer to continuous white light	Accumulation of	
	Chlorophyll (%)	LHCP (%)
None	100	100
2 days light–dark cycles	134	103
4 days light–dark cycles	20	18
4 days dark	70	76
6–10 days dark	<10	<5

Eight-day-old etiolated pea seedlings were exposed to 2 min white light every 2 h for up to 10 days, and then transferred to continuous white light for 4 days. The chlorophyll and LHCP contents of the third leaf pair were assayed, and expressed as a percentage of the levels found in plants which were transferred directly from darkness to continuous illumination.

2 h for up to 10 days) to maintain their capacity to synthesize chlorophyll on transfer to continuous illumination. The results of such an experiment are shown in Table 5. Etiolated pea leaves exposed to light–dark cycles for 6 days or more completely lose their capacity to accumulate further chlorophyll under continuous illumination, and even after 4 days of intermittent illumination the capacity to green is severely reduced compared with the capacity seen after 0–2 days of light–dark cycles. This result argues against the existence of a pre-programmed target for chlorophyll accumulation and suggests that it is the duration of the synthesis of chlorophyll or chlorophyll-binding proteins that limits the extent of chlorophyll accumulation. This conclusion has also been reached for bean leaves by Akoyunoglou & Argyroudi-Akoyunoglou (1978).

Since LHCP synthesis is readily detectable by labelling with [^{35}S]methionine in leaves that are incapable of further chlorophyll accumulation (J. Bennett, unpublished results), it would appear to be the duration of chlorophyll synthesis that is the primary limitation to chlorophyll accumulation.

Conclusions

1. The light-harvesting chlorophyll a/b complex is a multicomponent unit whose pigment components are synthesized within the chloroplast, but whose polypeptide components are encoded in nuclear DNA and synthesized in the cytoplasm in precursor form prior to uptake into the developing plastid.

2. The chlorophyll and protein components of LHC are under quite different photocontrols. Maximal rates of chlorophyll synthesis require continuous illumination to excite each successive protochlorophyllide molecule bound to protochlorophyllide reductase. In contrast, LHCP mRNA levels are controlled by phytochrome, probably through the regulation of nuclear gene transcription (see Gallagher *et al.* and Tobin *et al.*, this volume). Maximal rates of LHCP synthesis require only one or several pulses of red light to activate phytochrome. In both angiosperms and green algae, there appears to be no mechanism for tightly co-ordinating the synthesis of chlorophyll with the synthesis of LHCP. Excess LHCP that is unable to ligate with chl a and chl b is prevented from accumulating by a breakdown mechanism operating within the thylakoid membranes. Such a post-translational control of LHC accumulation eliminates the need for a regulatory mechanism involving communication between the developing plastid and the nucleocytoplasmic compartment.

3. Stabilization of LHC within the thylakoid membranes requires not only the synthesis of chl a and chl b but also some additional event associated with chloroplast maturation (possibly the sequestration of LHC units in the appressed membranes of the grana lamellae).

4. Topics on which further research is required include (i) the differential transcription of different LHCP genes, (ii) the mode of action of phytochrome in regulating transcription of LHCP genes in angiosperm nuclei, (iii) the mechanism by which light regulates LHCP gene transcription in green algae, (iv) the proteolytic processing of pre-LHCP and the proteolytic breakdown of mature LHCP, (v) the ligation of chl *a* and chl *b* to LHCP, (vi) the synthesis of chl *b* from chl *a*, (vii) the stabilization of LHCP against degradation in mature chloroplasts, (viii) the characterization of factors determining the duration of chlorophyll synthesis in mature leaves, and (ix) the identification of the basic endogenous controls that dictate which cells in a higher plant will synthesize chlorophyll and LHCP on illumination.

We wish to thank the Science and Engineering Research Council for financial support of this research.

References

Adamson, H. & Hiller, R. G. (1981). Chlorophyll synthesis in the dark in angiosperms. In *Photosynthesis V: Chloroplast Development*, ed. G. Akoyunoglou, pp. 213–21. Balaban International Science Services, Philadelphia.

Akoyunoglou, G. & Argyroudi-Akoyunoglou, J. H. (1972). CO_2-assimilation by etiolated bean leaves exposed to intermittent light. In *Proc. 2nd int. Congr. Photosynthesis*, vol. 3, ed. G. Forti, M. Avron & A. Melandri, pp. 2427–36. Dr W. Junk, The Hague.

Akoyunoglou, G. & Argyroudi-Akoyunoglou, J. H. (1978). Control of thylakoid growth in *Phaselous vulgaris. Pl. Physiol.*, **61**, 834–7.

Anderson, J. M., Waldron, J. C. & Thorne, S. W. (1978). Chlorophyll–protein complexes of spinach and barley thylakoids. Spectral characterization of six complexes resolved by an improved electrophoretic procedure. *FEBS Lett.*, **92**, 227–33.

Apel, K. (1977). The light-harvesting chlorophyll *a/b*–protein complex of the green alga *Acetabularia mediterranea*. Isolation and characterization of two subunits. *Biochim. biophys. Acta*, **462**, 390–402.

Apel, K. (1979). Phytochrome-induced appearance of mRNA activity for the apoprotein of the light-harvesting chlorophyll *a/b* protein of barley (*Hordeum vulgare*). *Eur. J. Biochem.*, **97**, 183–8.

Apel, K. & Kloppstech, K. (1978). The plastid membranes of barley (*Hordeum vulgare*). Light-induced appearance of mRNA coding for the apoprotein of the light-harvesting chlorophyll *a/b* protein. *Eur. J. Biochem.*, **85**, 581–8.

Apel, K. & Kloppstech, K. (1980). The effect of light on the biosynthesis of the light-harvesting chlorophyll *a/b* protein. Evidence for the requirement of chlorophyll *a* for the stabilization of the apoprotein. *Planta*, **150**, 426–30.

Argyroudi-Akoyunoglou, J. H. & Akoyunoglou, G. (1979). The chlorophyll–protein complexes of the thylakoids in greening plastids of *Phaseolus vulgaris*. *FEBS Lett.*, **104**, 78–84.

Argyroudi-Akoyunoglou, J. H., Akoyunoglou, A., Kalosakos, K. & Akoyunoglou, G. (1982). Reorganization of the photosystem II unit in developing thylakoids of higher plants after transfer to darkness. *Pl. Physiol.*, **70**, 1242–8.

Argyroudi-Akoyunoglou, J. H., Kondylaki, S. & Akoyunoglou, G. (1976). Growth of grana from primary thylakoids in *Phaseolus vulgaris*. *Pl. Cell Physiol.*, **17**, 939–54.

Armond, P. A., Arntzen, C. J., Briantais, J.-M. & Vernotte, C. (1976). Differentiation of chloroplast lamellae. *Archs. Biochem. Biophys.*, **175**, 54–63.

Beale, S. I. (1978). δ-Aminolevulinic acid in plants: its biosynthesis, regulation and role in plastid development. *A. Rev. Pl. Physiol.*, **29**, 95–120.

Bellemare, G., Bartlett, S. G. & Chua, N.-H. (1982). Biosynthesis of chlorophyll *a/b*-binding polypeptides in wild-type and the chlorina f 2 mutant of barley, *J. biol. Chem.*, **257**, 7762–7.

Bennett, J. (1981). Biosynthesis of the light-harvesting chlorophyll *a/b* protein. Polypeptide turnover in darkness. *Eur. J. Biochem.*, **118**, 61–70.

Bennett, J. (1983). Regulation of photosynthesis by reversible phosphorylation of the light-harvesting chlorophyll *a/b* complex. *Biochem. J.*, **212**, 1–13.

Bennett, J., Jenkins, G. I. & Hartley, M. R. (1984). Differential regulation of the accumulation of the light-harvesting chlorophyll *a/b* complex and ribulose bisphosphate carboxylase/oxygenase in greening pea leaves. *J. cell. Biochem.* (In press.)

Bennett, J., Markwell, J. P., Skrdla, M. & Thornber, J. P. (1981). Higher plant chlorophyll *a/b* protein complexes: studies on the phosphorylated apoproteins. *FEBS Lett.*, **131**, 325–30.

Block, M. A., Joyard, J. & Douce, R. (1980). Site of synthesis of geranylgeraniol derivatives in intact spinach chloroplasts. *Biochim. biophys. Acta*, **631**, 210–19.

Buchanan, B. B. (1980). The role of light in the regulation of chloroplast enzymes. *A. Rev. Pl. Physiol.*, **31**, 341–74.

Burke, J. J., Ditto, C. L. & Arntzen, C. J. (1978). Involvement of the light-harvesting complex in cation regulation of excitation energy distribution in chloroplasts. *Archs. Biochem. Biophys.*, **187**, 252–63.

Camm, E. L. & Green, B. R. (1980). Fractionation of thylakoid membranes with the detergent octyl-β-D-glucopyranoside. Resolution of chlorophyll–protein complex II into two chlorophyll–protein complexes. *Pl. Physiol.*, **66**, 428–32.

Camm, E. & Green, B. (1983). Relationship between the two minor chlorophyll *a* protein complexes and the photosystem II reaction centre. *Biochim. biophys. Acta*, **724**, 291–3.

Castelfranco, P. A. & Beale, S. I. (1983). Chlorophyll biosynthesis: Recent advances and areas of current interest. *A. Rev. Pl. Physiol.*, **34**, 241–78.

Chereskin, B. M. & Castelfranco, P. A. (1982). Effects of iron and O_2 on chlorophyll biosynthesis. II. Observations on the biosynthetic pathway in isolated etiochloroplasts. *Pl. Physiol.*, **68**, 112–16.

Chua, N.-H. & Blomberg, F. (1979). Immunochemical studies of thylakoid membrane polypeptides from spinach and *Chlamydomonas reinhardtii*. A modified procedure for crossed immunoelectrophoresis of sodium dodecyl sulfate–protein complexes. *J. biol. Chem.*, **254**, 215–23.

Chua, N.-H. & Schmidt, G. W. (1978). *In vitro* synthesis, transport and assembly of ribulose 1,5-bisphosphate carboxylase subunits. In *Photosynthetic Carbon Assimilation*, ed. H. W. Siegelman & G. Hind, pp. 325–47. Plenum Publishing Company, New York.

Coruzzi, G., Broglie, R., Cashmore, A. & Chua, N.-H. (1983). Nucleotide sequences of two pea cDNA clones encoding the small subunit of ribulose 1,5-bisphosphate carboxylase and the major chlorophyll *a/b* binding thylakoid polypeptide. *J. biol. Chem.*, **258**, 1399–1402.

Cuming, A. C. & Bennett, J. (1981). Biosynthesis of the light-harvesting chlorophyll *a/b* protein. Control of messenger RNA activity by light. *Eur. J. Biochem.*, **118**, 71–80.

Ellis, R. J. (1981). Chloroplast proteins: synthesis, transport and assembly. *A. Rev. Pl. Physiol.*, **32**, 111–37.

Fuesler, T. P., Wright, L. A. & Castelfranco, P. A. (1981). Properties of magnesium chelatase in greening etioplasts. Metal ion specificity and effect of substrate concentrations. *Pl. Physiol.*, **67**, 246–9.

Gallagher, T. F. & Ellis, R. J. (1982). Light-stimulated transcription of genes for two chloroplast polypeptides in isolated pea leaf nuclei. *EMBO J.*, **1**, 1493–8.

Gassman, M. & Bogorad, L. (1967). Control of chlorophyll production in rapidly greening bean leaves. *Pl. Physiol.*, **42**, 774–80.

Giddings, T. H., Withers, N. W. & Staehelin, L. A. (1980). Supramolecular structure of stacked and unstacked regions of the photosynthetic membranes of *Prochloron* sp., a prokaryote. *Proc. natn. Acad. Sci. USA*, **77**, 352–6.

Giles, A. B., Grierson, D. & Smith, H. (1977). *In vitro* translation of messenger RNA from developing bean leaves. Evidence for the existence of stored messenger RNA and its light-induced mobilization into polysomes. *Planta*, **136**, 31–6.

Gollmer, I. & Apel, K. (1983). The phytochrome-controlled accumulation of mRNA sequences encoding the light-harvesting chlorophyll *a/b* protein of barley (*Hordeum vulgare*). *Eur. J. Biochem.*, **133**, 309–13.

Gough, S. P., Girnth, C. & Kannangara, C. G. (1981). δ-Aminolevulinate synthesis in greening barley. I. Regulation. In *Photosynthesis V: Chloroplast Development*, ed. G. Akoyunoglou, pp. 107–16. Balaban International Science Services, Philadelphia.

Green, B. & Camm, E. (1982). The nature of the light-harvesting complex as defined by sodium dodecyl sulfate polyacrylamide gel electrophoresis. *Biochim. biophys. Acta*, **681**, 256–62.

Green, B. R., Camm, E. L. & van Houten, J. (1982). The chlorophyll–protein complexes of *Acetabularia*. A novel chlorophyll *a/b* complex which forms oligomers. *Biochim. biophys. Acta*, **681**, 248–55.

Griffiths, W. T. (1978). Reconstitution of chlorophyllide formation by isolated etioplast membranes. *Biochem. J.*, **174**, 681–92.

Grossman, A. R., Bartlett, S. G., Schmidt, G. W., Mullet, J. E. & Chua, N.-H. (1982). Optimal conditions for post-translational uptake of proteins by isolated chloroplasts. *In vitro* synthesis and transport of plastocyanin, ferredoxin-NADP$^+$ oxidoreductase and fructose-1,6-bisphosphatase. *J. biol. Chem.*, **257**, 1558–63.

Haworth, P., Watson, J. & Arntzen, C. J. (1983). The detection, isolation and characterization of a light-harvesting complex which is specifically associated with photosystem I. *Biochim. biophys. Acta*, **724**, 151–8.

Henriques, F. & Park, R. (1976). Development of the photosynthetic unit in lettuce. *Proc. natn. Acad. Sci. USA*, **73**, 4560–4.

Hoober, J. K., Millington, R. & D'Angelo, L. P. (1980). Structural similarities between the major polypeptides of thylakoid membranes from *Chlamydomonas reinhardtii*. *Archs. Biochem. Biophys.*, **202**, 221–34.

Hoober, J. K., Marks, D. B., Keller, B. J. & Margulies, M. M. (1982). Regulation of accumulation of the major thylakoid polypeptides in *Chlamydomonas reinhardtii* y-1 at 25 °C and 38 °C. *J. Cell Biol.*, **95**, 552–8.

Hoober, J. K. & Stegeman, W. J. (1976). Kinetics and regulation of synthesis of the major polypeptides of thylakoid membranes in *Chlamydomonas reinhardtii* y-1 at elevated temperatures. *J. Cell Biol.*, **70**, 326–37.

Jenkins, G. I., Gallagher, T. F., Hartley, M. R., Bennett, J. & Ellis, R. J. (1984). Photoregulation of gene expression during chloroplast biogenesis. In *Advances in Photosynthesis Research*, ed. C. Sybesma, vol. IV, pp. 863–72. Junk, The Hague.

Jenkins, G. I., Hartley, M. R. & Bennett, J. (1983). Photoregulation of chloroplast development: transcriptional, translational and post-translational controls? *Phil. Trans. R. Soc. Lond. Ser. B*, **303**, 419–31.

Kannangara, C. G., Gough, S. P. & von Wettstein, D. (1978). The biosynthesis of δ-aminolevulinate and chlorophyll and its genetic regulation. In *Chloroplast Development*, ed. G. Akoyunoglou, pp. 147–60. Elsevier/North-Holland, Amsterdam.

Kirk, J. T. O. (1974). The relation of chlorophyll synthesis to protein synthesis in the growing thylakoid membrane. *Port. Acta biol.*, **14**, 127–52.

Kirk, J. T. O. & Tilney-Bassett, R. A. (1978). *The Plastids*, 2nd edn. Elsevier/North-Holland, Amsterdam.

Leong, T.-Y. & Anderson, J. M. (1983). Changes in composition and function of thylakoid membranes as a result of photosynthetic adaptation of chloroplasts from pea plants grown under different light conditions. *Biochim. biophys. Acta*, **723**, 391–9.

Lichtenthaler, H.-K., Burkard, G., Kuhn, G. & Prenzel, U. (1981). Light-induced accumulation and stability of chlorophylls and chlorophyll-proteins during chloroplast development in radish seedlings. *Z. Naturf.*, **36c**, 421–30.

Lütz, C., Röper, U., Beer, N. S. & Griffiths, T. (1981). Subetioplast localization of the enzyme NADPH:protochlorophyllide oxidoreductase. *Eur. J. Biochem.*, **118**, 347–53.

Machold, O., Simpson, D. J. & Møller, B. L. (1979). Chlorophyll–proteins of thylakoids from wild-type and mutants of barley (*Hordeum vulgare*). *Carlsberg Res. Commun.*, **44**, 235–54.

Markwell, J. P., Reinman, S. & Thornber, J. P. (1978). Chlorophyll–protein complexes from higher plants: a procedure for improved stability and fractionation. *Archs. Biochem. Biophys.*, **190**, 136–41.

Melis, A. & Harvey, G. W. (1981). Regulation of photosystem stoichiometry, chlorophyll *a* and chlorophyll *b* content and relation to chloroplast ultrastructure. *Biochim. Biophys. Acta*, **637**, 138–45.

Mohr, H. (1977). Phytochrome and chloroplast development. *Endeavour (New Series)*, **1**, 107–14.

Mullet, J. E., Baldwin, T. O. & Arntzen, C. J. (1981). A mechanism for chloroplast thylakoid adhesion mediated by the chlorophyll *a/b* light-harvesting complex. In *Photosynthesis III. Structure and Molecular Organization of the Photosynthetic Membrane*, ed. G. Akoyunoglou, pp. 577–82. Balaban International Science Services, Philadelphia.

Mullet, J. E., Burke, J. J. & Arntzen, C. J. (1980). A developmental study of photosystem I peripheral chlorophyll proteins. *Pl. Physiol.*, **65**, 823–7.

Oelze-Karow, H., Kasemir, H. & Mohr, H. (1978). Control of chlorophyll *b* formation by phytochrome and a threshold level of chlorophyllide *a*. In *Chloroplast Development*, ed. G. Akoyunoglou, pp. 787–92. Elsevier/North-Holland, Amsterdam.

Ogawa, T., Obata, F. & Shibata, K. (1966). Two pigment proteins in spinach chloroplasts. *Biochim. biophys. Acta*, **112**, 223–34.

Rüdiger, W., Benz, J. & Guthoff, C. (1980). Detection and partial characterization of activity of chlorophyll synthetase in etioplast membranes. *Eur. J. Biochem.*, **109**, 193–200.

Ryrie, I. J., Anderson, J. M. & Goodchild, D. J. (1980). The role of the light-harvesting chlorophyll *a/b*–protein complex in chloroplast membrane stacking. Cation-induced aggregation of reconstituted proteoliposomes. *Eur. J. Biochem.*, **107**, 345–54.

Schmidt, G. W., Bartlett, S. G., Grossman, A. R. & Chua, N.-H. (1981). Biosynthetic pathway of two polypeptide subunits of the light-harvesting chlorophyll *a.b* protein complex. *J. Cell Biol.*, **91**, 468–78.

Schwarz, H. P. & Kloppstech, K. (1982). Effects of nuclear gene mutation on the structure and function of plastids in peas. The light-harvesting chlorophyll *a/b* protein. *Planta*, **155**, 116–23.

Sheperd, H. S., Ledoigt, G. & Howell, S. H. (1983). Regulation of light-harvesting chlorophyll-binding protein (LHCP) mRNA accumulation during the cell cycle in *Chlamydomonas reinhardtii*. *Cell*, **32**, 99–107.

Slovin, J. P. & Tobin, E. M. (1982). Synthesis and turnover of the light-harvesting chlorophyll *a/b*–protein in *Lemna gibba* grown with intermittent red light: possible translational control. *Planta*, **154**, 465–72.

Smith, S. M. & Ellis, R. J. (1979). Processing of small subunit precursor of ribulose bisphosphate carboxylase and its assembly into whole enzyme are stromal events. *Nature*, **278**, 662–4.

Stiekema, W. J., Wimpee, C. F., Silverthorne, J. & Tobin, E. M. (1983). Phytochrome control of the expression of two nuclear genes encoding chloroplast proteins in *Lemna gibba* L. G-3. *Pl. Physiol.*, **72**, 717–24.

Tanaka, A. & Tsuji, H. (1982). Calcium-induced formation of chlorophyll *b* and light-harvesting chlorophyll *a/b* protein complex in cucumber cotyledons in the dark. *Biochim. biophys. Acta*, **680**, 265–70.

Thomson, W. W. & Whatley, J. M. (1980). Development of non-green plastids. *A. rev. Pl. Physiol.*, **31**, 375–94.

Thornber, J. P., Gregory, R. P. F., Smith, C. A. & Bailey, J. L. (1967). Nature of the chloroplast lamellae. I. Preparation and some properties of two chlorophyll–protein complexes. *Biochemistry*, **6**, 391–6.

Thornber, J. P. & Highkin, H. R. (1974). Composition of the photosynthetic apparatus of normal barley and a mutant lacking chlorophyll *b*. *Eur. J. Biochem.*, **41**, 109–16.

Thornber, J. P., Markwell, J. P. & Reinman, S. (1979). Plant chlorophyll–protein complexes: recent advances. *Photochem. Photobiol.*, **29**, 1205–16.

Tobin, E. M. (1981). Phytochrome-mediated regulation of messenger RNAs for the small subunit of ribulose 1,5-bisphosphate carboxylase and the light-harvesting chlorophyll *a/b*–protein in *Lemna gibba*. *Pl. molec. Biol.*, **1**, 35–51.

Vaessen, R. T. M. J., Kreike, J. & Groot, G. S. P. (1981). Protein transfer to nitrocellulose filters. A simple method for quantification of single proteins in complex mixtures. *FEBS Lett.*, **124**, 193–6.

Vernon, L. P., Klein, S., White, F. G., Shaw, E. R. & Mayne, B. C. (1972). Properties of a small photosystem II particle obtained from spinach chloroplasts. In *Proc. 2nd int. Congr. on Photosynthesis Research*, ed. G. Forti, M. Avron & A. Melandri, vol. I, pp. 801–2. Dr W. Junk, The Hague.

Viro, M. & Kloppstech, K. (1982). Expression of genes for plastid membrane proteins in barley under intermittent light conditions. *Planta*, **154**, 18–23.

Wang, W.-Y., Boynton, J. E. & Gillham, N. W. (1977). Genetic control of chlorophyll biosynthesis: effect of increased δ-aminolevulinic acid synthesis on the phenotype of the y-1 mutant of *Chlamydomonas reinhardtii*. *Molec. gen. Genet.*, **152**, 7–12.

Wellburn, A. R. (1982). Bioenergetic and ultrastructural changes associated with chloroplast development. *Int. Rev. Cytol.*, **80**, 133–91.

Wessels, J. S. C., van Alphen-van Waveren, O. & Voorn, G. (1973). Isolation and properties of particles containing the reaction center complex of photosystem II from spinach chloroplasts. *Biochim. biophys. Acta*, **292**, 741–52.

Westhoff, P., Alt, J., Nelson, N., Bottomley, W., Büneman, H. & Herrmann, R. G. (1983). Genes and transcripts for the P700 chlorophyll *a* apoprotein and subunit 2 of the photosystem I reaction center complex from spinach thylakoid membranes. *Pl. molec. Biol.*, **2**, 95–108.

Whitfeld, P. R. & Bottomley, W. (1983). Organization and structure of chloroplast genes. *A. Rev. Pl. Physiol.*, **34**, 279–310.

Withers, N. W., Alberte, R. S., Lewin, R. A., Thornber, J. P., Britton, G. & Goodwin, T. W. (1978). Photosynthetic unit size, carotenoids and chlorophyll-protein composition of *Prochloron* sp., a prokaryotic green alga. *Proc. natn. Acad. Sci. USA*, **75**, 2301–5.

**R. DOUCE, A. J. DORNE, M. A. BLOCK
AND J. JOYARD**

The chloroplast envelope and the origin of chloroplast lipids

The extent of our knowledge of the role of polar lipids in chloroplast membranes is embarrassingly scanty. Polar lipids serve as the hydrophobic barrier in the membrane, but is this the sole function of these molecules? If it is, then why are the chloroplast polar lipids made in such diversity? The chloroplast acyl lipids, which contain high amounts of polyunsaturated fatty acids, comprise three glycolipids (monogalactosyldiacylglycerol or MGDG; digalactosyldiacylglycerol or DGDG; sulfoquinovosyldiacylglycerol or sulfo-lipid) and two phospholipids (phosphatidylglycerol and phosphatidylcholine), (see Fig. 1). Specific polar lipids may be required for the activity of some chloroplast membrane-bound enzymes. Unfortunately, the exact nature of polar lipid interactions with plastid membrane proteins is not clearly understood. For example, it is not known if all integral membrane proteins in the asymmetrical lipid bilayer of chloroplast membranes are surrounded by a boundary layer of relatively immobilized lipid (annulus) which does not participate in the ordered fluid phase transition.

Many of the present studies on the chloroplast polar lipids seek to understand their origin and precise biosynthetic routes. Analyses of membranes from higher plant plastids have shown that the capacity to assemble galactolipids is confined to limiting membranes of the organelles; i.e. the plastid envelope membranes, in which the complete galactolipid biosynthetic machinery has been demonstrated. Other experiments have been interpreted as indicating that some particular lipids (phospholipids) or at least parts of lipids (polyunsaturated fatty acids) located in thylakoids or envelopes have to be imported from their site of biosynthesis (endoplasmic reticulum?) into plastid membranes.

In this article, we wish to discuss the origin of the chloroplast polar lipids and the key role of envelope membranes as an active site of membrane biogenesis.

Fig. 1. Structure of some of the major chloroplast glycerolipids. There are significant differences in the positional distribution of fatty acids in plant galactolipids (reviewed by Heinz, 1977; Douce & Joyard, 1980). The most obvious specificity found in MGDG is that 16:3 (when present) is localized at the C-2 position of the glycerol backbone. Plants which are characterized by the occurrence of 16:3 together with 18:3 in their MGDG are called '16:3 plants'. In '18:3 plants', the polyunsaturated fatty acid components of galactolipids are almost exclusively 18:3. In DGDG and in sulfolipid, there are no polyunsaturated C_{16} fatty acids, and 16:0 is found at the C-1 and/or at the C-2 position of the glycerol backbone; this distribution is closely related to the position of the plant in taxonomy (see Heinz, 1977). For instance, in the green alga *Ulva*, 16:0 is almost exclusively at the C-2 position of DGDG glycerol backbone whereas it is at the C-1 position in higher plants such as *Vicia* or *Zea*. In 16:3 plants (such as spinach), 16:0 seems to be localized in appreciable proportion at the C-2 position of DGDG and sulfolipid (see Siebertz *et al.*, 1979). Chloroplast phosphatidylglycerol is characterized by a specific fatty acid, *trans*-Δ3-hexadecenoic acid, located almost exclusively at the C-2 position of the glycerol backbone (Haverkate & Van Deenen, 1965). In etioplasts, proplastids, etc., phosphatidylglycerol does not contain this fatty acid (see Dubacq & Trémolières, 1983). Finally, 16:0 in phosphatidylcholine is located almost exclusively at the C-1 position of the glycerol backbone.

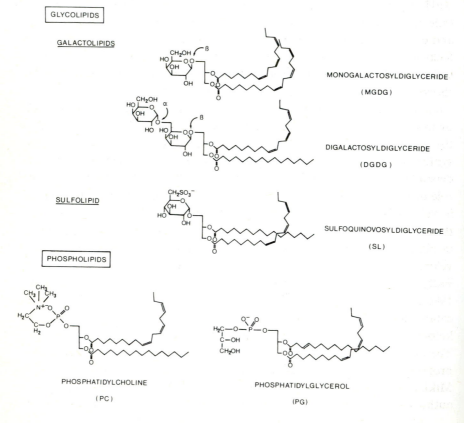

GLYCOLIPIDS

GALACTOLIPIDS

MONOGALACTOSYLDIGLYCERIDE
(MGDG)

DIGALACTOSYLDIGLYCERIDE
(DGDG)

SULFOLIPID

SULFOQUINOVOSYLDIGLYCERIDE
(SL)

PHOSPHOLIPIDS

PHOSPHATIDYLCHOLINE
(PC)

PHOSPHATIDYLGLYCEROL
(PG)

The plastid envelope: site of phosphatidic acid and diacylglycerol synthesis

Thioester derivatives of fatty acids (either CoA or acyl-carrier protein (ACP) derivatives of palmitate, $C_{16:0}$, and oleate, $C_{18:1}$) are utilized for the acylation of *sn*-glycerol-3-phosphate which provides the C_3 backbone for the chloroplast polar lipids.

Fatty acid synthesis

Numerous investigations have revealed that isolated chloroplasts from various leaves possess the complete machinery (palmitate synthetase, elongase and desaturase) for biosynthesis of the hydrocarbon chains of fatty acids ($C_{16:0}$, $C_{18:0}$ and $C_{18:1}$) (reviewed by Stumpf, 1980). For example, Roughan, Holland & Slack (1979a; Roughan et al., 1979b), using spinach chloroplasts isolated by techniques yielding preparations with high O_2-evolving activity, measured incorporation rates as high as 1500 nmol of acetate per hour per milligram of chlorophyll in buffered sorbitol solutions containing $NaHCO_3$ and [^{14}C]acetate. The same is true for *Narcissus* corona chromoplasts (Kleinig & Liedvogel, 1978) and avocado mesocarp plastids, cauliflower buds and castor bean endosperm proplastids (Zilkey & Canvin, 1972; Weaire & Kekwick, 1975). Moreover, Nothelfer, Barckhaus & Spener (1977) with isolated soybean cells, and Vick & Beevers (1978) with castor bean endosperm, showed that proplastids were the unique site of fatty acid synthesis in etiolated plant cells. Ohlrogge, Kuhn & Stumpf (1979) clearly demonstrated, by using antibodies raised against purified spinach ACP, that all the ACP of the leaf cell could be attributed to chloroplasts. Since all plant extracts that synthesize fatty acids from acetyl-CoA and malonyl-CoA are completely dependent upon the presence of ACP ($M_r \sim 10000$ Daltons), and since the role of ACP as the thioester component of the primary elongation substrate is well established, these results suggest that in the leaf cell, chloroplasts are the sole site of the synthesis *de novo* of C_{16} and C_{18} fatty acids. Consequently, in plant cells, plastids are the source of fatty acids for membranes of developing organelles and serve as precursors of glycolipids and phospholipids. Fatty acid synthetase is localized in the stromal phase of plastids (Stumpf, 1980). Seven different partial reactions are involved in overall synthesis catalysed by acetyltransferase, malonyltransferase, β-ketoacylsynthetase, β-ketoacylreductase, dehydrase, $\Delta2$–3-enoylreductase and palmitoyltransferase. The enzymes involved have been purified and their physical and substrate properties determined (Caughey & Kekwick, 1982; Høj & Mikkelsen, 1982; Mikkelsen & Høj, 1982; Shimakata & Stumpf, 1982a, 1982b, 1982c). All these authors have shown that the chloroplast fatty acid synthetase consists of separable components and therefore is analogous to prokaryotic enzymes. The system converts acetyl-CoA and malonyl-CoA (which is derived from acetyl-CoA) into palmitoyl-ACP via ACP derivatives. Palmitoyl-ACP is the preferred substrate for the elongation enzyme which converts it to stearoyl-ACP (Jaworski, Goldschmidt & Stumpf, 1974). Stearoyl-ACP is then desaturated to oleoyl-ACP (Jacobson, Jaworski & Stumpf, 1974) in the stroma. Consequently, in the stromal phase all the chloroplast fatty acids (16:0, 18:0, and 18:1) are attached to ACP during their biosynthesis *de novo* and during their elongation and desaturation to oleoyl-ACP. Spinach leaf ACP has been purified, and a partial amino acid sequence has been determined (Matsumura & Stumpf, 1968) which shows that the region surrounding the prosthetic group is identical to that of ACP from *Escherichia coli*. Acyl-ACP can be converted to acyl-CoA by a switching system involving acyl-ACP thioesterase and one or several acyl-CoA synthetase(s) (reviewed by Stumpf, 1980). Thus, Ohlrogge, Shine & Stumpf (1978) have characterized an acyl-ACP thioesterase in avocado mesocarp extracts. This enzyme, which was free of acyl-CoA thioesterase activity, has a pH optimum at 9.5 and a molecular weight of 70 000–80 000. Substrate specificities studies have shown that lauroyl-ACP, myristoyl-ACP, and palmitoyl-ACP are slowly hydrolyzed, whereas oleoyl-ACP is rapidly hydrolyzed, to free fatty acids. The other component of the switching system which allows acyl transfer from ACP to CoA derivatives in chloroplasts is the long-chain acyl-CoA synthetase which catalyses the following reaction:

$$\text{fatty acid} + \text{ATP} + \text{CoASH} \rightarrow \text{acyl-CoA} + \text{AMP} + P \sim P_i,$$

and is specifically associated with the chloroplast envelope membranes (Joyard & Douce, 1977; Roughan & Slack, 1977; Kleinig & Liedvogel, 1978; Joyard & Stumpf, 1981). In marked contrast, acetyl-CoA synthetase is concentrated in the stroma (Roughan & Slack, 1977; Joyard & Stumpf, 1981). The purified envelope membranes were also shown to be the site of a very active acyl-CoA thioesterase activity (E.C. 3.1.2.2) (Bertrams & Heinz, 1980; Joyard & Stumpf, 1980). This enzyme has also been found in chromoplasts (Liedvogel *et al.*, 1978). Detailed analysis of the acyl-CoA thioesterase and synthetase have shown a different fatty acid specificity for each enzyme: the thioesterase is more active with medium chain fatty acids than with long chain fatty acids (Joyard & Stumpf, 1980); conversely, the acyl-CoA synthetase is more active with long chain fatty acids (Joyard & Stumpf, 1981). In these circumstances, as the stroma is the sole site of fatty acid synthesis *de novo* in the cell (Ohlrogge *et al.*, 1979), it is possible that both acyl-CoA thioesterase and synthetase in the envelope membranes may be involved in the transport of oleic acid from the stromal phase of the plastids to the cytosol compartment of the cell (Joyard & Stumpf, 1980, 1981). To clarify the interacting roles of acyl-CoA thioesterase and acyl-CoA synthetase in leaf glycerolipid metabolism, we have studied the precise localization of these enzymes on the envelope membranes from spinach chloroplasts: the acyl-CoA synthetase and acyl-CoA thioesterase are respectively located on the outer and inner membrane of the envelope (Dorne *et al.*, 1982*b*; Block *et al.*, 1983*c*). The presence of acyl-CoA synthetase on the outer envelope membrane readily explains why acyl-CoA synthesis by isolated chloroplasts is stimulated by ATP and CoA (reviewed by Roughan & Slack, 1982). In addition, this result supports the view that the envelope acyl-CoA synthetase could be involved in fatty acid export outside the chloroplast (Douce & Joyard, 1980). Finally, it is likely that fatty acids, synthesized within the plastids, and destined for further metabolism (phospholipid metabolism) in the cytosol, are exported as acyl-CoA, as suggested by several authors (Roughan *et al.*, 1979*a*, 1979*b*, 1980; Stumpf *et al.*, 1980; Drapier *et al.*, 1982; Roughan & Slack, 1982). Another consequence of the localization of the envelope acyl-CoA synthetase and acyl-CoA thioesterase respectively on the outer and inner membrane is that acyl-CoA molecules are probably not the true physiological acyl donors for galactolipid synthesis on the inner envelope membrane (Block *et al.*, 1983*c*). This point will be discussed later.

Phosphatidic acid synthesis

Phosphatidic acid is formed by acylation of a water soluble precursor, *sn*-glycerol-3-phosphate. The primary source of this component is the cytosol (reviewed by Heinz, 1977).

Studies of Renkonen & Bloch (1969) showed, for the first time, that cell-free extracts of *Euglena gracilis* catalyze the transfer of acyl groups from thioesters of ACP or CoA to galactolipids (MGDG). This reaction was stimulated by *sn*-glycerol-3-phosphate. This experiment strongly suggested, but did not prove, that the diacylglycerol was formed by the acylation of *sn*-glycerol-3-phosphate followed by the dephosphorylation of phosphatidic acid (Kornberg–Pricer pathway). Direct support for this pathway was given by the observation that intact and purified plastids (chloroplasts and etioplasts) isolated from spinach and maize leaves incorporated label from *sn*-[^{14}C]glycerol-3-phosphate into MGDG (Douce & Guillot-Salomon, 1970). It has been found that chloroplasts from *Euglena* and spinach, chromoplasts from daffodil and amyloplasts from potato tubers contain two acyltransferases (reviewed by Douce & Joyard, 1980).

The first acyltransferase catalyses the acylation of *sn*-glycerol-3-phosphate to lysophosphatidic acid:

$$sn\text{-glycerol-3-phosphate} + \text{acyl-ACP} \rightarrow \text{lysophosphatidic acid} + \text{ACP}.$$
$$\text{(acyl-CoA)} \qquad\qquad\qquad\qquad \text{(CoA)}$$

Although several cellular compartments contain this enzyme, the plastid acyltransferase is unique since it is recovered as a soluble enzyme in the chloroplast extract (Bertrams & Heinz, 1976; Joyard & Douce, 1977). However, this enzyme is probably loosely bound to the inner face of the inner envelope membrane and becomes detached during the course of envelope preparation. Furthermore, localization of the soluble acyltransferase in the intermembrane space of the chloroplast envelope cannot be excluded. This soluble acyltransferase has been purified from the chloroplast extract by a factor of about 1000 (Bertrams & Heinz, 1979) and was shown by the same authors to be present in plastids as two isomeric forms having different isoelectric points (6.35 and 6.70, respectively). However, these two forms could not be differentiated further since they presented similar kinetic properties (Bertrams & Heinz, 1979). The pH optimum for lysophosphatidic acid synthesis was around 7.0 (Bertrams & Heinz, 1979; Joyard, 1979; Joyard, Chuzel & Douce, 1979). Furthermore, Bertrams & Heinz (1979) have shown that the soluble acyltransferase cannot use dihydroxyacetone phosphate, which is synthesized within the chloroplast by the enzymes of the Benson–Calvin cycle. Regardless of the acyl-CoA offered, the enzyme possesses a high positional specificity for acylation of C-1 of the glycerol backbone (Bertrams & Heinz, 1979; Joyard, 1979; Joyard *et al.*, 1979). In addition, during incubation of the soluble *sn*-glycerol-3-phosphate acyltransferase in presence of a mixture of palmitoyl- and oleoyl-CoA, the enzyme preferentially used oleic acid (Bertrams & Heinz, 1979). Using [^{14}C]acetate as a substrate for fatty acid synthesis and incorporation into envelope lipids, Joyard *et al.* (1979) demonstrated that lysophosphatidic acid thus formed contains C_{18} fatty acids at the C-1 position of the glycerol backbone. However, until recently, it was not clear whether the true substrate for lysophosphatidic acid was acyl-CoA or acyl-ACP thioesters. For instance, Bertrams & Heinz (1979) have shown that the soluble acyltransferase can use acyl-ACP as well as acyl-CoA. It is possible that the membrane environment and the acyl-CoA and/or acyl-ACP concentrations could strongly influence the preference for saturated or unsaturated acyl-CoA and/or acyl-ACP thioesters as substrate. Indeed, double-labeling experiments conducted with [^{14}C]oleoyl-ACP and [^{3}H]oleoyl-CoA by Frentzen *et al.* (1982, 1983) demonstrated that acyl-ACP rather than acyl-CoA is the physiological acyl donor for lysophosphatidic acid in chloroplasts. In addition, enzymes from pea and spinach gave similar results.

The second acyltransferase catalyses the acylation of lysophosphatidic acid to phosphatidic acid:

$$\text{lysophosphatidic acid} + \text{acyl-ACP} \rightarrow \text{phosphatidic acid} + \text{ACP}.$$
$$\text{(acyl-CoA)} \hspace{5cm} \text{(CoA)}$$

This enzyme is firmly and specifically bound to envelope membranes (Joyard & Douce, 1977). Neither thylakoids nor stroma contain this enzyme. Joyard *et al.* (1979), using [^{14}C]acetate as substrate, have shown that C_{16} fatty acids synthesized in the stromal phase of chloroplasts are preferentially transferred in envelope membranes at the C-2 position of lysophosphatidic acid which already has C_{18} fatty acids at the C-1 position. Frentzen *et al.* (1982, 1983) demonstrated that the envelope-bound acyltransferase from pea and spinach possesses a high selectivity for palmitoyl groups. However, lysophosphatidic acid, having at the C-1 position of the glycerol either oleic or palmitic acids, was used as a substrate by the acyltransferase (Frentzen *et al.*, 1982, 1983). Again, it was not clear whether *in vivo* the membrane-bound acyltransferase uses acyl-ACP and/or acyl-CoA thioesters. For instance, Roughan & Slack (1982) have suggested that 16:0-ACP could be a substrate for one of the chloroplast acyltransferases. However, in contrast to observations on *Euglena* (Renkonen & Bloch, 1969), Shine, Mancha & Stumpf (1976) have shown that in spinach chloroplasts, the acyl-ACP thioesters do not function as acyl donors. It is clear that the results obtained depend strongly on the nature of cofactors added to the medium and on the concentration of the different substrates used. Finally, Frentzen *et al.* (1982, 1983) have demonstrated that when [^{14}C]palmitoyl-ACP and [^{3}H]palmitoyl-CoA are offered to membrane-bound acyltransferase, lysophosphatidic acid is acylated with palmitic acid deriving mostly from ACP thioesters. These results are in good agreement with the localization of the acyl-CoA synthetase on the outer envelope membrane (Dorne *et al.*, 1982*b*; Block *et al.*, 1983*c*).

From these observations, we can conclude that the two acyltransferases associated with envelope membranes play a very important role in the synthesis of a diacylglycerol backbone having a specific fatty acid composition, i.e. with 18:1 and 16:0 respectively at the C-1 and C-2 positions. This conclusion is valid for 16:3 plants (such as spinach) as well as for 18:3 plants (such as pea). It is interesting to note that some galactolipid molecules from 16:3 plants and some other polar lipid molecules which are characteristic from plastids (such as sulfolipid and phosphatidylglycerol) have also a backbone with C_{18} fatty acids and C_{16} fatty acids, respectively, at the C-1 and C-2 position of the glycerol moiety. However, it is not yet clear if phosphatidic acid molecules having C_{18} fatty acids at both C-1 and C-2 positions of the glycerol moiety can be synthesized through the Kornberg–Pricer pathway, which is associated with the envelope membranes.

Diacylglycerol synthesis

Phosphatidic acid synthesized within envelope membranes by the two acyltransferases is hydrolyzed fairly rapidly into diacylglycerol by a phosphatidic acid phosphatase (Joyard & Douce, 1977, 1979; Fig. 2):

$$\text{phosphatidic acid} \rightarrow \text{diacylglycerol} + P_i.$$

The chloroplast envelope phosphatidic acid phosphatase is unique: it is strongly bound to envelope membranes and appears not to need divalent cations. In addition, this phosphatase has a pH optimum of 9.0, making it very different from all the other phosphatidic acid phosphatases described so far (Douce & Joyard, 1980). Finally, it is interesting to note that only endogenous (membrane-bound) phosphatidic acid, actively synthesized from *sn*-glycerol-3-phosphate by envelope membranes, can be hydrolyzed by the phosphatidic acid phosphatase (Fig. 2). We have shown that exogenously added phosphatidic acid is not directly accessible to the enzyme (Joyard & Douce, 1979), probably because of the inherent difficulties in presenting water-insoluble substrates to the enzyme. Joyard & Douce (1979) have

Fig. 2. Diacylglycerol and monogalactosyldiacylglycerol (MGDG) appearance during the course of endogenous phosphatidic acid (PA) hydrolysis by isolated spinach chloroplast envelope membranes (*a*) without UDP-galactose and (*b*) with UDP-galactose (UDP-gal), 1 mmol l⁻¹. Envelope membranes were loaded with [¹⁴C]phosphatidic acid by incubation of envelope membranes and stroma in presence of *sn*-[¹⁴C]glycerol-3-phosphate (Joyard & Douce, 1977). Envelope membranes were then separated and phosphatidic acid phosphatase assayed as described by Joyard & Douce (1979). This experiment demonstrates that diacylglycerol synthesized on envelope membranes is the substrate for galactosylation enzymes. (Reproduced with permission from Joyard & Douce, 1979.)

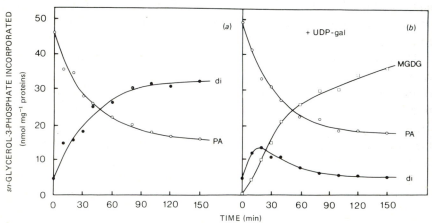

suggested that the envelope phosphatidic acid phosphatase could play a key role in the formation of diacylglycerol species involved in galactolipid synthesis. Indeed, Frentzen *et al.* (1982, 1983) and Gardiner & Roughan (1983) have demonstrated that chloroplasts from 18:3 plants have a much lower capacity to accumulate diacylglycerol than those from 16:3 plants, owing to a less active phosphatidic acid phosphatase in envelope membranes from 18:3 plants. For instance, the turnover rate of plastid phosphatidic acid was 20–30-fold faster in spinach than in pea (Gardiner & Roughan, 1983). This difference may explain why 16:3 plants, which are able to synthesize phosphatidic acid having 16:0/18:1 pairing, do not contain the corresponding diacylglycerol in MGDG. In addition, the origin of diacylglycerol species having C_{18} fatty acids at each C-1 and C-2 position of the glycerol backbone is not clear, although it is the substrate for galactosylation enzymes in 16:3 plants as well as in 18:3 plants. For instance, this diacylglycerol backbone having two C_{18} fatty acids could have been formed outside the chloroplast (Heinz & Roughan, 1982, 1983; Roughan & Slack, 1982; Williams, Khan & Mitchell, 1982). However, the postulated role of phosphatidylcholine (PC) in providing the C_{18}/C_{18} backbone to chloroplasts is not yet clear since the outer envelope membrane accumulates PC in spinach (Dorne *et al.*, 1982*b*; Block *et al.*, 1983*b*) as well as in pea (Cline *et al.*, 1981), an observation which does not fit with a precursor–product relationship between PC and galactolipids. In addition, isolated envelope membranes from spinach are devoid of the phospholipase C type activity which would yield diacylglycerol from hydrolysis of PC (Douce & Joyard, 1979).

It should be mentioned that diacylglycerol having polyunsaturated fatty acids (16:3/18:3 and 18:3/18:3 pairings) can be formed in isolated envelope membranes (Van Besouw & Wintermans, 1978; Dorne *et al.*, 1982*a*, 1982*b*). As we shall see later, this diacylglycerol results from the action of a galactolipid:galactolipid galactosyltransferase localized on the outer surface of the outer envelope membrane and probably not involved directly in galactolipid synthesis.

Therefore, the envelope phosphatidic acid phosphatase remains the only enzyme described so far which is involved directly in the synthesis of diacylglycerol molecules available for galactolipid synthesis. It is interesting to note that, in spinach chloroplasts, the phosphatidic acid phosphatase is localized on the inner envelope membrane (Block *et al.*, 1983*b*). Such a localization has major implications for the problem of galactolipid synthesis within chloroplasts.

The plastid envelope: a site of galactolipid synthesis

Localization of galactolipids

Galactolipids are the major polar lipids in photosynthetic tissues. Galactolipids clearly appear to be concentrated within the plastid membranes (reviewed by Douce & Joyard, 1980). Douce, Holtz & Benson (1973) and Mackender & Leech (1974) have shown that both types of chloroplast membranes (envelope and thylakoids) contain galactolipids. Furthermore, using antibodies raised against galactolipids, Billecocq, Douce & Faure (1972) and Billecocq (1974) have demonstrated the occurrence of galactolipids in the outer envelope membrane. This was confirmed recently: outer and inner membranes of the envelope from purified spinach and pea leaf chloroplasts were prepared and their lipid composition determined (Cline *et al.*, 1981; Dorne *et al.*, 1982*b*; Block *et al.*, 1983*b*), the presence of MGDG and DGDG was detected in both membranes. In addition, using a modification of the

Fig. 3. Thin sections of part of a young spinach leaf cell, post-stained according to a modification (Carde *et al.*, 1982) of the method of Thiery (1967). A mild oxidation (30 min, 4 °C) of spinach leaves allows the localization of glycolipids within the cell (Carde *et al.*, 1982). Note that the plastid membranes (outer and inner envelope membranes, prolamellar body, thylakoids) containing galactolipids – and to a lesser extent the plasma membrane – are highly contrasted; the other cell membranes (nuclear and mitochondrial membranes, endoplasmic reticulum), however, are poorly contrasted.

CW, cell wall; N, nucleus; ne, nuclear envelope; M, mitochondria; er, endoplasmic reticulum; pe, plastid envelope; pm, plasma membrane. (Reproduced with permission from Carde *et al.*, 1982.)

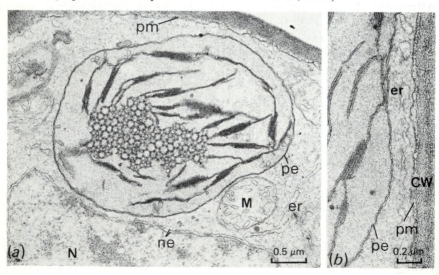

method of Thiery (1967) to localize glycolipids within the cell, Carde, Joyard & Douce (1982) demonstrated that, after a mild oxidation (30 min, 3 °C), plastid membranes, but not the other cell membranes, are strongly stained by the treatment (Fig. 3). The same result was obtained with an OsO_4-ferri(o)-cyanide mixture (Fig. 4). Finally, by comparing polar lipid composition of pure organelle membranes with that of etiolated and greening photosynthesizing tissues, Douce & Joyard (1980) have concluded that inside the plant cell, galactolipids are exclusively localized in plastid membranes.

Since (*a*) galactolipids are characteristic of plastid membranes, (*b*) envelope membrane structure develops before the thylakoids, and (*c*) envelope membranes contain the enzymes of the Kornberg–Pricer pathway which yield diacylglycerol, the substrate of galactosylation enzymes, it is reasonable to suggest that the plastid envelope is involved in the synthesis of galactolipids.

Galactolipid synthesis

The first studies on the biosynthesis of galactolipids in plants were carried out by Benson *et al.* (1958) on *Chlorella pyrenoidosa*, and by Kates (1960) on runner bean leaves. These workers found that $^{14}CO_2$ was rapidly incorporated into MGDG and DGDG and that the acyl residues were labeled more slowly than the polar head groups. The main advantage of using $^{14}CO_2$ to study lipid metabolism *in vivo* is that exposures may be terminated rapidly, and thus the start of a 'cold chase' is clearly defined. According to Ferrari & Benson (1961), MGDG was labeled first, and later the decline in its radioactivity was matched by a rise in that of DGDG. This was consistent with the suggestion that the biosynthesis of galactolipids in plants occurs by a stepwise addition of galactose residues to diacylglycerol from UDP-galactose (UDP-gal):

$$1,2\text{-diacyl-}sn\text{-glycerol} + \text{UDP-gal} \rightarrow \text{MGDG} + \text{UDP}$$
$$\text{MGDG} + \text{UDP-gal} \rightarrow \text{DGDG} + \text{UDP}.$$

Evidence in support of this biosynthetic pathway was obtained by Neufeld & Hall (1964) and Ongun & Mudd (1968), who found that isolated spinach chloroplasts catalyzed the transfer of galactose from UDP-gal to an endogenous acceptor, yielding MGDG, DGDG, and probably TGDG and TTDG. In contrast, there was greater incorporation of galactose into DGDG than into MGDG in *Euglena gracilis* chloroplasts (Chang & Kulkarni, 1970; Matson, Fei & Chang, 1970). A later report showed that the highest specific activity of UDP-gal incorporation into galactolipids was associated with the 40 000 and 100 000 *g* pellets, and the conclusion was that the activity was associated with the 'microsomal fraction' (Van Hummel, 1974). However, Helmsing & Barendese (1970) showed that preservation of the integrity of

Fig. 4. Thin sections of part of young spinach leaf cells post-fixed with OsO_4–ferricyanide mixtures (*a* to *c*) or OsO_4–ferrocyanide mixtures (*d* to *f*). Under these conditions, the plastid membranes and especially both the outer and inner membranes of the envelope, unlike the other cell membranes, are highly contrasted. Cytoplasmic and plastid ribosomes are no longer visible after ferrocyanide treatment (*d* to *f*). These results demonstrate that the chemical composition of plastid membranes is clearly distinct from that of other cell membranes (*a*, *c*, *e* and *f*). The thin sections were counterstained with lead citrate; (*b* and *d*), no additional contrast.

ga, Golgi apparatus; P, plastid; ST, starch; C, cytoplasm; th, thylakoid; V, vacuole; t, tonoplast; see also legend for Fig. 3. (Reproduced with permission from Carde *et al.*, 1982.)

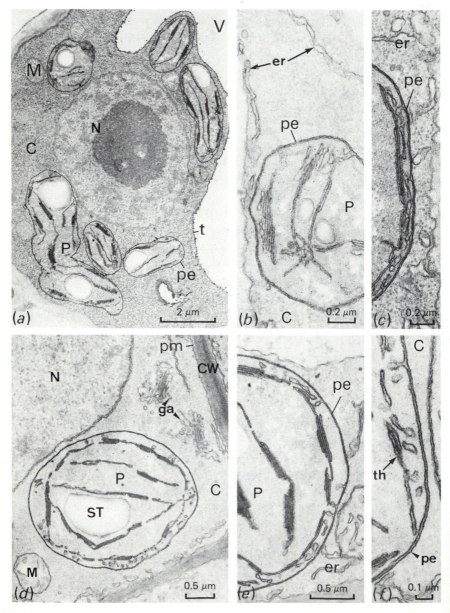

chloroplasts resulted in increased enzyme activity. It has been subsequently demonstrated unequivocally that the chloroplast envelope is the major site of UDP-gal incorporation into both MGDG and DGDG in leaf cells (Douce, 1974). As previously noted, the reason for confusion arising from earlier reports is the tendency for chloroplast envelopes to lyse during the isolation procedure. The supposed 'intact chloroplasts' used were stripped of their outer envelope membranes, and the 'microsomes', defined operationally as a membranous fraction which could be sedimented from a postmitochondrial supernatant at high speed (Fleischer & Kervina, 1974), were extensively contaminated by envelope membrane vesicles. In addition, DGDG synthetase seems to be relatively easily removed from the envelope membranes; hence much of its activity was found in the cytoplasmic fraction of the $100000\,g$ supernatant (Mudd, Van Vliet & Van Deenen, 1969; Siebertz & Heinz, 1977). The ability of the isolated chloroplast envelope to synthesize galactolipids was confirmed by other workers using various leaves from $C_{16:3}$ or $C_{18:3}$ plants, and cells of *Euglena* (reviewed by Douce & Joyard, 1980). Furthermore, Liedvogel & Kleinig (1976, 1977) have clearly demonstrated that the non-photosynthetic chromoplast inner membranes from the corona of *Narcissus pseudonarcissus* also contain galactolipid-synthesizing activities. The same is true for the envelope of amyloplasts from potato tubers (Fishwick & Wright, 1980). Significant incorporation of galactose into galactolipids from UDP-[^{14}C]gal has also been observed with avocado and cauliflower mitochondria (Ongun & Mudd, 1970) and 'microsome' preparations extracted from wheat roots or sycamore cell suspensions (Axelos & Péaud-Lenoël, 1978). However, the incorporation observed could be entirely attributable to contamination of the crude mitochondrial or 'microsomal' fraction by envelope membrane vesicles deriving from the fragile proplastids. The same observation has been made with 'microsomes' from spinach leaves (Joyard & Douce, 1976c). In carefully prepared envelope vesicles, the specific activity of the galactolipid-synthesizing enzyme is extremely high: 45 nmol mg protein^{-1} min^{-1} (Joyard & Douce, 1976a; Van Besouw & Wintermans, 1978). This activity exceeds the corresponding figure for total cell proteins by a factor of at least 100. Obviously, the reaction catalyzed is not the rate-limiting step of galactolipid synthesis because its activity *in vitro* is the highest of all the enzymes in the pathway. Moreover, the thylakoid fraction, practically devoid of envelope membrane vesicles, exhibits low galactolipid-synthesizing activity (Douce & Joyard, 1979). All these results demonstrate that, in plant cells, the plastid envelope and probably the inner membrane (Liedvogel & Kleinig, 1976) catalyze specifically the final steps in galactolipid synthesis. However, Cline & Keegstra (1983) have localized the UDP-gal:diacylglycerol galactosyltransferase on the outer envelope membrane from pea (and also from spinach, K. Keegstra, personal communication) chloroplasts. This is different from our own results (Dorne *et al.*, 1982*b*; Block *et al.*, 1983*b*): we have demonstrated that, in spinach chloroplasts, the UDP-gal:diacylglycerol galactosyltransferase is localized on the inner envelope membrane, together with the phosphatidic acid phosphatase (Fig. 5). The reasons for this difference are not yet clear and are under investigation.

It has also been demonstrated that two distinct enzymes responsible for the synthesis of MGDG and DGDG are associated with chloroplast envelope membranes (Joyard & Douce, 1976*a*). One is specific for the formation of the β-glycosidic bond and the other for the formation of the α-glycosidic bond

Fig. 5. Galactolipid synthesis by membrane fractions separated from spinach chloroplasts and enriched in outer envelope membrane (light fraction) and in inner envelope membrane (heavy fraction). We have obtained similar results with the fractions obtained according to our procedure (Block *et al.*, 1983*a*) or with the fractions obtained according to Cline *et al.* (1981). The different fractions separated on a continuous sucrose gradient (from 0.4 mol l^{-1} to 1.2 mol l^{-1}) were assayed in the presence of 1 mmol l^{-1} UDP-gal and 1 mg diacylglycerol ml^{-1}, according to Douce (1974). This experiment clearly demonstrates that UDP-gal:diacylglycerol galactosyl-transferase is localized in the membrane fraction which is enriched in inner envelope membrane.

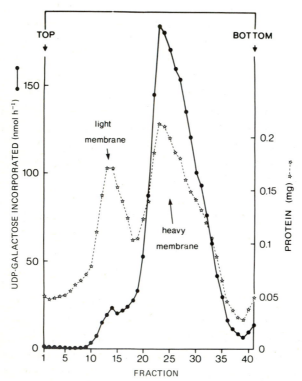

(see Fig. 1). The first enzyme, or UDP-gal: diacylglycerol galactosyltransferase, catalyses the synthesis of MGDG. In this case, UDP-gal is the galactosyl donor for the galactosylation of envelope diacylglycerol (Joyard & Douce, 1976b) (Fig. 2). The second enzyme is either a UDP-gal:MGDG galactosyltransferase (Ongun & Mudd, 1968) or a galactolipid: galactolipid galactosyltransferase (Van Besouw & Wintermans, 1978):

$$MGDG + MGDG \rightarrow DGDG + diacylglycerol.$$

According to Van Besouw & Wintermans (1978), the second enzyme (a galactosidase?) does not require the presence of UDP-gal. The galactosyl transfer proceeds by direct exchange of galactosyl groups between molecules of galactolipids. The large amounts of diacylglycerol found in isolated envelope membranes are probably a result of the activity of this interlipid galactosyltransferase during the course of envelope purification (Siebertz *et al.*, 1979). Dorne *et al.* (1982a, 1982b) have demonstrated that the fusion of the outer and inner envelope membrane during the swelling of intact plastids is responsible for the stimulation of diacylglycerol production in isolated envelope membranes. On the other hand, the results of Williams, Khan & Leung (1975) obtained with *Vicia faba* leaves support the view that DGDG is formed by galactosylation of MGDG. However, this does not take place by a rapid two-enzyme system reaction but more slowly in two phases: galactosylation of a pool of newly formed MGDG, and a more random galactosylation of MGDG. We suggest that both reactions occur on the spinach chloroplast envelope (Dorne *et al.*, 1982a, 1982b). The first galactosylation enzyme has its pH optimum above 7.5. In contrast, the second galactosylation enzyme has its maximum activity at about pH 6.5 (Joyard & Douce, 1976a). Triton X-100 (0.9% by volume) is a strong inhibitor of the second galactosylation enzyme but is almost without effect on the first galactosylation enzyme (Heinz *et al.*, 1978). The sulfhydryl nature of the galactosyltransferase has been established in spinach chloroplast preparations (Chang, 1970).

What is the physiological significance of the galactolipid: galactolipid galactosyltransferase? Up to now, it is has not been possible to give an answer to this question. Recently, we have shown that treatment of isolated intact chloroplasts with thermolysin completely inactivates the galactolipid: galactolipid galactosyltransferase. Since this treatment hydrolyses proteins located on the outer face of the outer membrane (Joyard *et al.*, 1983), it is clear that this enzymatic activity is topologically distinct from the galactolipid synthesis pathway localized on the inner envelope membrane and linked to the Kornberg–Pricer enzymes. Thus, it is unlikely that the galactolipid: galactolipid galactosyltransferase could be involved in galactolipid synthesis. Although numerous reports regarding plant enzymes catalysing headgroup incorporation into galactolipids now exist, no extensive purifications of these enzymes have been reported.

Origin of polyunsaturated fatty acids

Questions about the origin of plastid polyunsaturated fatty acids still remain. Several hypotheses have been proposed in which the endoplasmic reticulum plays a fundamental role in the desaturation of oleic acid to linoleic and linolenic acids (see Douce & Joyard, 1980, for review). Four points can be made in this connection:

(1) Some blue-green algae, suspected to be the progenitors of plastids, contain polyunsaturated fatty acids (reviewed by Douce & Joyard, 1979).

(2) Joyard *et al.* (1979) demonstrated for the first time that, with *sn*-glycerol-3-phosphate and UDP-galactose in the incubation medium, the labeled saturated and mono-unsaturated fatty acids synthesized *de novo* from [^{14}C]acetate in the chloroplast stroma could be rapidly incorporated into diacylglycerol and MGDG in the envelope membranes. Siebertz *et al.* (1980), with radioactive chloroplast membranes (thylakoid and envelope membranes) isolated from spinach leaves labeled with $^{14}CO_2$, demonstrated that the newly made MGDG (containing essentially 18:1/16:0 and 18:2/16:0 pairings) formed a very small independent pool which did not disappear into the large pre-existing MGDG pool rich in polyunsaturated fatty acids. With increasing lengths of time, more label appeared in unsaturated species. They indicated also that oligoene species of MGDG (i.e. 18:1/16:0 pairing) were derived rather directly from newly made oligoene species of diacylglycerol in the envelope membranes. In addition, these authors found that oligoene species of DGDG were derived from newly made MGDG, whereas hexaene DGDG (i.e. 18:3/18:3 pairing) was synthesized by galactosylation of pre-existing hexaene MGDG. Consequently, all these results strongly suggest that desaturation occurs *in vivo* after the formation of galactolipids (reviewed by Douce & Joyard, 1980, and Roughan & Slack, 1982).

(3) Roughan, Mudd, McManus & Slack (1979) have shown that oleate at C-1 position of the glycerol backbone of MGDG was desaturated to linoleate and linolenate when isolated spinach chloroplasts were incubated in presence of [^{14}C]acetate, *sn*-glycerol-3-phosphate and UDP-gal. This result was then extended to other 16:3 plants (Heinz & Roughan, 1982, 1983). In addition, palmitate at the C-2 position of the glycerol backbone of MGDG was desaturated to hexadecatrienoate (Heinz & Roughan, 1982, 1983).

(4) In 18:3 plants, which probably have a pathway for galactolipid synthesis different from that of 16:3 plants (Heinz & Roughan, 1982; Williams *et al.*, 1982), desaturation of oleate at C-1 position of MGDG can

also occur (Heinz & Roughan, 1983). It is possible that the quinones found in envelope membranes (Lichtenthaler *et al.*, 1981) play an important role in the mechanism of desaturation. This mechanism is completely unknown. So far, the direct desaturation of galactolipid molecules has only been deduced from changing labeling patterns of fatty acids in galactolipids, and awaits a direct demonstration. Unfortunately all attempts which have been made *in vitro* in order to look at the direct desaturation of MGDG molecular species containing mono-unsaturated fatty acids have failed (Heinz *et al.*, 1979). However, it is very likely that the exogenous added galactolipids, in marked contrast to the thioesters of CoA, which behave as powerful detergents, are not directly accessible to the membrane desaturases.

These findings suggest, in our opinion, that in 16:3 plants as well as in 18:3 plants, there is no need to postulate that chloroplast fatty acids are desaturated on the endoplasmic reticulum. The study of the mechanism of desaturation has thus far been hindered by the lack of a proper assay for the desaturase activity.

Conclusions

The results presented here show that the envelope is a site of assembly of the three parts of the galactolipid molecules (galactose, glycerol, fatty acids) and it is clear that the envelope contains the complete array of galactolipid biosynthetic machinery (Fig. 6). In addition we are convinced that *in vivo* most of the ACP in the stroma is operating near the internal surface of the inner membrane of the envelope. This system must operate in close proximity to all the enzymes involved in galactolipid synthesis and localized in the envelope membranes. The acyl-ACP thioesters are used for *sn*-glycerol-3-phosphate acylation, and constitute a link between the fatty acid synthetase and the galactolipid enzyme machinery. We believe also that *in vivo* all the intermediates of this complex biosynthetic pathway, such as lysophosphatidic acid, phosphatidic acid and probably diacylglycerol are never released but are metabolized by a coupled series of membrane-bound enzymes (Douce & Joyard, 1980). In support of these suggestions, unexchangeability and inaccessibility of enzyme-bound intermediates have been demonstrated during lipid biosynthesis in erythrocyte membranes (Hirata & Axelrod, 1978). The accumulation of these intermediates *in vitro* strongly suggests that the very fragile complex involved in galactolipid synthesis is partially dislocated during the course of the preparation of chloroplast membranes, and consequently the kinetic links between all enzymes are partially broken (see Douce & Joyard, 1980).

Synthesis of galactolipids following the pathway described in Fig. 6 has only been clearly demonstrated for molecules having the characteristic fatty acid pairing 16:3/18:3. However, most of the galactolipid molecules, in both 16:3 and 18:3 plants, have a different fatty acid pairing, i.e. 18:3/18:3. According to several authors (Roughan, 1975; Williams, Watson & Leung, 1976; Roughan et al., 1980; Drapier et al., 1982; Heinz & Roughan, 1982,

Fig. 6. Isolated intact chloroplasts are able to synthesize galactolipids, sulfolipid and phosphatidylglycerol. The backbone for these polar lipids probably comes from phosphatidic acid and diacylglycerol synthesized on the chloroplast envelope. The enzymes involved are (1) acyl-ACP: sn-glycerol-3-phosphate acyltransferase, which is responsible for the formation of lysophosphatidic acid; (2) acyl-ACP: monoacylglycerol-3-phosphate acyl-transferase, which is responsible for the formation of phosphatidic acid; (3) phosphatidic acid phosphatase, which yields diacylglycerol. MGDG is then synthesized by galactosylation of diacylglycerol on the inner envelope membrane. These three steps have been clearly demonstrated on envelope membranes whereas the other reactions (desaturation (d) of polar lipids, sulfolipid and phosphatidylglycerol synthesis) occur in isolated, intact chloroplasts and could take place on envelope membranes. Phosphatidyl-glycerol synthesis (7) probably involves the formation of intermediates such as CDP-diacylglycerol (5) and phosphatidylglycerol-3-phosphate (6). Sulfo-lipid synthesis (8) probably requires the formation of a sulfoquinovosyl donor which is still unknown. This scheme is based on observations that, in isolated intact chloroplasts, galactolipid, sulfolipid and phosphatidyl-glycerol molecules can be synthesized which have the same diacylglycerol backbone (see text). The origin of the polar lipids having C_{18} fatty acids at each C-1 and C-2 position is not yet clear, but could involve extrachloroplastic compartments.

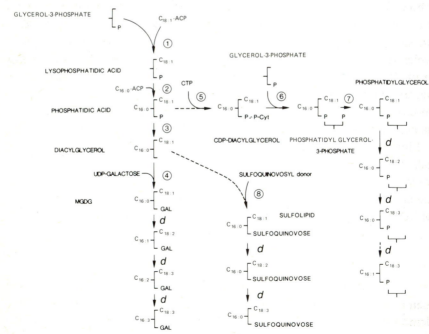

1983; Dubacq, Drapier & Trémolières, 1983), there is cooperation between the endoplasmic reticulum and the chloroplast in the formation of the diacylglycerol backbone having the correct fatty acid composition, i.e. with C_{18} fatty acids at each position of the glycerol. Unfortunately, the mechanisms which could supply diacylglycerol from the endoplasmic reticulum to the envelope membranes, where both galactosylation of diacylglycerol for the synthesis of galactolipids and desaturation of galactolipid molecules occur, are completely unknown.

Thylakoids seem to lack the enzymes for the synthesis of galactolipids; consequently during the course of thylakoid biogenesis massive transport of galactolipids must occur between the inner layer of the inner envelope membrane and the growing thylakoids. According to Williams, Simpson & Chapman (1979) this exchange occurs very rapidly. A possible solution to this problem may lie either in galactolipid-exchange proteins localized in the stromal space, or in membrane flow, as proposed by Morré & Mollenhauer (1976). In other words, the inner membrane of the envelope behaves as a 'generating element', the 'end product' being the thylakoid membranes. However, during the course of the synthesis of chloroplast membrane lipids and chlorophyll in synchronous culture of *Chlamydomonas reinhardtii*, Beck & Levine (1977) could not distinguish between insertion of new components, such as galactolipids, into very small regions of the membrane where membrane growth was occurring, and truly random insertion throughout the whole chloroplast membrane. At the moment, the mechanisms of transfer are not sufficiently well established to permit a conclusion regarding the physiological significance of any route. One of the most interesting questions which remain to be answered is how galactolipid synthesis is regulated. For example, it will be interesting to determine whether or not this lipid exchange occurs simultaneously with specific proteins and/or pigments such as carotenoids. We now need to establish more precisely just how the controls of galactolipid synthesis operate.

Finally, it is very likely that all the enzymes involved in fatty acids and galactolipid synthesis are encoded by nuclear DNA. The plastids of a barley mutant which contains no plastid ribosomes are surrounded by a double membrane system (Börner, Schumann & Hagemann, 1976) and contain MGDG as well as DGDG (Dorne *et al.*, 1982c).

Origin of other chloroplast lipids
Sulfolipid

As described above for galactolipids, careful analyses of purified membrane systems and of etiolated and photosynthesizing tissues have clearly demonstrated that sulfolipid is localized solely within plastid membranes

(Douce & Joyard, 1980). Furthermore, the use of specific antibodies raised against sulfolipid (Billecocq, 1975), and analyses of purified outer envelope membrane (Cline *et al.*, 1981; Dorne *et al.*, 1982b; Block *et al.*, 1983b) have shown that the outer envelope membrane, as well as the other plastid membranes, contains sulfolipid. Localization of sulfolipid in plastids is also supported by labeling experiments from Haas *et al.* (1980). These authors demonstrated that application of $^{35}SO_4^{2-}$ to Percoll-purified chloroplasts resulted in a light-dependent labeling of a lipid component which was identified as sulfolipid. In addition, mesophyll protoplasts from ^{35}S-labeled oat primary leaves were gently disrupted and separated into organelles by sucrose gradient centrifugation. Labeled sulfolipid was located almost exclusively in the chloroplasts. Consequently, from these data, Haas *et al.* (1980) rightly concluded that the final assembly steps in the biosynthesis of sulfolipid are confined to the chloroplasts. This was shown directly by Mudd & Sparace (1981); using purified intact chloroplasts, they succeeded in incorporating (although at a low rate) both radioactive sulfate and radioactive acetate into sulfolipid. However, despite numerous studies the precise biosynthetic route for the production of sulfolipid remains uncertain; thus it is very difficult to localize precisely within the chloroplast the different enzymes involved in sulfolipid biosynthesis. Since diacylglycerol synthesis occurs only in envelope membranes (owing to the presence of the enzymes of the Kornberg–Pricer pathway), it is tempting to suggest that sulfolipid synthesis occurs also on envelope membranes (Fig. 6). So far, the evidence is still lacking.

Finally, as already shown for galactolipids, it is very likely that all the enzymes involved in sulfolipid synthesis are coded for by nuclear DNA, since plastid ribosome-deficient barley mutants contain sulfolipid (Dorne *et al.*, 1982c).

Phospholipids

Plastid membranes contain phospholipids in proportions lower than those in other cell membranes (Douce & Joyard, 1979, 1980). Furthermore, phosphatidylethanolamine, a major phospholipid present in large amounts in all cell membranes, is entirely absent from plastid membranes (Douce & Joyard, 1979, 1980). In fact, the two major plastid phospholipids are phosphatidylcholine (PC) and phosphatidylglycerol (PG); the former predominates in the envelope and the latter in the thylakoids (Douce *et al.*, 1973; Mackender & Leech, 1974). The presence of large amounts of PC in envelope membranes reflects the chemical nature of the outer envelope membrane (Cline *et al.*, 1981; Dorne *et al.*, 1982b; Block *et al.*, 1983b). PG in plastids is unique in that it contains trans-Δ3-hexadecenoic acid esterified almost exclusively at the C-2 position of the glycerol-3-phosphate backbone (Haverkate & Van Deenen, 1965). It is interesting to note that envelope membranes, as well as thylakoids, contain this specific fatty acid (Siebertz *et al.*, 1979).

The origin of PC by the nucleotide pathway has been conclusively demonstrated in spinach leaves (reviewed by Mudd, 1980, and Moore, 1982):

$$\text{CDP-choline} + sn\text{-1,2-diacylglycerol} \rightarrow \text{phosphatidylcholine} + \text{CMP}.$$

Similarly, Marshall & Kates (1972) provided evidence that the biosynthesis of PG, which involves the intermediate formation of phosphatidylglycerophosphate from CDP-diacylglycerol and *sn*-glycerol-3-phosphate, occurs in spinach leaves (reviewed by Mudd, 1980, and Moore, 1982);

$$sn\text{-glycerol-3-phosphate} + \text{CDP-diacylglycerol} \rightarrow$$
$$3\text{-}sn\text{-phosphatidyl-1}'\text{-}sn\text{-glycerol-3-phosphate} + \text{CMP}.$$

Subcellular localization studies of PC- (Marshall & Kates, 1974) and PG-synthetases (Marshall & Kates, 1972) revealed that the fraction with the highest specific activity and largest proportion of the total activity is the $40\,000\text{--}100\,000 \times g$ 'microsomal' pellet, after removal of the larger cell organelles such as chloroplasts ($4000 \times g$ pellet) and mitochondria ($10\,000 \times g$ pellet). These results suggest that plastid phospholipid synthesis occurs in the endoplasmic reticulum. Unfortunately, as well as fragments of lamellar and cristae membranes from chloroplasts and mitochondria, the 'microsomal' fraction contains numerous heterogeneous vesicles derived from the microbodies, dictyosomes and endoplasmic reticulum. It also contains chloroplast envelope vesicles. Therefore, although PC- and PG-synthetases are called 'microsomal' activities, it should be emphasized that they may not be located solely (or mostly) in the endoplasmic reticulum. Joyard & Douce (1976*a*) have demonstrated that the purified chloroplast envelopes from spinach leaves, although containing diacylglycerol (Joyard & Douce, 1976*b*), are devoid of CDP-choline:diacylglycerol phosphorylcholine transferase activity indicating that PC is probably synthesized outside the chloroplasts. All these results strongly suggest, but do not prove, that the chloroplast membrane phospholipids may be synthesized by the endoplasmic reticulum system and then incorporated into chloroplast envelope membranes. However, Chammaï (1980) has demonstrated that the envelope isolated from *Euglena* chloroplasts[1] contains the complete array of enzymes involved in the PG biosynthetic

[1] In the brown algae and chromophyta, the chloroplasts are surrounded by four membranes. The inner pair represents the double chloroplast envelope. Traditionally, the outer pair of membranes is referred to as chloroplast endoplasmic reticulum and forms a loosely fitting sac surrounding the chloroplast. In addition, one characteristic of the Dinophyceae and Euglenophyceae is that their chloroplasts are surrounded by three membranes. An interesting discussion about the possible origin of their chloroplasts is presented by Whatley, John & Whatley (1979).

machinery. In support of this conclusion, Mudd & DeZacks (1981) found that highly intact chloroplasts isolated from spinach leaves incorporate *sn*-glycerol-3-phosphate into PG. The omission of ATP, CoA, bicarbonate and acetate decreased the incorporation. According to Mudd & DeZacks, PG synthesis is linked to fatty acid synthesis in the stromal phase and phosphatidic acid and diacylglycerol synthesis in envelope membranes. However, CDP-diacylglycerol was not detected among the synthesized products, but CTP increased the proportion of PG synthesized from [^{14}C]acetate (Sparace & Mudd, 1982*a*, 1982*b*). PG molecules synthesized from radioactive *sn*-glycerol-3-phosphate or acetate were digested by lipases and analyzed by Sparace & Mudd (1982*a*, 1982*b*). The data indicate that the PG obtained is the product of synthesis *de novo*, and rule out the possibilities of PG synthesis by transphosphatidylation catalyzed by phospholipase D or acylation of lyso derivatives (Sparace & Mudd, 1982*b*). Furthermore, Sparace & Mudd (1982*b*) observed that approximately 30% of the C_{18} fatty acids at the C-1 position of the molecule was linolenate. This result is very interesting since it is another demonstration that desaturation of polar lipids occurs in isolated, purified chloroplasts. The fatty acid pattern of the newly synthesized PG molecule is similar to that of phosphatidic acid molecules synthesized in isolated envelope membranes (Joyard *et al.*, 1979; Frentzen *et al.*, 1982, 1983). Therefore, we suggest that the phosphatidic acid molecules synthesized by the enzymes of the Kornberg–Pricer pathway located in the envelope (Joyard & Douce, 1977) can be used for PG synthesis within the chloroplast (and probably on the envelope membranes) as proposed in Fig. 6. The results obtained by Sparace & Mudd (1982*a*, 1982*b*) demonstrate that phospholipids can be synthesized outside the endoplasmic reticulum (ER) thus attributing to that membrane system less importance than had been previously suggested.

If chloroplast PC is really synthesized in the ER, we must imagine a direct transfer of PC molecules between the ER and the chloroplast envelope. A possible solution to this problem may lie in either membrane flow, as proposed by Morré, Merrit & Lembi (1971) or phospholipid-exchange protein(s) located in the cytoplasm (reviewed by Mazliak & Kader, 1980).

Structural continuity between various cellular membranes, particularly between the ER and the outer envelope membrane (Morré & Mollenhauer, 1974), if correct, would mean that phospholipids synthesized in the ER could diffuse laterally within a continuous membrane network to the outer envelope membrane. In other words, the ER behaves as a 'generating element', the 'end product' being the envelope membranes and subsequently the thylakoids. However, one must imagine a specific mechanism in order to explain why phospholipids could diffuse from ER to the outer envelope membrane and galactolipids could not diffuse from the outer envelope membrane to ER.

Membrane flow between these two membrane systems should lead to a uniform lipid composition. Obviously, this is not the case (Douce & Joyard, 1979); indeed, analyses of the lipid composition of the outer envelope membrane from spinach (Dorne *et al.*, 1982*b*; Block *et al.*, 1983*a*) and from pea (Cline *et al.*, 1981) clearly demonstrated that this membrane system differs from ER (Douce & Joyard, 1981). In addition, observations of an intimate association between ER and the chloroplast envelope are so exceptional that it is highly unlikely that this is the main method by which PC molecules enter chloroplasts.

The second hypothesis to explain a possible transfer of phospholipids to plastids is based on the existence of a phospholipid-exchange protein (reviewed by Mazliak & Kader, 1980). However, Stuhne-Sekalec & Stanacev (1980) have demonstrated spontaneous (protein-independent) transfer of phosphatidic acid, PC and diacylglycerol from 'microsomes' to mitochondrial membranes by a mechanism which involves a close contact between the membrane-donor and the membrane-acceptor.

It is clear that further work is needed in order to understand the mechanism which prevents intermixing of polar lipids between ER and plastids and allows the specific accumulation of PC in envelope membranes.

Conclusions

The purification of envelope membranes from spinach chloroplasts paved the way for the analysis of the chemical composition and functions of this membrane system. The results obtained clearly demonstrate that the envelope membranes from all the plastids isolated so far contain all the enzymes of the Kornberg–Pricer pathway and the complete array of galactolipid biosynthetic machinery (Douce & Joyard, 1980).

Although most current research activities concerned with chloroplast glycerolipids continue at the descriptive level, important questions relating to the regulation of enzyme activity are beginning to be addressed. Strong evidence indicates that the biosynthesis of MGDG is asymmetric, occurring on the inner surface of the inner membrane of the chloroplast envelope. Synthesis in such an asymmetric location focuses attention on transmembrane movement and the sorting of polar lipids within the envelope and thylakoids.

Another aspect of the regulation of the synthesis of glycerolipids concerns the cooperation between ER and organelles. Most authors give the leading role to the ER. We do not believe this to be correct. Plastids seem to provide all the fatty acids for all membrane lipids; in addition, they contain all the enzymes involved in the synthesis of phosphatidic acid and diacylglycerol, and they are able to synthesize their major polar lipids (galactolipids). This could also be the case for PG, the major chloroplast phospholipid, and for sulfolipid. A better understanding of the interactions between the organelles (plastids but also mitochondria) and the ER requires the use of highly characterized cellular fractions. Unfortunately, it has so far proved impossible to prepare pure ER fractions. Experiments with well-characterized envelope fractions have clearly demonstrated that, in a plant cell, the envelope is a major site of membrane synthesis. This membrane system is very dynamic, since it is involved in both the transport of plastid proteins synthesized on free cytoribosomes and the synthesis of plastid components such as carotenoids, prenylquinones and galactolipids (Douce & Joyard, 1981). Consequently, we must discard the view that the ER provides all the membrane material for all the plant cell membranes.

Finally, an overriding problem in chloroplast biogenesis concerns the regulation of the types and quantities of glycerolipids synthesized within chloroplasts. An approximate mixture of products must be synthesized to meet the several, and often simultaneous, demands for membrane biogenesis. Greater understanding of the roles of chloroplast glycerolipids and the regulation of their synthesis within green leaf cells should come with enzyme purification and the raising of monospecific antibodies, the development of specific inhibitors, and the selection and characterization of mutant cell lines defective in glycerolipid synthetic enzymes.

References

Axelos, M. & Péaud-Lenoël, C. (1978). Glycosyl transfers from UDP-sugars to lipids of plant membranes: identification and specificity of the transferases. *Biochimie*, **60**, 35–44.

Beck, J. C. & Levine, R. P. (1977). Synthesis of chloroplast membrane lipids and chlorophyll in synchronous cultures of *Chlamydomonas reinhardtii*. *Biochim. biophys. Acta*, **489**, 360–9.

Benson, A. A., Wiser, R., Ferrari, R. A. & Miller, J. A. (1958). Photosynthesis of galactolipids. *J. Am. chem. Soc.*, **80**, 4740.

Bertrams, M. & Heinz, E. (1976). Experiments on enzymatic acylation of *sn*-glycerol-3-phosphate with enzyme preparations from pea and spinach leaves. *Planta*, **132**, 161–8.

Bertrams, M. & Heinz, E. (1979). Soluble, isomeric forms of glycerophosphate acyltransferase in chloroplasts. In *Recent Advances in the Biochemistry and Physiology of Plant Lipids*, ed. L. A. Appelqvist & C. Liljenberg, pp. 139–44. Elsevier/North-Holland Biomedical Press, Amsterdam.

Bertrams, M. & Heinz, E. (1980). Long chain acyl-coenzyme A thioesters as substrates in glycerolipid biosynthesis of chloroplasts. In *Biogenesis and Functions of Plant Lipids*, ed. P. Mazliak, P. Benveniste, C. Costes & R. Douce, pp. 67–72. Elsevier/North-Holland Biomedical Press, Amsterdam.

Billecocq, A. (1974). Structure des membranes biologiques: localisation des galactosyldiglycérides dans les chloroplastes au moyen des anticorps spécifiques. II. Etude en microscopie électronique à l'aide d'un marquage à la péroxydase. *Biochim. biophys. Acta*, **352**, 245–51.

Billecocq, A. (1975). Structure des membranes biologiques: localisation du sulfoquinovosyldiglycéride dans les diverses membranes des chloroplastes au moyen des anticorps spécifiques. *Annls Immunol. (Institut Pasteur)*, **126C**, 337–52.

Billecocq, A., Douce, R. & Faure, M. (1972). Structure des membranes biologiques: localisation des galactosyldiglycérides dans les chloroplastes au moyen des anticorps spécifiques. *C.r. hebd. Séanc. Acad. Sci. Paris*, **275**, 1135–7.

Block, M. A., Dorne, A.-J., Joyard, J. & Douce, R. (1983a). Preparation and characterization of membrane fractions enriched in outer and inner envelope membranes from spinach chloroplasts. I. Electrophoretic and immunochemical analyses. *J. biol. Chem.*, **258**, 13273–80.

Block, M. A., Dorne, A.-J., Joyard, J. & Douce, R. (1983b). Preparation and characterization of membrane fractions enriched in outer and inner envelope membranes from spinach chloroplasts. II. Biochemical characterization. *J. biol. Chem.* **258**, 13281–6.

Block, M. A., Dorne, A.-J., Joyard, J. & Douce, R. (1983c). The acyl-CoA synthetase and acyl-CoA thioesterase are located on the outer and inner membrane of the chloroplast envelope, respectively. *FEBS Lett.*, **153**, 377–81.

Block, M. A., Dorne, A.-J., Joyard, J. & Douce, R. (1983d). The phosphatidic acid phosphatase of the chloroplast envelope is located on the inner envelope membrane. *FEBS Lett.*, **164**, 111–15.

Börner, T., Schumann, B. & Hagemann, R. (1976). Biochemical studies on a plastid ribosome-deficient mutant of *Hordeum vulgare*. In *Genetics and Biogenesis of Chloroplasts and Mitochondria*, ed. Th. Bücher, W. Neupert, W. Sebald & S. Werner, pp. 41–8. Elsevier/North-Holland Biomedical Press, Amsterdam.

Carde, J.-P., Joyard, J. & Douce, R. (1982). Electron microscopic studies of envelope membranes from spinach plastids. *Biol. Cell*, **44**, 315–24.

Caughey, I. & Kekwick, R. G. O. (1982). The characteristics of some components of the fatty acid synthetase system in the plastids from the mesocarp of avocado (*Persea americana*) fruit. *Eur. J. Biochem.*, **123**, 553–61.

Chammaï, A. (1980). Métabolisme des phospholipides chez *Euglena gracilis* en culture synchrone. Biosynthèse du phosphatidylglycérol. Thèse de Doctorat d'Etat, Université de Strasbourg.

Chang, S. S. (1970). Sulfhydryl nature of galactosyl transfer enzymes of spinach chloroplasts. *Phytochemistry*, **9**, 1947–8.

Chang, S. S. & Kulkarni, N. D. (1970). Enzymatic reactions of galactolipid synthesis with a soluble, sub-chloroplast fraction from *Spinacia oleracea*. *Phytochemistry*, **9**, 927–34.

Cline, K., Andrews, J., Mersey, B., Newcomb, E. H. & Keegstra, K. (1981).

Separation and characterization of inner and outer envelope membranes of pea chloroplasts. *Proc. natn. Acad. Sci. USA*, **78**, 3595–9.

Cline, K. & Keegstra, K. (1983). Galactosyltransferases involved in galactolipid biosynthesis are located on the outer membrane of pea chloroplast envelopes. *Pl. Physiol.*, **71**, 366–72.

Dorne, A.-J., Block, M. A., Joyard, J. & Douce, R. (1982*a*). The galactolipid:galactolipid galactosyltransferase is located on the outer surface of the outer membrane of the chloroplast envelope. *FEBS Lett.*, **145**, 30–4.

Dorne, A.-J., Block, M. A., Joyard, J. & Douce, R. (1982*b*). Studies on the localization of enzymes involved in galactolipid metabolism in chloroplast envelope membranes. In *Biochemistry and Metabolism of Plant Lipids*, ed. J. F. G. M. Wintermans & P. J. C. Kuiper, pp. 153–64. Elsevier Biomedical Press, Amsterdam.

Dorne, A.-J., Carde, J.-P., Joyard, J., Börner, T. & Douce, R. (1982*c*). Polar lipid composition of a plastid ribosome-deficient barley mutant. *Pl. Physiol.*, **69**, 1467–70.

Douce, R. (1974). Site of biosynthesis of galactolipids in spinach chloroplasts. *Science*, **183**, 852–3.

Douce, R. & Guillot-Salomon, T. (1970). Sur l'incorporation de la radio-activité du *sn*-glycérol 3-phosphate-^{14}C dans le monogalactosyldiglycéride des plastes isolés. *FEBS Lett.*, **11**, 121–6.

Douce, R., Holtz, R. B. & Benson, A. A. (1973). Isolation and properties of the envelope of spinach chloroplasts. *J. biol. Chem.*, **248**, 7215–22.

Douce, R. & Joyard, J. (1979). Structure and function of the plastid envelope. *Adv. bot. Res.*, **7**, 1–116.

Douce, R. & Joyard, J. (1980). Plant galactolipids. In *The Biochemistry of Plants*, vol. 4, *Lipids: Structure and Function*, ed. P. K. Stumpf & E. E. Conn, pp. 321–62. Academic Press, New York.

Douce, R. & Joyard, J. (1981). Does the plastid envelope derive from the endoplasmic reticulum? *Trends biochem. Sci.*, **6**, 237–9.

Douce, R. & Joyard, J. (1982). Purification of the chloroplast envelope. In *Methods in Chloroplast Molecular Biology*, ed. M. Edelman, R. Hallick & N. H. Chua, pp. 239–56. Elsevier/North-Holland Biomedical Press, Amsterdam.

Drapier, D., Dubacq, J.-P., Trémolières, A. & Mazliak, P. (1982). Cooperative pathway for lipid biosynthesis in young pea leaves: oleate exportation from chloroplasts and subsequent integration into complex lipids of added microsomes. *Pl. Cell Physiol.*, **23**, 125–35.

Dubacq, J.-P., Drapier, D. & Trémolières, A. (1983). Synthesis of polyunsaturated fatty acids by a mixture of chloroplasts and microsomes from spinach leaves: evidence for two distinct pathways of the biosynthesis of trienoic acids. *Pl. Cell Physiol.*, **24**, 1–9.

Dubacq, J.-P. & Trémolières, A. (1983). Occurrence and function of phosphatidylglycerol containing Δ3-*trans*-hexadecenoic acid in photosynthetic lamellae. *Physiol. Veg.*, **21**, 293–312.

Ferrari, R. A. & Benson, A. A. (1961). The path of carbon in photosynthesis of the lipids. *Archs Biochem. Biophys.*, **93**, 185–92.

Fishwick, M. J. & Wright, A. J. (1980). Isolation and characterization of amyloplast envelope membranes from *Solanum tuberosum*. *Phytochemistry*, **19**, 55–9.

Fleischer, S. & Kervina, M. (1974). Subcellular fractionation of rat liver. *Meth. Enzym.*, **31**, 6–41.

Frentzen, M., Heinz, E., McKeon, T. A. & Stumpf, P. K. (1982). De-novo biosynthesis of galactolipid molecular species by reconstituted enzyme systems from chloroplasts. In *Biochemistry and Metabolism of Plant Lipids*, ed. J. F. G. M. Wintermans & P. J. C. Kuiper, pp. 141–52. Elsevier Biomedical Press, Amsterdam.

Frentzen, M., Heinz, E., McKeon, T. A. & Stumpf, P. K. (1983). Specificities and selectivities of glycerol 3-phosphate acyltransferase and monoacyl-glycerol 3-phosphate acyltransferase from pea and spinach chloroplasts. *Eur. J. Biochem.*, **129**, 629–36.

Gardiner, S. E. & Roughan, P. G. (1983). Relationship between fatty-acyl composition of digalactosylglycerol and turnover of chloroplast phosphatidate. *Biochem. J.*, **210**, 949–52.

Haas, R., Siebertz, H. P., Wrage, K. & Heinz, E. (1980). Localization of sulfolipid labelling within cells and chloroplasts. *Planta*, **148**, 238–44.

Haverkate, F. & Van Deenen, L. L. M. (1965). Isolation and chemical characterization of phosphatidylglycerol in spinach leaves. *Biochim. biophys. Acta*, **106**, 78–92.

Heinz, E. (1977). Enzymatic reactions in galactolipid biosynthesis. In *Lipids and Lipid Polymers in Higher Plants*, ed. M. Tevini & H. K. Lichtenthaler, pp. 102–20. Springer-Verlag, Berlin.

Heinz, E., Bertrams, M., Joyard, J. & Douce, R. (1978). Demonstration of an acyltransferase activity in chloroplast envelopes. *Z. Pfl. Physiol.*, **87**, 325–31.

Heinz, E. & Roughan, P. G. (1982). *De novo* synthesis, desaturation and acquisition of monogalactosyldiacylglycerol by chloroplasts from '16:3'- and '18:3'-plants. In *Biochemistry and Metabolism of Plant Lipids*, eds. J. F. G. M. Wintermans & P. J. C. Kuiper, pp. 169–82. Elsevier Biomedical Press, Amsterdam.

Heinz, E. & Roughan, P. G. (1983). Similarities and differences in lipid metabolism of chloroplasts isolated from 18:3 and 16:3 plants. *Pl. Physiol.*, **72**, 273–9.

Heinz, E., Siebertz, H. P., Linsheid, M., Joyard, J. & Douce, R. (1979). Investigations on the origin of diglyceride diversity in leaf lipids. In *Advances in the Biochemistry and Physiology of Plant Lipids*, ed. L. A. Appelqvist & C. Liljenberg, pp. 99–120. Elsevier/North-Holland Biomedical Press, Amsterdam.

Helmsing, P. J. & Barendese, G. W. M. (1970). Incorporation of UDP-galactose-^{14}C by spinach chloroplasts. *Acta bot. neer.*, **19**, 567–71.

Hirata, F. & Axelrod, J. (1978). Enzymatic synthesis and rapid translocation of phosphatidylcholine by two methyltransferases in erythrocyte membranes. *Proc. natn. Acad. Sci. USA*, **75**, 2348–52.

Høj, P. B. & Mikkelsen, J. D. (1982). Partial separation of individual enzyme activities of an ACP-dependent fatty acid synthetase from barley chloroplasts. *Carlsberg Res. Commun.*, **47**, 119–41.

Jacobson, B. S., Jaworski, J. G. & Stumpf, P. K. (1974). Stearyl-acyl carrier protein desaturase from spinach chloroplasts. *Pl. Physiol.*, **54**, 484–6.

Jaworski, J. G., Goldschmidt, E. E. & Stumpf, P. K. (1974). Properties of the palmityl acyl carrier protein:stearyl acyl carrier protein elongation system in maturing safflower seed extracts. *Archs. Biochem. Biophys.*, **163**, 769–76.

Joyard, J. (1979). L'enveloppe des chloroplastes. Thèse de Doctorat d'Etat, Université Scientifique et Médicale de Grenoble.

Joyard, J., Billecocq, A., Bartlett, S. G., Block, M. A., Chua, N.-H. & Douce, R. (1983). Localisation of polypeptides to the cytosolic side of the outer envelope membrane of spinach chloroplasts. *J. biol. Chem.*, **258**, 10000–6.

Joyard, J., Chuzel, M. & Douce, R. (1979). Is the chloroplast envelope a site of galactolipid synthesis? Yes!. In *Recent Advances in the Biochemistry and Physiology of Plant Lipids*, ed. L. A. Appelqvist & C. Liljenberg, pp. 181–6. Elsevier/North-Holland Biomedical Press, Amsterdam.

Joyard, J. & Douce, R. (1976a). Préparation et activités enzymatiques de l'enveloppe des chloroplastes d'épinard. *Physiol. Veg.*, **14**, 31–8.

Joyard, J. & Douce, R. (1976b). Mise en évidence et rôle des diacylglycérols de l'enveloppe des chloroplastes d'épinard. *Biochim. biophys. Acta*, **424**, 125–31.

Joyard, J. & Douce, R. (1976c). L'enveloppe des chloroplastes est-elle capable de synthétiser la phosphatidylcholine? *C.r. hebd. Séanc. Acad. Sci. Paris*, **282**, 1515–18.

Joyard, J. & Douce, R. (1977). Site of synthesis of phosphatidic acid and diacylglycerol in spinach chloroplasts. *Biochim. biophys. Acta*, **486**, 273–85.

Joyard, J. & Douce, R. (1979). Characterization of phosphatidate phosphohydrolase activity associated with chloroplast envelope membranes. *FEBS Lett.*, **102**, 147–50.

Joyard, J. & Stumpf, P. K. (1980). Characterization of an acyl-Coenzyme A thioesterase associated with the envelope of spinach chloroplasts. *Pl. Physiol.*, **65**, 1039–43.

Joyard, J. & Stumpf, P. K. (1981). Synthesis of long-chain acyl-CoA in chloroplast envelope membranes. *Pl. Physiol.*, **67**, 250–6.

Kates, M. (1960). Chromatographic and radioisotopic investigations of the lipid components of runner bean leaves. *Biochim. biophys. Acta*, **41**, 315–28.

Kleinig, H. & Liedvogel, B. (1978). Fatty acid synthesis by isolated chromoplasts from the daffodil: [^{14}C]-acetate incorporation and distribution of labelled acids. *Eur. J. Biochem.*, **83**, 499–505.

Lichtenthaler, H. K., Prenzel, U., Douce, R. & Joyard, J. (1981). Localization of prenylquinones in the envelope of spinach chloroplasts. *Biochim. biophys. Acta*, **641**, 99–105.

Liedvogel, B. & Kleinig, H. (1976). Galactolipid synthesis in chromoplast internal membranes of the daffodil. *Planta*, **129**, 19–21.

Liedvogel, B. & Kleinig, H. (1977). Lipid metabolism in chromoplast membranes from the daffodil: glycosylation and acylation. *Planta*, **133**, 249–53.

Liedvogel, B., Kleinig, H., Thompson, J. A. & Falk, H. (1978). Chromoplasts of *Tropaeolum majus* L.: lipid synthesis in whole organelles and sub-fractions. *Planta*, **141**, 303–9.

Mackender, R. O. & Leech, R. M. (1974). The galactolipid, phospholipid and fatty acid composition of the chloroplast envelope membranes of *Vicia faba* L. *Pl. Physiol.*, **53**, 496–502.

Marshall, M. O. & Kates, M. (1972). Biosynthesis of phosphatidylglycerol by cell-free preparations from spinach leaves. *Biochim. biophys. Acta*, **260**, 558–70.

Marshall, M. O. & Kates, M. (1974). Biosynthesis of nitrogenous phospho-lipids in spinach leaves. *Can. J. Biochem./J. canad. Biochim.*, **52**, 469–82.

Matson, R. S., Fei, M. & Chang, S. B. (1970). Comparative studies of biosynthesis of galactolipids in *Euglena gracilis* Strain Z. *Pl. Physiol.*, **45**, 531–2.

Matsumura, S. & Stumpf, P. K. (1968). Fat metabolism in higher plants. XXXV. Partial primary structure of spinach acyl carrier protein. *Archs. Biochem. Biophys.*, **125**, 932–41.

Mazliak, P. & Kader, J. C. (1980). Phospholipid-exchange systems. In *The Biochemistry of Plants*, vol. 4, *Lipids: Structure and Function*, ed. P. K. Stumpf, pp. 283–300. Academic Press, New York.

Mikkelsen, J. D. & Høj, P. B. (1982). Separation of component enzyme activities of fatty acid synthetase from barley chloroplasts. In *Biochemistry and Metabolism of Plant Lipids*, ed. J. F. G. M. Wintermans & P. J. C. Kuiper, pp. 13–20. Elsevier Biomedical Press, Amsterdam.

Moore, T. S. (1982). Phospholipid biosynthesis. *A. Rev. Pl. Physiol.*, **33**, 235–59.

Morré, D. J., Merritt, W. D. & Lembi, C. A. (1971). Connections between mitochondria and endoplasmic reticulum in rat liver and onion stem. *Protoplasma*, **73**, 43–9.

Morré, D. J. & Mollenhauer, H. H. (1974). The endomembrane concept: a functional integration of endoplasmic reticulum and Golgi apparatus. In *Dynamic Aspects of Plant Ultrastructure*, ed. A. W. Robards, pp. 84–137. McGraw-Hill, London.

Morré, D. J. & Mollenhauer, H. H. (1976). Interactions among cytoplasm, endomembranes, and the cell surface. In *Transports in Plants*, vol. 3, ed. C. R. Stocking & U. Heber, pp. 288–344. Springer Verlag, Berlin and New York.

Mudd, J. B. (1980). Phospholipid biosynthesis. In *The Biochemistry of Plants*, vol. 4, *Lipids: Structure and Function*, ed. P. K. Stumpf, pp. 250–82. Academic Press, New York.

Mudd, J. B. & Sparace, S. A. (1981). Phosphatidylglycerol and sulfolipid synthesis in chloroplasts. In *XIII International Botanical Congress*, *Sydney, Australia*, Abstracts, p. 17.

Mudd, J. B. & DeZacks, R. (1981). Synthesis of phosphatidylglycerol by chloroplasts from leaves of *Spinacia oleracea* L. (spinach). *Archs. Biochem. Biophys.*, **209**, 584–91.

Mudd, J. B., Van Vliet, H. H. D. M. & Van Deenen, L. L. M. (1969).

Biosynthesis of galactolipids by enzyme preparation from spinach leaves. *J. Lipid Res.*, **10**, 623–30.

Neufeld, E. F. & Hall, C. W. (1964). Formation of galactolipids by chloroplasts. *Biochem. biophys. Res. Commun.*, **14**, 503–8.

Nothelfer, H. G., Barckhaus, R. H. & Spener, F. (1977). Localization and characterization of the fatty acid synthesizing system in cells of *Glycine max* (soybean) dispersion cultures. *Biochim. biophys. Acta*, **489**, 370–80.

Ohlrogge, J. B., Kuhn, D. A. & Stumpf, P. K. (1979). Subcellular localization of acyl carrier protein in leaf protoplasts of *Spinacia oleracea*. *Proc. natn. Acad. Sci. USA*, **76**, 1194–8.

Ohlrogge, J. B., Shine, W. E. & Stumpf, P. K. (1978). Characterization of plant acyl-ACP and acyl-CoA hydrolases. *Arch. Biochem. Biophys.*, **189**, 382–91.

Ongun, A. & Mudd, J. B. (1968). Biosynthesis of galactolipids in plants. *J. biol. Chem.*, **243**, 1558–66.

Ongun, A. & Mudd, J. B. (1970). The biosynthesis of steryl glucosides in plants. *Pl. Physiol.*, **45**, 255–62.

Renkonen, O. & Bloch, H. (1969). Biosynthesis of monogalactocyl diglycerides in photoauxotrophic *Euglena gracilis*. *J. biol. Chem.*, **244**, 4899–903.

Roughan, P. G. (1975). Phosphatidylcholine: donor for C18 fatty acids for glycerolipid synthesis. *Lipids*, **10**, 609–14.

Roughan, P. G., Holland, R. & Slack, C. R. (1979*a*). On the control of long-chain fatty acid and synthesis in isolated intact spinach (*Spinacia oleracea*) chloroplasts. *Biochem. J.*, **184**, 193–202.

Roughan, P. G., Holland, R. & Slack, C. R. (1980). The role of chloroplasts and microsomal fractions in polar lipid synthesis from [^{14}C]-acetate by cell-free preparations from spinach (*Spinacia oleracea*) leaves. *Biochem. J.*, **188**, 17–24.

Roughan, P. G., Mudd, J. B., McManus, T. T. & Slack, C. R. (1979*b*). Linoleic and α-linolenate synthesis by isolated spinach (*Spinacia oleracea*) chloroplasts. *Biochem. J.*, **184**, 571–4.

Roughan, P. G. & Slack, C. R. (1977). Long-chain acyl-coenzyme A synthetase activity of spinach chloroplasts is concentrated in the envelope. *Biochem. J.*, **162**, 457–9.

Roughan, P. G. & Slack, C. R. (1982). Cellular organization of glycerolipid metabolism. *A. Rev. Pl. Physiol.*, **33**, 97–132.

Shimakata, T. & Stumpf, P. K. (1982*a*). Fatty acid synthetase of *Spinacia oleracea* leaves. *Pl. Physiol.*, **69**, 1257–62.

Shimakata, T. & Stumpf, P. K. (1982*b*). The procaryotic nature of the fatty acid synthetase of developing *Carthamus tinctorius* L. (safflower) seeds. *Archs. Biochem. Biophys.*, **217**, 144–54.

Shimakata, T. & Stumpf, P. K. (1982*c*). Purification and characterizations of β-hydroxyacyl-[acyl-carrier-protein]-dehydrase, and enoyl-[acyl-carrier protein]-reductase from *Spinacia oleracea* leaves. *Archs. Biochem. Biophys.*, **218**, 77–91.

Shine, W. E., Mancha, M. & Stumpf, P. K. (1976). The function of acyl-thioesterases in the metabolism of acyl-coenzyme A and acyl-acyl carrier protein. *Archs. Biochem. Biophys.*, **172**, 110–16.

Siebertz, H. P. & Heinz, E. (1977). Labelling experiments on the origin of hexa- and octa-decatrienoic acids in galactolipids from leaves. *Z. Pfl. Physiol.*, **32**, 193–205.

Siebertz, H. P., Heinz, E., Joyard, J. & Douce, R. (1980). Labelling *in vivo* and *in vitro* of molecular species of lipids from chloroplast envelopes and thylakoids. *Eur. J. Biochem.*, **108**, 177–85.

Siebertz, H. P., Heinz, E., Linscheid, M., Joyard, J. & Douce, R. (1979). Characterization of lipids from chloroplast envelopes. *Eur. J. Biochem.*, **101**, 429–38.

Sparace, S. A. & Mudd, J. B. (1982*a*). Studies on chloroplast lipid metabolism: stimulation of phosphatidylglycerol biosynthesis and analysis of the radioactive lipid. In *Biochemistry and Metabolism of Plant Lipids*, ed. J. F. G. M. Wintermans & P. J. C. Kuiper, pp. 111–19. Elsevier Biomedical Press, Amsterdam.

Sparace, S. A. & Mudd, J. B. (1982*b*). Phosphatidylglycerol synthesis in spinach chloroplasts: characterization of the newly synthesized molecule. *Pl. Physiol.*, **70**, 1260–4.

Stuhne-Sekalec, L. & Stanacev, N. Z. (1982). On the mechanism of spontaneous transfer of lipids from isolated microsomal to mitochondrial membranes. *Can. J. Biochem./J. Canad. Biochim.*, **60**, 137–43.

Stumpf, P. K. (1980). Biosynthesis of saturated and unsaturated fatty acids. In *The Biochemistry of Plants*, vol. 4, *Lipids: Structure and Function*, ed. P. K. Stumpf, pp. 177–204. Academic Press, New York.

Stumpf, P. K., Kuhn, D. N., Murphy, D. J., Pollard, M. R., McKeon, T. & MacCarty, J. (1980). Oleic acid, the central substrate. In *Biogenesis and Function of Plant Lipids*, ed. P. Mazliak, P. Benveniste, C. Costes & R. Douce, pp. 3–10. Elsevier/North-Holland Biomedical Press, Amsterdam.

Thiery, J. P. (1967). Mise en évidence des polysaccharides sur coupes fines en microscopie électronique. *J. Microscopie*, **6**, 987–1017.

Van Besouw, A. & Wintermans, J. F. G. M. (1978). Galactolipid formation in chloroplast envelopes. I. Evidence for two mechanisms in galactosylation. *Biochim. biophys. Acta*, **529**, 44–53.

Van Hummel, H. C. (1974). Some observations on biosynthesis and ratio regulation of mono- and di-galactosyl diglycerides by chloroplasts and other subcellular fractions from spinach leaves. *Z. Pfl. Physiol.*, **71**, 228–41.

Vick, B. & Beevers, H. (1978). Fatty acid synthesis in endosperm of young castor bean seedlings. *Pl. Physiol.*, **62**, 173–8.

Weaire, P. J. & Kekwick, R. G. O. (1975). The synthesis of fatty acids in avocado mesocarp and cauliflower bud tissue. *Biochem. J.*, **146**, 425–37.

Whatley, J. M., John, P. & Whatley, F. R. (1979). From extracellular to intracellular: the establishment of mitochondria and chloroplasts. *Proc. R. Soc. Lond. Ser. B*, **204**, 165–87.

Williams, J. P., Khan, M. & Leung, S. P. K. (1975). Biosynthesis of digalactosyl diglyceride in *Vicia faba* leaves. *J. Lipid Res.*, **16**, 61–6.

Williams, J. P., Khan, M. & Mitchell, K. (1982). Galactolipid biosynthesis in *Brassica napus* and *Vicia faba*: a comparison of lipid biosynthesis in 16:3- and 18:3-plants. In *Biochemistry and Metabolism of Plant Lipids*, ed. J. F. G. M. Wintermans & P. J. C. Kuiper, pp. 183–6. Elsevier Biomedical Press, Amsterdam.

Williams, J. P., Simpson, E. E. & Chapman, D. J. (1979). Galactolipid synthesis in *Vicia faba* leaves. III. Site(s) of galactosyl transferase activity. *Pl. Physiol.*, **63**, 669–73.

Williams, J. P., Watson, G. R. & Leung, S. P. K. (1976). Galactolipid synthesis in *Vicia faba* leaves. II. Formation and desaturation of long chain fatty acids in phosphatidylcholine, phosphatidylglycerol, and the galactolipids. *Pl. Physiol.*, **57**, 179–84.

Zilkey, B. F. & Canvin, D. T. (1972). Localization of oleic acid biosynthesis in the proplastids of developing castor bean endosperm. *Can. J. Bot./ J. Canad. Bot.*, **50**, 323–6.

W. RÜDIGER AND J. BENZ

Synthesis of chloroplast pigments

Pigmentation is one of the outstanding characteristics of chloroplasts. The term 'chloroplast' suggests its green colour, which arises predominantly from chlorophylls (greenish-blue) and carotenoids (orange-yellow). These are the main pigments of green plants (including green algae). Brown algae accumulate large amounts of special carotenoids (e.g. fucoxanthin) in the chloroplasts. In addition to chlorophyll *a* and carotenoids, chloroplasts of red algae also contain red and blue pigments and the biliproteins phycoerythrin and phycocyanin. The quantities of all the pigments are such that they contribute considerably to the general pigmentation of the whole plant. The reason for this can be deduced from their function: as 'mass pigments' they harvest sunlight, which is used for photosynthesis.

The light-harvesting function requires a well-defined spatial arrangement of the pigments. On the one hand, the absorption of light should be as effective as possible (maximal absorptive cross-section); this requires a certain minimum distance between the single pigment molecules. On the other hand, the absorbed light energy has to be transferred to the reaction centre with high efficiency. This requires a certain maximal distance between the single pigment molecules which must not be exceeded. Both of these requirements are fulfilled by combination of the pigments with either intrinsic or extrinsic thylakoid membrane proteins. The connection between the spatial arrangement of chromophores and their energy transfer properties has been most successfully elucidated with phycobiliproteins, which occur in aggregates (phycobilisomes) on the surface of the thylakoid membrane (for review see Gantt, 1981).

All the various above-mentioned 'mass pigments' are derived from just two biosynthetic pathways, the tetrapyrrole and the isoprenoid (Table 1). Because both pathways are necessary for chlorophyll biosynthesis, we will first look at how the pathways are interconnected. Specific aspects of the biosynthesis of carotenoids and phycobilins will be treated briefly thereafter.

Chlorophyll biosynthesis
Tetrapyrrole moiety

Early experiments with *Chlorella* mutants (Granick, 1948 *a*, 1948 *b*) suggested that the tetrapyrrole moiety of chlorophyll is formed via protoporphyrin, the direct precursor of haem. Numerous investigations on the enzyme systems confirmed that the first steps of haem and chlorophyll biosynthesis are basically identical. Excellent reviews on the enzymology of chlorophyll biosynthesis are available (Bogorad, 1976; Schneider, 1980; Battersby *et al.*, 1980). Therefore, only some points of recent interest are mentioned here.

The first specific precursor of tetrapyrroles is 5-aminolaevulinic acid or ALA. This can be produced either from succinyl-CoA and glycine (Shemin pathway) or from glutamate (C_5 pathway). The latter pathway (Beale, 1978) strongly predominates in chlorophyll formation (Oh-hama *et al.*, 1982; Porra, Klein & Wright, 1983), although both pathways can operate in algae and

Table 1. *Biosynthetic pathways of chloroplast pigments*

| | % carbon derived from | |
Pigment types	Tetrapyrrole pathway	Isoprenoid pathway
Chlorophylls	64	36
Carotenoids	—	100
Phycobilins	100	—

Fig. 1. Tetrapyrrole pathway of chlorophyll biosynthesis.

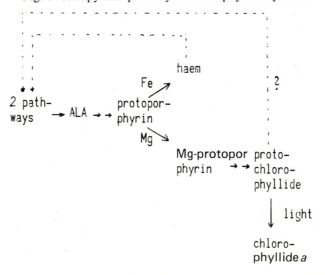

higher plants (Klein & Senger, 1978*a*, 1978*b*; Foley, Dzelzhalns & Beale, 1982). Interestingly, the ALA-forming enzyme system seems to be subject to feedback inhibition by haem (Lascelles, 1968; Gough & Kannangara, 1979; Chereskin & Castelfranco, 1982) and possibly by protochlorophyllide (protochlide) holochrome (Bogorad, 1976; Kasemir, 1979) (see Fig. 1).

After insertion of magnesium into protoporphyrin, a number of modifications of side-chains led to protochlide. These reactions can take place in the dark. We shall return to protochlide accumulation in the dark later. The subsequent, light-dependent reduction to chlorophyllide (chlide *a*) will be treated in detail by Griffiths and Oliver, this volume; this reaction is shown in Fig. 2.

The reaction steps can also be studied with photosynthetic bacteria because bacteriochlorophyllide synthesis follows the same route as chlide formation, but is extended by some more steps. This fact (among others) has been taken as evidence that evolution proceeded from chlorophyll-containing to bacteriochlorophyll-containing thylakoids (Olson, 1981).

The modifications of side-chains are, to some extent, independent of one other, which can give rise to parallel biosynthetic pathways (Rüdiger, 1969). Early reduction of one vinyl group leads to 'monovinyl protoporphyrin' derivatives (Belanger & Rebeiz, 1982), whereas omission of this reduction can lead to 'divinyl' protochlide 1*a* (Jones, 1968), which on photoreduction yields the chlide 2*a* (Duggan & Rebeiz, 1982) (see Fig. 2). The corresponding

Fig. 2. Light-dependent reduction of protochlorophyllide to chlorophyllide *a*.

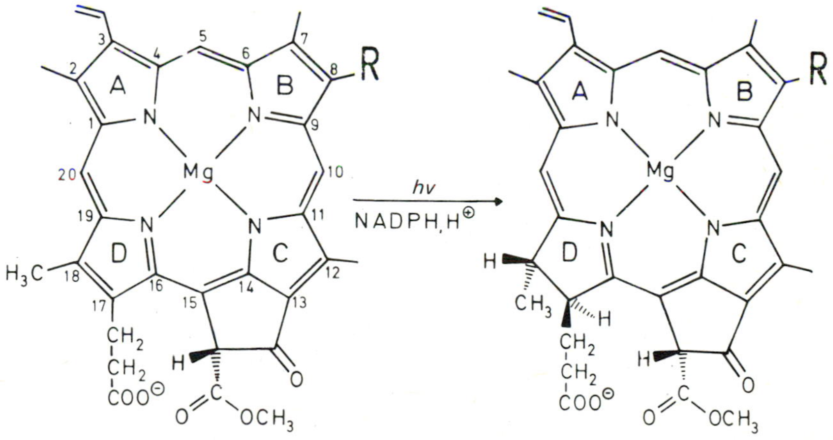

1 : R = ethyl
1*a*: R = vinyl

2 : R = ethyl
2*a*: R = vinyl

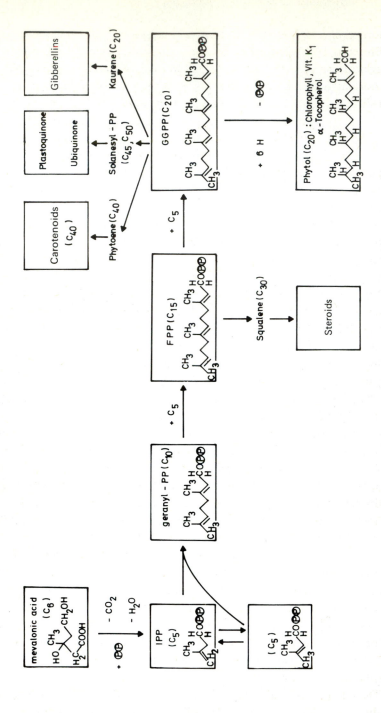

Fig. 3. Pathways of isoprenoid biosynthesis. For abbreviations see text.

chlorophyll *a* has recently been identified in a mutant of *Zea mais* (Bazzaz & Brereton, 1982). Because evidence from fluorescence emission and excitation suggests the presence of such (and possibly still other) chlorophyll species also in normal higher plants (Belanger & Rebeiz, 1980; Rebeiz *et al.*, 1980), parallel chlorophyll pathways may be a general rule (cf. Rebeiz *et al.*, 1981).

Isoprenoid moiety

The isoprenoid moiety of the chlorophylls of algae and higher plants is phytol. It is generally agreed that its biosynthesis follows the main route of isoprenoid synthesis (Fig. 3). The first specific precursor is mevalonic acid. Although the specific incorporation of labelled mevalonic acid into the phytol moiety of chlorophyll in greening barley seedlings was demonstrated 20 years ago (Fischer *et al.*, 1962), detailed knowledge on phytol formation has only recently been obtained.

Isoprenoids are formed from the C_5 precursor isopentenyldiphosphate (IPP) in such a way that one double-bond is retained per C_5 unit, i.e. the C_{15} compound farnesyldiphosphate (FPP) contains three, while the C_{20} compound (geranylgeranyldiphosphate) contains four, double-bonds. Prenyltransferase activity giving rise to GGPP has been demonstrated in plastids either indirectly (Benz & Rüdiger, 1981*a*; Kreuz & Kleinig, 1981) or directly (Block, Joyard & Douce, 1980). The hydrogenation of GGPP to phytol requires six hydrogen-equivalents. Such a hydrogenase activity has recently been demonstrated in chloroplasts (Soll & Schultz, 1981; Soll *et al.*, 1983). An alternative pathway is the hydrogenation of GG after incorporation into chlide *a* (Rüdiger *et al.*, 1976; Schoch, Lempert & Rüdiger, 1977; Schoch, 1978). The latter pathway predominates in greening, etiolated seedlings (Fig. 4). In this case, the intermediates, dihydrogeranylgeraniol and tetra-hydrogeranylgeraniol, have been characterized (Rüdiger *et al.*, 1976; Schoch & Schäfer, 1978). Hydrogen-equivalents are derived from NADPH (Benz, Wolf & Rüdiger, 1980*b*; Soll & Schultz, 1981).

The reason for this hydrogenation, i.e. the necessity to synthesize phytol-containing chlorophyll$_P$ instead of chlorophyll$_{GG}$, is not clear. Phytol is chemically much more stable than geranylgeraniol. It can be speculated that this increased stability of chlorophyll$_P$ over chlorophyll$_{GG}$ has been an advantage in evolution. It should be noted, however, that some bacterio-chlorophylls contain alcohols different from phytol, e.g. farnesol, GG, or tetrahydrogeranylgeraniol, an isomer of the intermediate found in higher plants (Steiner *et al.*, 1981).

The manner of regulation of the isoprenoid pathway is still unknown. The finding of a soluble pool of GGPP in etiolated tissue (Benz, Hampp & Rüdiger, 1981; Benz, Haser & Rüdiger, 1983; Benz, Fischer & Rüdiger, 1983) and the easy penetration of GGPP through the etioplast membrane suggest that this compound may be a feed-back regulator. A membrane-bound phytol pool is discussed below (see Fig. 6).

Esterification of chlorophyllide

Although chlorophyllase has been assumed by many authors to be active in chlorophyll formation (reviews: Bogorad, 1976; Schneider, 1980), the detection of another enzyme activity, termed 'chlorophyll synthetase' (Rüdiger *et al.*, 1977; Rüdiger, Benz & Guthoff, 1980), supports the view that different enzymes are responsible for the biosynthesis and for the degradation of chlorophyll. Chlorophyll synthetase has been found in etioplasts (Rüdiger *et al.*, 1980), in chloroplasts (Block *et al.*, 1980; Soll & Schultz, 1981; Soll *et al.*, 1983), and in chlorophyll-free chromoplasts (Kreuz & Kleinig, 1981). It is localized in prothylakoid fractions of *Avena sativa* etioplasts (Lütz, Benz & Rüdiger, 1981) and in the stromal fraction of chromoplasts (Kreuz & Kleinig, 1981). It is probably a peripheral membrane protein. Chlorophyll deficiency in a mutant of *Arabidopsis thaliana* L. has been demonstrated to be the result of partial deficiency of chlorophyll synthetase (Benz, Kranz & Rüdiger, 1980*a*).

The enzyme specificity was tested with regard to both its isoprenoid and its tetrapyrrole substrates (Table 2; Rüdiger *et al.*, 1980; Benz & Rüdiger, 1981*b*; Soll *et al.*, 1983). The best isoprenoid substrate is GGPP in etioplasts

Fig. 4. Precursors of chlorophyll in greening etiolated seedlings. Chlide = chlorophyllide; Chl = chlorophyll; DHGG = dihydrogeranylgeraniol; THGG = tetrahydrogeranylgeraniol.

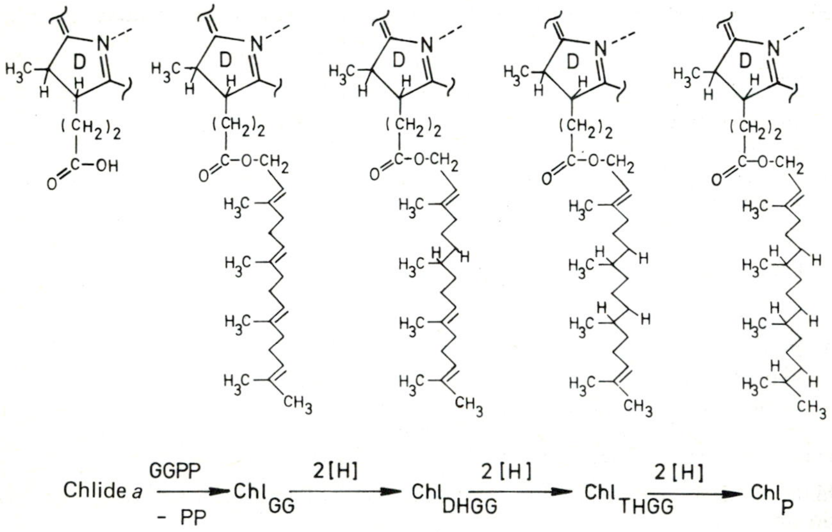

and phytyldiphosphate in chloroplasts (Soll *et al.*, 1983). Because the latter can be formed in chloroplasts from GGPP (Soll & Schultz, 1981; Soll *et al.*, 1983), both compounds are presumably natural substrates. FPP is a poor substrate: the product chlorophyll$_F$ has only been found *in vitro*. Its lack *in vivo* means either that FPP cannot penetrate to the site of chlorophyll synthesis or that prenyltransferase, which forms GGPP from FPP and IPP, has such a high activity that FPP is very rapidly metabolized (Benz & Rüdiger, 1981*a*). The C_{10} compound geraniol (or its diphosphate) does not react with chlorophyll synthetase. Free alcohols are incorporated in the presence of ATP (Rüdiger *et al.*, 1980). The presence of isoprenoid kinase(s) deduced from these experiments was confirmed in experiments on prenyltransferase (Benz & Rüdiger, 1981*a*; see Fig. 5). As tetrapyrrole substrates, chlides *a* and *b* are equally suited (Table 2). This fact, together with the finding of a chlide *b* pool in pea chloroplasts (Aronoff, 1981), suggests that the transformation from the *a* to the *b* series takes place before esterification (cf. Oelze-Karow & Mohr, 1978). But this concept needs still further experimental evidence for verification. Modification of side-chains at rings A or C had little effect on the property of chlides as substrates for chlorophyll synthetase (Benz & Rüdiger, 1981*b*).

Table 2. *Substrate specificity of chlorophyll synthetase*

Broken etioplasts were incubated
(*a*) after illumination (formation of chlide) with various alcohols + ATP;
(*b*) in the dark with various chlides, solubilized with 0.1 % cholate, and 160 nmol GGPP.
Each sample contained 8 nmol pchlide (after Rüdiger *et al.*, 1980; Benz & Rüdiger, 1981*b*)

	Substrate	(nmol)	Esterification (%) of total chlide, increase over control
	Geranylgeraniol	13	45–50
	Phytol	15	48–53
(*a*)	Farnesol	16	7–12
	Geraniol	20–100	0
	n-Pentadecanol	50–250	0
	Chlide *a*	36.0	52
	Chlide *b*	41.1	53
	Acetylchlide *a*	41.0	52
(*b*)	Pyrochlide *a*	10.8	26
	Pheophorbide *a*	27.6	6
	Bacteriochlide *a*	20.5	4
	Pchlide	42.7	0.9

Chlide = chlorophyllide; pchlide = protochlorophyllide; bacteriochlide = bacteriochlorophyllide.

However, the hydrogenation state seems to be essential: neither bacterio-chlorophyllide (hydrogenated ring B) nor protochlide (unsaturated ring D) is a substrate. Removal of magnesium destroys the substrate property. Most of the properties of chlorophyll synthetase differ from those of chlorophyllase (see Table 3).

The chlorophyll synthetase reaction can be used as a tool for 'coupled enzyme tests' in membrane fractions. We were able for example to detect prenylkinase activity in etioplast preparations by this means (Fig. 5). IPP is incorporated into GGPP, and this into chlorophyll$_{GG}$. The specific radio-activity of chlorophyll$_{GG}$ is about three times that of the applied IPP, which is evidence of its self-condensation. Chlorophyll$_{GG}$ was also found in similar experiments with chloroplasts and chromoplasts; in the latter case, as expected, chlorophyll$_{GG}$ had nearly four times the specific activity of IPP (Block *et al.*, 1980; Kreuz & Kleinig, 1981). In our experiments, labelled IPP was applied together with unlabelled farnesol or geraniol in the presence of ATP (Fig. 5). Although the esterification of chlide was increased, the specific radioactivity of the formed chlorophyll$_{GG}$ decreased, as might be expected for a stoichiometric incorporation of the applied alcohols. This result was achieved only in the presence of ATP. The kinase which forms the diphosphates of geraniol and farnesol is certainly different from chlorophyll synthetase because chlorophyll synthetase does not react with geraniol + ATP (Rüdiger *et al.*, 1980).

An interesting observation was made in the chlorophyll synthetase test with

Table 3. *Properties of chlorophyll synthetase and chlorophyllase*

Properties	Chlorophyll synthetase	Chlorophyllase
Activity in phosphate buffer	+ +	(−)
Activity in presence of detergents or organic solvents	−	+ +
Substrates		
Diphosphate derivates of isoprenoid alcohols	+ +	−
Free isoprenoid alcohols without ATP	−	+ +
Alcohols other than C_{20} isoprenoids	(−)	+ +
Pheophorbide, bacteriochlide	−	+ +
Pchlide	−	−
Chlide *a, b*	+ +	+ +

+ +, Activity present; (−), trace of activity; −, no activity.
Chlide = chlorophyllide; pchlide = protochlorophyllide; bacteriochlide = bacterio-chlorophyllide.

FPP (Table 4). Etioplast membranes contain a pool of phytol precursors which are incorporated into chlide immediately after photoconversion of protochlide to chlide. This pool consists of all four C_{20} isoprenoids leading to the four known chlorophylls (see Fig. 4). The main product is chlorophyll$_P$ (8% of photoconverted pigment in the controls), whereas the other three chlorophylls are minor products (together 9%, giving a total of 17% esterification, see Table 4). Incubation with FPP leads to an increase of esterification (up to 43% of photoconverted pigment), which results only partially from the formation of chlorophyll$_F$ (11%), but mainly from

Fig. 5. Some enzyme activities in broken illuminated etioplasts from oat. Detection of chlorophyll synthetase and isoprenoid kinase. Incubation was with [^{14}C]IPP alone or in addition to G (=geraniol+ATP) or F (=farnesol+ATP).
    1. Esterification as percentage of chlide formed by phototransformation.
2. Distribution of chlorophylls (chl) esterified with various alcohols as percentage of total esterified pigment; each column from left to right is chl$_{GG}$, chl$_{DHGG}$, chl$_{THGG}$, chl$_P$ (see Fig. 4). 3. Specific radioactivity in chl$_{GG}$, chl$_{DHGG}$, chl$_{THGG}$ and chl$_P$.

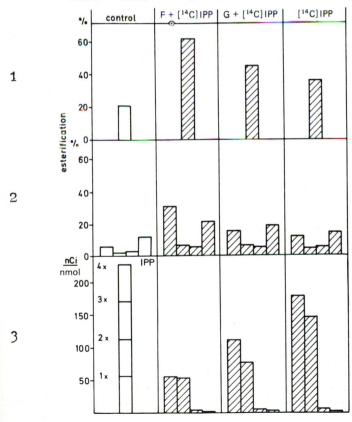

chlorophyll$_P$ (30%). After application of [^{14}C]FPP, chlorophyll$_F$ is labelled but chlorophyll$_P$ is not. This means that the latter is formed from the endogenous phytol pool. Furthermore, the increase of chlorophyll$_P$ after incubation with FPP means that the endogenous phytol pool can be 'mobilized' by FPP. It can also be mobilized by GGPP but to a smaller extent (Fig. 6). The smaller effect of GGPP can perhaps be explained by the substrate specificity of chlorophyll synthetase: GGPP is a better substrate than FPP and can, therefore, effectively compete with the endogenous phytol compound. The nature of this phytol compound is not yet known. We presume that it is similar to phytyldiphosphate because it can be mobilized by FPP, but not

Table 4. *Products of esterification after incubation of membranes with farnesyldiphosphate*

Etioplast membranes were illuminated and incubated with varying amounts of [^{14}C]farnesyldiphosphate. The product chlorophyll$_F$ (Chl$_F$) was labelled, the product chlorophyll$_P$ (Chl$_P$) was unlabelled. All numbers are given as percentage of chlide formed by the illumination (after Benz, 1980).

[^{14}C]farnesyldiphosphate (nmol per nmol chlide)	Products of esterification (% of chlide)			
	Total	(>control)	Chl$_F$ (labelled)	Chl$_P$ (unlabelled)
0 (control)	17	(−)	0	8
2	20	(+3)	3	10
5	29	(+12)	5	14
10	41	(+24)	8	20
20	59	(+42)	10	29
90	60	(+43)	11	30

Chlide = chlorophyllide.

Fig. 6. Detection of membrane-bound phytol pool. PPP = compound which behaves like phytyldiphosphate but is membrane-bound.

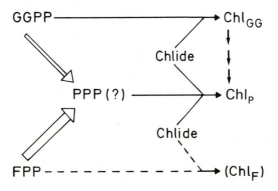

by farnesol with or without ATP. The compound is, however, membrane-bound whereas phytyldiphosphate itself should be water-soluble. It could, however, bind specifically to membrane proteins. Soll & Schultz (1981) used organic solvents to extract phytyldiphosphate from chloroplast preparations.

Carotenoid biosynthesis

Carotenoids have been under investigation for a long time. Many details of their biosynthesis are, therefore, already known (Davies, 1980; Goodwin, 1980). We wish to restrict ourselves here to aspects of our current research.

The biosynthesis of carotenoids is identical with that of phytol (of chlorophyll and other compounds) up to the formation of GGPP. The condensation of two molecules of GGPP yields phytoene, which leads to the various carotenoids by a number of dehydrogenation and cyclization steps. Because all of these reactions should take place in the plastids it is reasonable to look for labelling of carotenoids after application of labelled GGPP or other precursors to plastid preparations.

Our results with broken, illuminated etioplasts (etiochloroplasts) from oat (Table 5) demonstrate that a number of compounds are formed from GGPP. The main product (29% of organic phase) is free GG, which is presumably formed by phosphatase action. Phosphatases are present in the soluble and the particulate fraction of etioplasts (J. Benz and G. Jung, unpublished results). Large amounts of free GG have also been found after incubation of chloroplasts (Block *et al.*, 1980) or chromoplasts (Kreuz & Kleinig, 1981) with IPP, and chloroplasts with mevalonic acid (Buggy, Britton & Goodwin, 1974). The product formed in the second largest amount in our experiment is chlorophyll$_{GG}$ (24%). In terms of yield, the next most important product (17%) is a colourless compound which on saponification gives free GG. We presume that it is a fatty acid ester of GG. Fatty acid esters of phytol have been described before (Fischer & Bohn, 1957; Csupor, 1970; Liljenberg, 1977).

Of special interest is the incorporation of GGPP (11%) into a yellow pigment which co-migrates with β-carotene. It is clearly separated from other yellow pigments and from phytoene. No significant amount of label was found in the position of phytoene. We suggest that our system of broken etiochloroplasts, in which carotenoid biosynthesis has been started by a short illumination, is a good system in which to study biosynthetic reactions of the pathway from phytoene to β-carotene. Cell-free plastid systems which have been described earlier incorporate isoprenoid precursors mainly into phytoene (e.g. Graebe, 1968; Buggy *et al.*, 1974), into lycopene (Subbarayan *et al.*, 1970), and into violaxanthene (Costes *et al.*, 1979), but not into β-carotene.

Incorporation of GGPP and other precursors into β-carotene has been achieved, however, with cell-free preparations from a cyanobacterium (Clarke *et al.*, 1982) and with chromoplast preparations (Beyer, Kreuz & Kleinig, 1980; Camara & Brangeon, 1981; Kreuz & Kleinig, 1981; Kreuz *et al.*, 1982).

Biliprotein biosynthesis

Biosynthesis of phycobilins, the tetrapyrrole chromophores of biliproteins, is apparently identical with that of chlorophylls up to the step of protoporphyrin. This has been deduced mainly from the specific incorporation of ALA into phycocyanobilin in the red alga *Cyanidium caldarium* (Troxler & Brown, 1970; Troxler, Brown & Köst, 1978 *a*; Troxler, Kelly & Brown, 1978 *b*) and in protoplasts of the cyanobacterium *Anacystis nidulans* (Cosner & Troxler, 1978). The specific incorporation of haem into phycocyanobilin (Brown *et al.*, 1981; Schuster *et al.*, 1983) demonstrates that phycobilins are members of the iron branch and not of the magnesium branch of tetrapyrrole biosynthesis (see Fig. 1). The chlorophyll was not labelled in these experiments. That the ring of the plant porphyrin macrocyclus opens in the same way as the mammalian compound has indirectly been deduced from the stoichio-

Table 5. *Isotope distribution in the organic phase after incubation of broken, illuminated etioplasts with GGPP*

Samples containing 10 nmol l^{-1} chlorophyll *a* were incubated with 100 nmol l^{-1} GGPP according to Rüdiger *et al.*, 1980. The organic phase which contained 17% of the applied radioactivity was separated on a sugar column with petroleum ether/n-propanol (99.5:0.5, v/v) yielding fractions *a*, *b*, *c* which were then further separated by thin-layer chromatography. Numbers are percentage radioactivity of the organic phase (data from Benz, 1980).

Fraction of sugar column		*a*			*b*		*c*	
Radioactivity		41			46		13	
Thin-layer chromatography		a_1	a_2	a_3	b_1	b_2	c_1	c_2
Radioactivity recovered		20	11	1.5	18	24	11	6
Radioactivity in GG after saponification		17	—	1	14	24	—	—
Identification		free GG	a_2*	?	GG- ester	Chl$_{GG}$	?	?

* a_2 co-migrates with β-carotene in three solvent systems on silicagel G:
  (1) CCl$_4$/acetone (9:1, v/v), R_F = 0.75–0.80.
  (2) Benzene/ethylacetate/methanol (75:20:5, v/v), R_F = 0.85–0.90.
  (3) Petroleum ether/diethylether (99.5:5, v/v), R_F = 0.17–0.20.
GG = geranylgeraniol; Chl = chlorophyll.

metric formation of carbon monoxide and phycocyanobilin (Troxler, 1972; Troxler & Dokos, 1973).

The effective incorporation of ALA into the tetrapyrrole pigments in *Cyanidium caldarium* facilitates biosynthetic studies. We found recently the whole series from octacarboxylic (uro-) to dicarboxylic (proto-) porphyrins formed from ALA in this organism (Bauer, 1980). The corresponding porphyrinogens can be considered as intermediates in chlorophyll and phycobilin biosynthesis. Among the bile pigments formed from ALA, biliverdin (Köst & Benedikt, 1982) is of especial interest. This compound is the expected first product of the opening of the haem ring. It is believed to be the direct precursor of phycocyanobilin (Fig. 7).

Contrary to the binding of chlorophylls and carotenoids, phycobilins are covalently linked to cysteine residues of the protein moiety of biliproteins. The nature and stereochemistry of the thioether linkage has been worked out using both chemical and spectroscopic methods (Rüdiger, 1980). The connection of phycobilins with the apoprotein via this thioether bond is presumably the last step of biliprotein biosynthesis (Fig. 7). Because not all cysteine residues of the peptide chains are linked to phycobilins (Frank *et al.*, 1978; Freidenreich, Apell & Glazer, 1978; Offner *et al.*, 1981; Troxler *et al.*, 1981), and because the reaction of free phycocyanobilin with thiols is not a straightforward reaction (Klein & Rüdiger, 1979), it can be concluded that the formation of the thioether linkage is under enzymatic control. Neither this enzyme nor the apoprotein has so far been detected or characterized.

Fig. 7. Presumed pathway of phycocyanin biosynthesis from biliverdin (*left*) via phycocyanobilin (*bottom*) to phycocyanin (*right*).

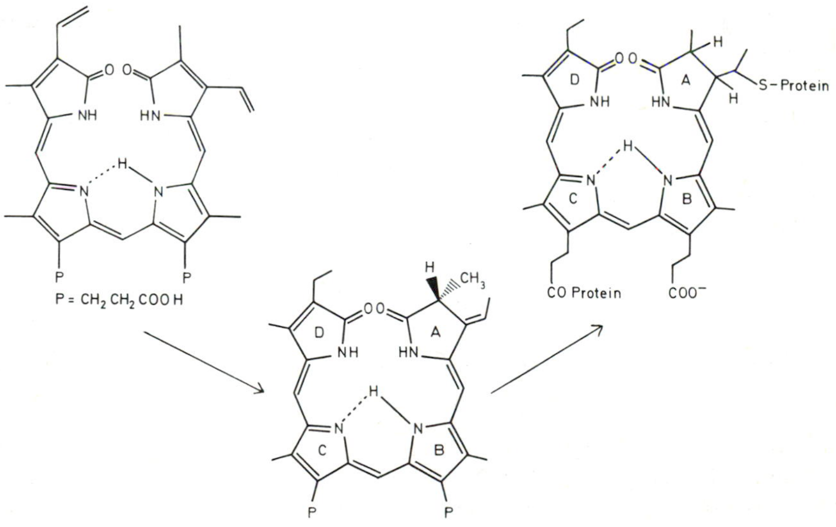

Control of pigment synthesis by light

The 'mass pigments' of chloroplasts function to harvest light energy, so they are not needed in the dark. It is, therefore, not surprising that many (but not all) photosynthetic organisms have developed mechanisms for the regulation of pigment biosynthesis by light.

The best-known example is perhaps *chlorophyll* biosynthesis in angiosperms. Biosynthesis without light stops at protochlide, which is accumulated during prolonged growth of seedlings in the dark. This condition ('etiolated') has very often been used experimentally. One has to be aware that this is a special physiological situation; extrapolations from etiolated to green tissues have to be made with caution.

Two points at which light regulation operates are known in chlorophyll biosynthesis: protochlide reductase (see Griffiths and Oliver, this volume), which needs stoichiometric amounts of light quanta, and the phytochrome system which operates after a red light pulse. The latter (as P_{fr}) increases the capacity of the chlorophyll pathway at the level of ALA synthesis (Mohr, 1980), but also induces the synthesis of the apoprotein of the light-harvesting chlorophyll a/b protein complex (Apel, 1979; Apel & Kloppstech, 1980). Whether the presence of this protein – which can be considered as one of the receptor proteins for esterified chlorophyll molecules – influences the biosynthesis of chlorophyll is still unknown. In many algae, the blue-light receptor cryptochrome substitutes for phytochrome in its action on chlorophyll biosynthesis. Regulation of chlorophyll biosynthesis does not involve light control of chlorophyll synthetase or of the hydrogenase which forms phytol. This can be shown by assaying the corresponding reactions in the dark with etioplast membranes *in vitro* (Benz & Rüdiger, 1981b) or with etiolated leaves *in vivo* after infiltration of chlide (Vezitskii & Walter, 1981), and has also been concluded from irradiation experiments with intact plants (Kasemir & Prehm, 1976). The light-dependency of phytol accumulation (Steffens *et al.*, 1976) is an indirect regulation of protochlide reductase because no appreciable amounts of free phytol are accumulated. Excess phytol which accumulates as fatty acid ester presumably comes from photodestruction of newly formed chlorophyll (Steffens *et al.*, 1976).

Light control of *carotenoid* biosynthesis has been comprehensively reviewed (Rau & Schrott, 1979; Harding & Shropshire, 1980; Rau, 1980, 1983). Although strict photoregulation seems to be restricted to some fungi and bacteria, light also stimulates carotenoid biosynthesis in algae and higher plants. It is not yet known exactly which steps of carotenoid biosynthesis are under light control. But it is evident from the foregoing results on esterification of chlide that carotenoid-specific light control should occur at steps after

GGPP. A rather simple means of control could be the availability of those thylakoid proteins to which the carotenoids must be bound. Light control of synthesis of these proteins would then control carotenoid accumulation. Such a dependency cannot be reversed: membrane assembly and chlorophyll accumulation are still possible when carotenoid synthesis is blocked by the herbicide SAN 9789 (Frosch *et al.*, 1979).

Photoregulation of *biliprotein* synthesis is best known as 'chromatic adaption' in many cyanobacteria, in which either phycoerythrin or phyco-cyanin is preferentially accumulated according to the colour of the light (Bogorad, 1975; Tandeau de Marsac, Castets & Cohen-Bazire, 1980). But light is also required for phycocyanin synthesis in *Cyanidium caldarium*, where it induces the formation of ALA (Schneider & Bogorad, 1979). However phycocyanobilin is formed in the dark after application of ALA (Troxler *et al.*, 1978*b*), so it is clear that regulation of phycocyanin synthesis can be overcome by exogenous ALA. However, because no phycocyanin is formed under these conditions, light must also somehow induce either the formation of the apoprotein or the linkage between phycocyanobilin and the apoprotein. How this may be achieved has still to be elucidated.

References

Apel, K. (1979). Phytochrome-induced appearance of mRNA activity for the apoprotein of the light-harvesting *a/b* protein. *Eur. J. Biochem.*, **97**, 183–8.

Apel, K. & Kloppstech, K. (1980). The effect of light on the biosynthesis of the light-harvesting chlorophyll *a/b* protein. *Planta*, **150**, 426–30.

Aronoff, S. (1981). Chlorophyllide *b*. *Biochem. Biophys. Res. Commun.*, **102**, 108–12.

Battersby, A. R., Fookes, C. J. R., Matcham, G. W. J. & McDonald, E. (1980). Biosynthesis of the pigments of life: formation of the macrocycle. *Nature*, **285**, 17–21.

Bauer, M. (1980). Aufnahme von Aminolaevulinsäure und Ausscheidung von Tetrapyrrolen durch die Rotalge *Cyanidium caldarium*. *Zulassungs-sarbeit*, pp. 1–76. Botanisches Institut, Universität München.

Bazzaz, M. B. & Brereton, R. G. (1982). 4-Vinyl-4-desethyl chlorophyll *a*: a new naturally occurring chlorophyll. *FEBS Lett.*, **138**, 104–8.

Beale, S. I. (1978). δ-Aminolevulinic acid in plants: its biosynthesis, regulation and role in plastid development. *A. Rev. Pl. Physiol.*, **29**, 95–120.

Belanger, F. C. & Rebeiz, C. A. (1980). Chloroplast biogenesis. 30. Chloro-phyll(ide) (E459F675) and chlorophyll(ide) (E449F675). The first detect-able products of divinyl and monovinyl protochlorophyll photoreduction. *Pl. Sci. Lett.*, **18**, 343–50.

Belanger, F. C. & Rebeiz, C. A. (1982). Chloroplast biogenesis. Detection of monovinyl magnesium-protoporphyrin monoester and other monovinyl magnesium-porphyrins in higher plants. *J. biol. Chem.*, **257**, 1360–71.

Benz, J. (1980). Untersuchungen zur Chlorophyll-Biosynthese in etiolierten Haferkeimlingen: Veresterung von Chlorophyllid *in vitro*. Ph.D. thesis. Universität München.

Benz, J., Fischer, I. & Rüdiger, W. (1983). Determination of phytyl diphosphate and geranylgeranyl diphosphate in etiolated oat seedlings (*Avena sativa* L.). *Phytochemistry*, **22**, 2801–4.

Benz, J., Hampp, R. & Rüdiger, W. (1981). Chlorophyll biosynthesis by mesophyll protoplasts and plastids from etiolated oat (*Avena sativa* L.) leaves. *Planta*, **152**, 54–8.

Benz, J., Haser, A. & Rüdiger, W. (1983). Changes in the endogenous pools of tetraprenyl diphosphates in etiolated oat seedlings after irradiation. *Z. Pfl. Physiol.*, **11**, 349–56.

Benz, J., Kranz, A. R. & Rüdiger, W. (1980*a*). Partial deficiency of chlorophyll synthetase in the mutant IM_{28} of *Arabidopsis thaliana*. *Arabidop. Inf. Ser.*, **17**, 33–8.

Benz, J. & Rüdiger, W. (1981*a*). Incorporation of 1-^{14}C-isopentenyldiphosphate, geraniol and farnesol into chlorophyll in plastid membrane fractions of *Avena sativa* L. *Z. Pfl. Physiol.*, **102**, 95–100.

Benz, J. & Rüdiger, W. (1981*b*). Chlorophyll biosynthesis: various chlorophyllides as exogenous substrates for chlorophyll synthetase. *Z. Naturforsch.*, **36c**, 51–7.

Benz, J., Wolf, Ch. & Rüdiger, W. (1980*b*). Chlorophyll biosynthesis: hydrogenation of geranyl-geraniol. *Pl. Sci. Lett.*, **19**, 225–30.

Beyer, P., Kreuz, K. & Kleinig, H. (1980). β-Carotene synthesis in isolated chromoplasts from *Narcissus pseudonarcissus*. *Planta*, **150**, 435–8.

Block, M. A., Joyard, J. & Douce, R. (1980). Site of synthesis of geranylgeraniol derivatives in instant spinach chloroplasts. *Biochim. biophys. Acta*, **631**, 210–19.

Bogorad, L. (1975). Phycobiliproteins and complementary chromatic adaption. *A. Rev. Pl. Physiol.*, **26**, 369–401.

Bogorad, L. (1976). Chlorophyll biosynthesis. In *Chemistry and Biochemistry of Plant Pigments*, vol. 1, ed. T. W. Goodwin, pp. 64–148. Academic Press, London.

Brown, S. B., Holroyd, A. J., Troxler, R. F. & Offner, G. D. (1981). Bile pigment synthesis in plants. Incorporation of haem into phycocyanobilin and phycobiliproteins in *Cyanidium caldarium*. *Biochem. J.*, **194**, 137–47.

Buggy, M. J., Britton, G. & Goodwin, T. W. (1974). Terpenoid biosynthesis by chloroplasts isolated in organic solvents. *Phytochemistry*, **13**, 125–9.

Camara, B. & Brangeon, J. (1981). Carotenoid metabolism during chloroplast to chromoplast transformation in *Capsicum annuum* fruit. *Planta*, **151**, 359–64.

Chereskin, B. M. & Castelfranco, P. A. (1982). Effects of iron and oxygen on chlorophyll biosynthesis. II. Observations on the biosynthetic pathway in isolated etiochloroplasts. *Pl. Physiol.*, **68**, 112–16.

Clarke, I. E., Sandmann, G., Bramley, P. M. & Böger, P. (1982). Carotene biosynthesis with isolated photosynthetic membranes. *FEBS Lett.*, **140**, 203–6.

Cosner, J. C. & Troxler, R. F. (1978). Phycobiliprotein synthesis in protoplasts of the unicellular cyanophyte, *Anacystis nidulans. Biochim. biophys. Acta*, **519**, 474–88.

Costes, C., Burghoffer, C., Joyard, J., Block, M. & Douce, R. (1979). Occurrence and biosynthesis of violaxanthin in isolated spinach chloroplast envelope. *FEBS Lett.*, **103**, 17–21.

Csupor, L. (1970). Das Phytol in vergilbten Blättern. *Planta Med.*, **19**, 37–41.

Davies, B. H. (1980). Carotenoid biosynthesis. In *Pigments in Plants*, 2nd edn, ed. F.-C. Czygan, pp. 31–56. G. Fischer Verlag, Stuttgart and New York.

Duggan, J. W. & Rebeiz, C. A. (1982). Chloroplast biogenesis 37. Induction of chlorophyllide *a* (E459F675) accumulation in higher plants. *Pl. Sci. Lett.*, **24**, 27–37.

Fischer, F. G. & Bohn, H. (1957). Über die Bildung und das Vorkommen von Phytol. *Liebigs Annalen der Chemie*, **611**, 224–35.

Fischer, F. G., Märkl, G., Hönel, H. & Rüdiger, W. (1962). Bildung und Vorkommen von Phytol. III. Einbau von Essigsäure- und Mevalonsäure-[2-¹⁴C] in Chlorophyll, Sterine und Carotinoide von Gerstenkeimlingen. *Liebigs Annl Chem.*, **657**, 199–212.

Foley, T., Dzelzhalns, V. & Beale, S. I. (1982). δ-Aminolevulinic acid synthase of *Euglena gracilis*: regulation of activity. *Pl. Physiol.*, **70**, 219–26.

Frank, G., Sidler, W., Widmer, H. & Zuber, H. (1978). The complete amino sequence of both subunits of C-phycocyanin from the cyanobacterium *Matigocladus laminosus. Hoppe-Seyler's Z. physiol. Chem.*, **359**, 1491–507.

Freidenreich, P., Apell, G. S. & Glazer, A. N. (1978). Structural studies on phycobiliproteins. *J. biol. Chem.*, **253**, 212–19.

Frosch, S., Jabben, M., Bergfeld, R., Kleinig, H. & Mohr, H. (1979). Inhibition of carotenoid biosynthesis by the herbicide SAN 9789 and its consequences for the action of phytochrome on plastogenesis. *Planta*, **145**, 497–505.

Gantt, E. (1981). Phycobilisomes. *A. Rev. Pl. Physiol.*, **32**, 327–47.

Goodwin, T. W. (1980). *The Biochemistry of the Carotenoids*, 2nd edn, vol. 1. Chapman and Hall, London and New York.

Gough, S. P. & Kannangara, C. G. (1979). Biosynthesis of δ-aminolevulinate in greening barley leaves. III. The formation of δ-aminolevulinate in *trigina* mutants of barley. *Carlsberg Res. Commun.*, **44**, 403–16.

Graebe, J. E. (1968). Biosynthesis of kaurene, squalene and phytoene from mevalonate-(2-¹⁴C) in a cell-free system from pea fruits. *Phytochemistry*, **7**, 2003–20.

Granick, S. (1948*a*). Protoporphyrin *a* as a precursor of chlorophyll. *J. biol. Chem.*, **172**, 717–27.

Granick, S. (1948*b*). Magnesium protoporphyrin as a precursor of chlorophyll in *Chlorella. J. biol. Chem.*, **175**, 333–42.

Harding, R. W. & Shropshire, Jr, W. (1980). Photocontrol of carotenoid biosynthesis. *A. Rev. Pl. Physiol.*, **31**, 217–38.

Jones, O. T. G. (1968). Biosynthesis of chlorophylls. In *Porphyrins and Related Compounds*, ed. T. W. Goodwin, pp. 131–45. Academic Press, London and New York.

Kasemir, H. (1979). Control of chloroplast formation by light. *Cell Biol. int. Rep.*, **3**, 197–214.

Kasemir, H. & Prehm, G. (1976). Control of chlorophyll by phytochrome. III. Does phytochrome regulate the chlorophyllide esterification in mustard seedlings? *Planta*, **132**, 291–5.

Klein, G. & Rüdiger, W. (1979). Thioether formation of phycocyanobilin: a model reaction of phycocyanin biosynthesis. *Z. Naturfors.*, **34c**, 192–5.

Klein, O. & Senger, H. (1978a). Two biosynthetic pathways to δ-aminolevulinic acid in a pigment mutant of the green alga, *Scenedesmus obliquus*. *Pl. Physiol.*, **62**, 10–13.

Klein, O. & Senger, H. (1978b). Biosynthetic pathways to δ-ALA induced by blue light in the pigment mutant C-2A′ of *Scenedesmus obliquus*. *Photochem. Photobiol.*, **27**, 203–8.

Köst, H.-P. & Benedikt, E. (1982). Biliverdin IXa, intermediate and end product of tetrapyrrole biosynthesis. *Z. Naturf.*, **37c**, 1057–63.

Kreuz, K., Beyer, P. & Kleinig, H. (1982). The site of carotenogenic enzymes in chloroplasts from *Narcissus pseudonarcissus* L. *Planta*, **154**, 66–9.

Kreuz, K. & Kleinig, H. (1981). Chlorophyll synthetase in chlorophyll-free chromoplasts. *Pl. Cell Rep.*, **1**, 40–2.

Lascelles, J. (1968). The regulation of haem and chlorophyll synthesis. In *Porphyrins and Related Compounds*, ed. T. W. Goodwin, pp. 49–59. Academic Press, London.

Liljenberg, C. (1977). The occurrence of phytylpyrophosphate and acylester of phytol in irradiated dark grown barley seedlings and their role in the biosynthesis of chlorophyll. *Physiologia Pl.*, **39**, 101–5.

Lütz, C., Benz, J. & Rüdiger, W. (1981). Esterification of chlorophyllide in prolamellar body (PLB) and prothylakoid (PT) fractions from *Avena sativa* etioplasts. *Z. Naturfors.*, **36c**, 58–61.

Mohr, H. (1980). Light and pigments. In *Pigments in Plants*, 2nd edn, ed. F. C. Czygan, pp. 7–30. G. Fischer Verlag, Stuttgart and New York.

Oelze-Karow, H. & Mohr, H. (1978). Control of chlorophyll *b* biosynthesis by phytochrome. *Photochem. Photobiol.*, **27**, 189–93.

Offner, G. D., Brown-Mason, A. S., Ehrhardt, M. M. & Troxler, R. F. (1981). Primary structure of phycocyanin from the unicellular rhodophyte *Cyanidium caldarium*. *J. biol. Chem.*, **256**, 12167–75.

Oh-hama, T., Seto, H., Otake, N. & Miyachi, S. (1982). ^{13}C-NMR evidence for the pathway of chlorophyll biosynthesis in green algae. *Biochem. Biophys. Res. Commun.*, **105**, 647–52.

Olson, J. M. (1981). Evolution of photosynthetic reaction centers. *BioSystems*, **14**, 89–94.

Porra, R. J., Klein, O. & Wright, P. E. (1983). The proof by ^{13}C-NMR spectroscopy of the predominance of the C_5 pathway over the Shemin pathway in chlorophyll biosynthesis in higher plants and the formation of the methyl ester group of chlorophyll from glycine. *Eur. J. Biochem.*, **130**, 509–16.

Rau, W. (1980). Photoregulation of carotenoid biosynthesis: an example of photomorphogenesis. In *Pigments in Plants*, 2nd edn, ed. F.-C. Czygan, pp. 80–103. G. Fischer Verlag, Stuttgart and New York.

Rau, W. (1983). Photoregulation of carotenoid biosynthesis. In *Biosynthesis of Isoprenoid Compounds*, ed. J. W. Porter & S. L. Spurgeon, vol. 2, pp. 123–57. Wiley, New York.

Rau, W. & Schrott, E. L. (1979). Light-mediated biosynthesis in plants. *Photochem. Photobiol.*, **30**, 755–65.

Rebeiz, C. A., Belanger, F. C., Freyssinet, G. & Saab, D. G. (1980). Chloroplast biogenesis. XXIX. The occurrence of several novel chlorophyll *a* and *b* chromophores in higher plants. *Biochim. biophys. Acta*, **590**, 234–47.

Rebeiz, C. A., Belanger, F. C., McCarthy, S. A., Freyssinet, G., Duggan, J. X., Wu, S.-M. & Mattheis, J. R. (1981). Biosynthesis and accumulation of novel chlorophyll *a* and *b* chromophoric species in green plants. In *Photosynthesis V*, ed. G. Akoyunoglou, pp. 197–212. Balaban International Science Services, Philadelphia.

Rüdiger, W. (1969). Biosynthese von Chlorophyllen. *Chemiker-Zeitung*, **93**, 664–8.

Rüdiger, W. (1980). Plant biliproteins. In *Pigments in Plants*, 2nd edn, ed. F.-C. Czygan, pp. 314–51. G. Fischer Verlag, Stuttgart and New York.

Rüdiger, W., Benz, J. & Guthoff, C. (1980). Detection and partial characterization of activity of chlorophyll synthetase in etioplast membranes. *Eur. J. Biochem.*, **109**, 193–200.

Rüdiger, W., Benz, J., Lempert, U., Schoch, S. & Steffens, D. (1976). Hemmung der Phytolakkumulation mit Herbiziden. Geranylgeraniol und Dihydrogeranylgeraniol-haltiges Chlorophyll aus Weizenkeimlingen. *Z. Pfl. Physiol.*, **80**, 131–43.

Rüdiger, W., Hedden, P., Köst, H.-P. & Chapman, D. J. (1977). Esterification of chlorophyllide by geranylgeranyl pyrophosphate in a cell-free system from maize shoots. *Biochem. biophys. Res. Commun.*, **74**, 1268–72.

Schneider, Hj. (1980). Chlorophyll biosynthesis: enzymes and regulation of enzyme activities. In *Pigments in Plants*, 2nd edn, ed. F.-C. Czygan, pp. 237–307. G. Fischer Verlag, Stuttgart and New York.

Schneider, Hj. & Bogorad, L. (1979). Spectral response curves for the formation of phycobiliproteins, chlorophyll and δ-aminolevulinic acid in *Cyanidium caldarium*. *Z. Pfl. Physiol.*, **94**, 449–59.

Schoch, S. (1978). The esterification of chlorophyllide *a* in greening bean leaves. *Z. Naturfors.*, **33c**, 712–14.

Schoch, S., Lempert, U. & Rüdiger, W. (1977). On the last steps of chlorophyll biosynthesis: intermediates between chlorophyllide and phytol-containing chlorophyll. *Z. Pfl. Physiol.*, **83**, 427–36.

Schoch, S. & Schäfer, W. (1978). Tetrahydrogeranylgeraniol, a precursor of phytol in the biosynthesis of chlorophyll *a* – localization of the double bonds. *Z. Naturfors.*, **33c**, 408–12.

Schuster, A., Köst, H.-P., Rüdiger, W., Holroyd, J. A. & Brown, S. B. (1983). Incorporation of haem into phycocyanobilin in levulinic acid-treated *Cyanidium caldarium*. *Pl. Cell Rep.*, **2**, 85–7.

Soll, J. & Schultz, G. (1981). Phytol synthesis from geranylgeraniol in spinach chloroplasts. *Biochem. biophys. Res. Commun.*, **99**, 907–12.

Soll, J., Schultz, G., Rüdiger, W. & Benz, J. (1983). Hydrogenation of geranylgeraniol. Two pathways exist in spinach chloroplasts. *Pl. Physiol.*, **71**, 849–54.

Steffens, D., Blos, I., Schoch, S. & Rüdiger, W. (1976). Lichtabhängigkeit der Phytolakkumulation. *Planta*, **130**, 151–8.

Steiner, R., Schäfer, W., Blos, I., Wieschoff, H. & Scheer, H. (1981). Δ2,10-Phytadienol as esterifying alcohol of bacteriochlorophyll *b* from *Ectothiorhodospira halochloris*. *Z. Naturfors.*, **36c**, 417–20.

Subbarayan, C., Kushwaha, S. C., Suzue, G. & Porter, J. W. (1970). Enzymatic conversion of isopentenyl pyrophosphate-4-^{14}C and phytoene-^{14}C to acyclic carotenes by an ammonium sulfate-precipitated spinach enzyme system. *Archs. Biochem. Biophys.*, **137**, 547–57.

Tandeau de Marsac, N., Castets, A. M. & Cohen-Bazire, G. (1980). Wavelength modulation of phycoerythrin synthesis in *Synechocystis* sp. 6701. *J. Bacteriol.*, **142**, 310–14.

Troxler, R. F. (1972). Synthesis of bile pigments in plants. Formation of carbon monoxide and phycocyanobilin in wild-type and mutant strains of alga *Cyanidium caldarium*. *Biochemistry*, **11**, 4235–42.

Troxler, R. F. & Brown, A. (1970). Biosynthesis of phycocyanin in vivo. *Biochim. biophys. Acta*, **215**, 503–11.

Troxler, R. F., Brown, A. S. & Köst, H.-P. (1978*a*). Quantitative degradation of radiolabelled phycobiliproteins. *Eur. J. Biochem.*, **87**, 181–9.

Troxler, R. F. & Dokos, J. M. (1973). Formation of carbon monoxide and bile pigment in red and blue-green algae. *Pl. Physiol.*, **51**, 72–5.

Troxler, R. F., Ehrhardt, M. M., Brown-Mason, A. S. & Offner, G. D. (1981). Primary structure of phycocyanin from the unicellular rhodophyte *Cyanidium caldarium*. *J. biol. Chem.*, **256**, 12176–84.

Troxler, R. F., Kelly, P. & Brown, S. B. (1978*b*). Phycocyanobilin synthesis in the unicellular rhodophyte *Cyanidium caldarium*. *Biochem. J.*, **172**, 569–76.

Vezitskii, A. Yu. & Walter, G. (1981). Alcoholic component of chlorophyll *a* synthesized in darkness from infiltrated chlorophyllide in etiolated leaves of Rye (*Secale cereale* L.). *Photosynthetica*, **15**, 104–8.

W. T. GRIFFITHS AND R. P. OLIVER

Protochlorophyllide reductase – structure, function and regulation

Chlorophyll biosynthesis by higher plants is light-requiring, in contrast to the process in most algae and gymnosperms which can produce chlorophyll in darkness. Two steps in the chlorophyll biosynthetic pathway have now been identified as being light-mediated in higher plants, viz. the systems involved respectively with the synthesis of 5-aminolaevulinic acid (ALA) and with the reduction of protochlorophyllide (pchlide) to chlorophyllide (chlide). The synthesis of ALA is only indirectly influenced by light, with illumination required for actual formation or derepression of the 'ALA-synthetase' system. In contrast, pchlide reduction has an obligate direct requirement for light and as such has fascinated photobiologists for a long time. Not only is this process important as a key step in chlorophyll formation but it is also probably of immense significance to photomorphogenesis in higher plants.

From early chemical analyses, pchlide was established as the precursor of chlorophyll which accumulates in dark-grown plants, and which on illumination is converted to chlide by reduction of the C7–C8 double bond of the porphyrin (Fig. 1). Pchlide and chlide have very different light-absorbing properties, the former absorbing maximally at 623 nm in ether compared with chlide having a corresponding red absorption maximum in the same solvent at 663 nm. Despite the fact that *in vivo* the absorption maxima of the compounds are red-shifted to varying degrees, it is nevertheless still possible to observe the photoconversion reaction by spectroscopic analysis of illuminated etiolated leaves (Fig. 2). Extensive use of such techniques over the past 30 years or so has identified different spectral forms of pchlide and chlide *in vivo* displaying different roles in the photoconversion process. The most widely studied of these absorbing forms include the non-photoactive and photoactive form of pchlide with absorption maxima at approximately 630 and 638/650 nm respectively (Kahn, Boardman & Thorne, 1970); the early transient photoproduct chlide form, chlide 678 nm (Gassman, Granick & Mauzerall, 1968; Bonner, 1969); the first stable product formed from this in darkness, chlide 684 nm (Shibata, 1957); and finally chlide 672 nm, which is formed from chlide 684 by a slower dark reaction (see Fig. 2). Apart from identifying such pigment complexes and their sequential involvement in pchlide photoconversion, the spectroscopic work *in vivo* offered no information either on the nature of these complexes or on the mechanism of the photoconversion process. This early work did, however, lead to the suggestion of a 'photoenzyme' which functioned repeatedly during the photoconversion process. During these investigations *in vivo*, parallel studies were carried out on photoconversion *in vitro* to try to provide better characterisation of the process. In this respect, the first cell-free preparations retaining pchlide phototransformation activity were made by Smith (1952) and Krasnovsky & Kosobutskaya (1952), using etiolated leaves of barley or bean respectively as the starting material. This led to the introduction of the term pchlide-holochrome, by Smith (1952), to describe the photoactive pchlide complex found *in vitro* and presumably also occurring *in vivo*. Analysis of the preparation *in vitro* led to a molecular weight determination of $c. 0.5 \times 10^6$ for the protein of the holochrome and functional studies by Boardman (1962) led to the conclusion that only the native pchlide complex is required for photoconversion, with no evidence of a dialysable cofactor or reduced nucleotide as hydrogen donor for the pigment reduction. Progress in the isolation of the holochrome resulted in the purification of a type with molecular weight 0.3×10^6 by Schopfer & Siegelman (1968) from Triton X-100

Fig. 1. The phototransformation of protochlorophyllide to chlorophyllide. ALA = 5-aminolaevulinic acid.

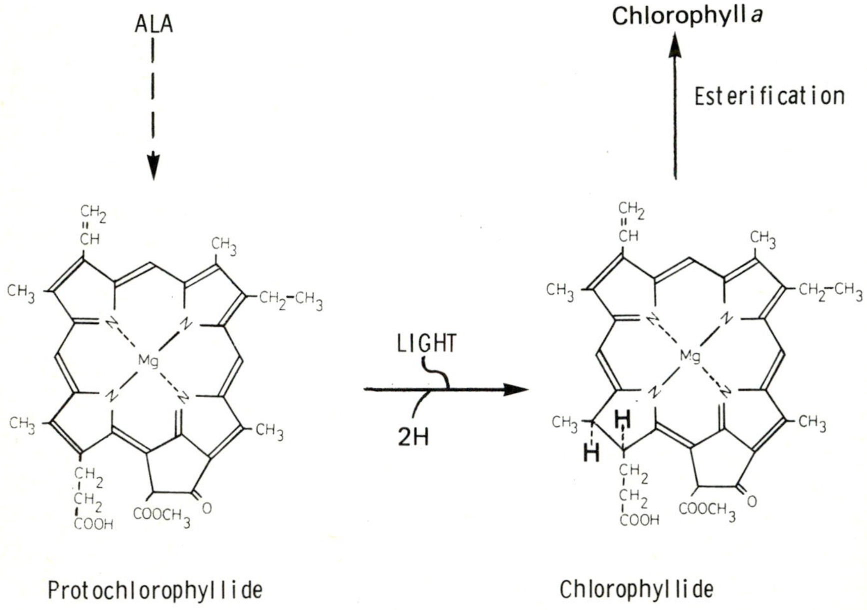

Protochlorophyllide

Chlorophyllide

extracts of etiolated bean leaves, and subsequently led to the identification of individual subunits (mol. wt 63000) of the complex using the natural detergent saponin (Henningsen & Kahn, 1971). The absence of energy transfer between pigment molecules in the saponin preparation indicated that each of the 63000 Dalton subunits bears a single chromophore. Despite this progress towards holochrome purification no unambiguous identification of the basic peptide of the holochrome subunit (e.g. by sodium dodecylsulphate–polyacrylamide gel electrophoresis (SDS–PAGE)) was achieved and inconsistent data on this point prevailed at this time in the literature (Canaani & Sauer, 1977; Guignery, Luzzati & Duranton, 1974; Redlinger & Apel, 1980).

Fig. 2. Spectral properties of pigment complexes in etiolated leaves before and after illumination. Spectra of 7-day-old etiolated oat leaves were recorded before (i), immediately after a brief (10 s) illumination (ii), and following incubation in darkness for a further 30 s (iii) and 20 min (iv). λ = wavelength.

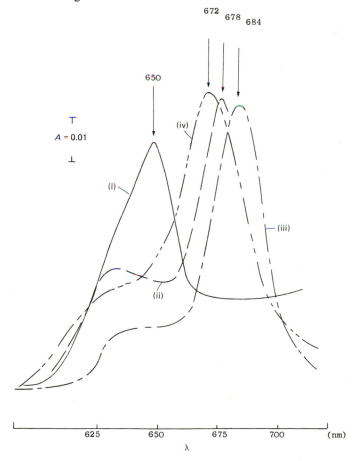

Again, while spectroscopic studies *in vivo* had adequately described the spectroscopic intermediates involved in the photoreduction process, no structural explanation of the various forms or sequence of events had emerged from this work. In particular, the source of hydrogen for the pchlide reduction remained a mystery. A new strategy, viz. the use of isolated etioplasts, had to be employed to provide the answers to these questions.

Phototransformation by isolated etioplasts

Etioplasts are chloroplast precursors formed from proplastids in plants grown for prolonged periods in darkness. As such they contain the pchlide-reducing system associated with the organellar membrane system. The original isolation of etioplasts still retaining pchlide phototransformation activity was carried out by Horton & Leech (1972). This technique has since been modified in our laboratory, and exploited both to reconstitute pchlide photoconversion *in vitro* and to identify the enzyme catalysing the reaction, NADPH:protochlorophyllide oxidoreductase (Griffiths, 1975, 1978; Oliver & Griffiths, 1980).

Protochlorophyllide reductase, a protein loosely bound to the inner membranes of etioplasts, catalyses pchlide photoreduction by forming a photoactive ternary complex with the two substrates pchlide and NADPH. On exposure of this complex to light, hydrogen transfer from the nucleotide to the porphyrin results, producing chlide. Using [³H]NADPH as reductant, a stoichiometric incorporation of ³H into chlide has been observed, thereby confirming the role of NADPH as hydrogen donor in the photoconversion process (Griffiths, 1981). It has been possible using enzyme-enriched purified membranes from etioplasts to reproduce *in vitro* all the spectral intermediates observed during photoconversion *in vivo* (see above), and to incorporate these into a rational scheme for the photoconversion process (Fig. 3). According to this scheme, non-photoactive pchlide in etioplast membranes is considered to be 'free' or non-specifically bound pigment. In this state the pigment, pchlide 630, absorbs at *c.* 628–630 nm, a form which can readily be reproduced *in vitro* merely by preparing an aqueous 'solution' of purified pchlide, e.g. in cholate suspension or in the detergent Triton X-100 (Griffiths, 1978). The pigment in this state can, in the presence of NADPH, bind to the enzyme protochlorophyllide reductase, resulting in a shift in absorption with formation of the photoactive pigment complex absorbing at 638 and 650 nm (pchlide 638/650); the complex reconstituted *in vitro* is spectroscopically indistinguishable from the native photoactive complex of etiolated leaves (Griffiths, 1978). A number of different chlide spectroscopic forms can also be produced during photoconversion *in vitro* under different conditions, and these correspond, at least spectroscopically, to the forms observed during pchlide photoconversion in whole leaves (Fig. 2). The production of these forms *in vitro* is best demonstrated by repeated scanning absorption spectra of recently illuminated membranes in a split-beam spectrophotometer. From the spectra, the chlide absorption maxima are estimated and these maxima are plotted as a function of time after illumination in Fig. 4. When unsupplemented membranes are illuminated to effect complete photoconversion, chlide absorbing at 678 nm is formed which over a period of about 30 min shifts to give a form absorbing maximally at 672 nm (Fig. 4). When this procedure is repeated with membranes supplemented with NADPH (0.5 mmol l^{-1} and a regenerating system) the initial photoproduct again absorbs at 678 nm but on subsequent incubation the absorption maximum shifts to longer wavelengths stabilising after approximately 45 min at 684 nm (Fig. 4). This 684 nm-absorbing chlide remains stable under these conditions i.e. 22 °C and in the presence of NADPH, for at least 16 h. Finally, when photoconversion is carried out in the presence of NADP$^+$ (0.5 mmol l^{-1}), a chlide species absorbing at 678 nm is formed as before, but now on incubation the absorption maximum shifts slightly within *c.* 5 min to give a chlide form absorbing maximally at 677 nm (Fig. 4). It should be mentioned that addition of the NADP$^+$-reducing system (4 mmol l^{-1} glucose-

Fig. 3. Formation of a ternary complex in the photoconversion of protochlorophyllide. The numbers refer to the different spectral forms of the pigments (see text).

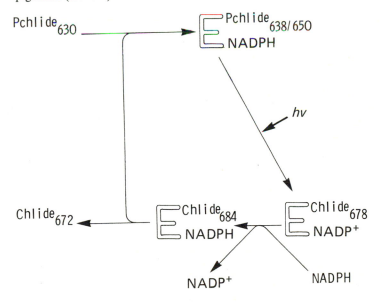

6-phosphate + 0.1 units glucose-6-phosphate dehydrogenase) to the NADP-incubated sample results in a red shift of this peak from 677 nm to approximately 681 nm (Fig. 4).

While these spectral forms generated *in vitro* appear qualitatively similar to the forms seen during photoconversion of pchlide in whole leaves (Fig. 2), differences in behaviour between the whole leaf forms and the corresponding complexes *in vitro* are apparent. Thus, while photoconversion *in vivo* results in only a transient appearance of chlide 678 (Fig. 2), a form related to this is always readily apparent immediately after photoconversion *in vitro*, irrespective of any additions (Fig. 4). In the presence of added NADP+, apart from undergoing a slight shift of *c.* 1 nm to lower wavelengths, this chlide 678 form is very stable *in vitro*. In view of this we suggest that chlide 678, the initial photoproduct, represents a ternary protochlorophyllide reductase-products complex i.e. [Enz–chlide–NADP+] participating in the photo-conversion process as outlined in Fig. 3.

In whole leaves, the chlide 678 form is rapidly transformed into chlide 684 (Fig. 2). In purified membranes a spectroscopically similar form can be

Fig. 4. Absorption maxima of chlide complexes produced on illumination of etiolated membranes *in vitro*. Suspensions of purified etioplast membranes were illuminated under different conditions as indicated and their absorption spectra recorded at intervals from which the chlide absorption maxima were estimated. At the point indicated (▲), endogenous NADP+ was reduced by addition of the reducing system glucose-6-phosphate (G6P) and its corresponding dehydrogenase (G6PDH).

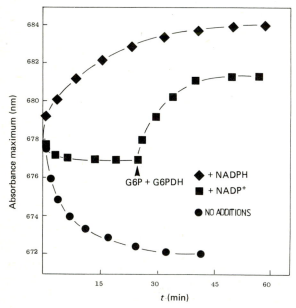

accumulated as a result of photoconversion in the presence of excess NADPH (Fig. 4 – see also El Hamouri & Sironval, 1980). On addition of the NADP$^+$-reducing system, i.e. glucose 6-phosphate and the corresponding dehydrogenase, to membranes containing the chlide 678 this is transformed to a species absorbing at 682 nm (Fig. 4) and related to the chlide 678 seen in whole leaves and in membranes photoconverted in the presence of added NADPH. We suggest, therefore, that the complex designated chlide 684 represents protochlorophyll reductase with bound chlide + NADPH, reduction of the nucleotide on the enzyme surface inducing a shift of the absorbance peak from 678 to 684 nm. The most widely studied of the chlide spectral shifts occurring *in vivo* is the so-called Shibata shift which results in the formation of a chlide species absorbing at 672 nm, chlide 672 (Fig. 2). In isolated membranes a chlide form absorbing at 672 nm is formed fairly rapidly (within *c.* 20 min; Fig. 4) of incubation of unsupplemented illuminated membranes in darkness. The progress of the Shibata shift can be readily studied by continuous monitoring of the absorbance of samples at 690 nm relative to that at 710 nm. At 690 nm a large difference in absorbance exists between the chlide 684 and chlide 672 forms whereas at 710 nm there is no specific pigment absorption. Using this technique it has been demonstrated (Oliver & Griffiths, unpublished) that the Shibata shift can be induced in isolated illuminated membranes by the addition of traces of detergents e.g. Triton X-100, thiol reagents (e.g. *N*-phenylmaleimide), which combine with the substrate-binding SH group on the reductase, and also by freezing and thawing treatment. Of more physiological significance, Fig. 5 indicates that exogenous pchlide, added to membranes which have been photoconverted in the presence of excess NADPH and thus contain chlide 684, will also experience a rapid Shibata shift, giving chlide 672. We have interpreted these data by proposing that chlide 672 corresponds to 'free' or non-enzyme bound chlide, with the Shibata shift reflecting production of such a form by dissociation of the chromophore from the enzyme. Further, the inference is that under normal conditions *in vivo* this dissociation is similarly induced by pchlide, initiating a turnover of the protochlorophyllide reductase as indicated in Fig. 3. There is ample evidence in the literature to support this suggestion; in particular, many workers have shown that if the endogenous pchlide level in plants is elevated by ALA feeding, this results in a marked stimulation of the rate of the Shibata shift (Gassman, 1973; Klockare & Sundquist, 1977; Walter, 1974). A barley mutant (*alb f*[17]), which contains no photoactive pchlide, does not show the Shibata shift, which further supports our suggestions (Henningsen & Thorne, 1974). Such a shift can, however, be induced in membranes from the mutant by pchlide addition (Oliver & Griffiths, unpublished). It should be apparent that according to our model (Fig. 3) the rate of pchlide photoconversion is governed by the availability of pchlide, which under normal conditions reflects the flux through the porphyrin pathway producing this pigment.

Identification of protochlorophyllide reductase

Central to the functioning of the scheme in Fig. 3 is the enzyme protochlorophyllide reductase, the identity of which had to be ascertained. It might be expected that this could be readily achieved by identification of the protein that binds the highly coloured pchlide. However, because of the insolubility of the enzyme, detergents have to be employed to solubilise the enzyme prior to study, and this invariably leads to dissociation of the pigment from the protein. Detergents can also, under certain conditions, produce artificial pigment–protein complexes, further complicating identification attempts (Thomas, Ryan & Levin, 1976). In the authors' laboratory (Oliver & Griffiths, 1980) unambiguous identification of the enzyme has been achieved by covalent labelling of the enzyme with a radioactive inhibitor.

The technique employed involved labelling of the substrate-binding thiol groups, exposed on illumination of the photoactive enzyme-substrate complex, with ^{3}H-labelled *N*-phenylmaleimide, followed by identification of the labelled peptide by SDS–PAGE and fluorography. The substrates protochlorophyllide

Fig. 5. Rate of the Shibata shift in isolated etioplast membranes under different conditions. Isolated etioplast membranes, supplemented with NADPH, were photoconverted (A) while absorbance changes Δ (690–710 nm) were continuously monitored in a dual wavelength spectrophotometer. At B 30 μl of CH_3OH were added, followed by three successive aliquots of pchlide in 25 μl CH_3OH at C.

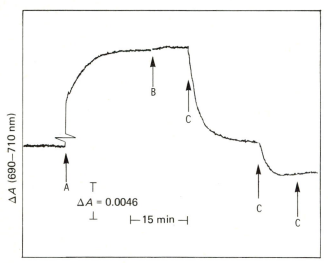

and NADPH were initially added to an etioplast membrane preparation to protect the essential thiol groups of the enzyme. Following this, the remaining thiol groups in the sample were made to react with unlabelled *N*-phenylmaleimide. Excess of the unlabelled thiol reagent and substrates was then removed by gel filtration, *N*-[³H]phenylmaleimide added, and the thiol group in the active site of the reductase exposed to the reagent by illumination, which resulted in specific radioactive covalent labelling of the enzyme. Under these conditions, increasing the illumination results in an increased photoconversion of the enzyme–substrates complex which should lead to specific increased labelling of the enzyme. Further, increasing photoconversion under these conditions should also lead to the simultaneous increase in inhibition of the reductase activity, because of reaction of the exposed thiol groups on the enzyme to the radioactive thiol reagent. The validity of the latter prediction is illustrated in Fig. 6, which shows that under these experimental conditions, i.e. photoconversion in the presence of *N*-[³H]phenylmaleimide, there is a direct relationship between the degree of photoconversion and inhibition of the reductase.

The essential results leading to identification of the reductase in this experiment are shown in the fluorograph (Fig. 7). At the sides we have included for comparison the stained polypeptide pattern of the original gel prior to the fluorographic treatment. This pattern illustrates the large number of different polypeptides present in the experimental material. The developed

Fig. 6. Correlation between photoconversion in the presence of *N*-phenylmaleimide and inhibition of protochlorophyllide reductase. Purified etioplast membranes containing substrate-protected protochlorophyllide reductase were photoconverted in the presence of *N*-phenylmaleimide. During the photoconversion, samples were removed and assayed for reductase activity.

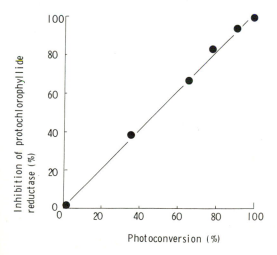

fluorograph (Fig. 7) shows: (i) incubation of the membranes in darkness with the radioactive inhibitor for up to 14 min results in only slight incorporation of ^{3}H into polypeptides (tracks 1 and 2); (ii) flash illumination of the incubated membranes gives a rapid incorporation of ^{3}H into a closely running doublet of polypeptides of *c.* mol. wt 35000 and 37000 (tracks 4 to 11). Furthermore, the extent of incorporation appears qualitatively to correlate with the percentage photoconversion resulting from the illumination, i.e. both appear to saturate after 10 min of flash illumination. Bearing in mind the rationale behind this experiment, it is apparent that these two polypeptides possess thiol groups that can be protected from *N*-phenylmaleimide by NADPH and protochlorophyllide and, as such, qualify as the polypeptides of the enzyme protochlorophyllide reductase. We have yet to explain why the two polypeptides both qualify for identification as protochlorophyllide reductase in these experiments, but should add that when the technique is

Fig. 7. Light-dependent labelling of etioplast membranes by *N*-[^{3}H]phenyl-maleimide. Substrate-protected *N*-phenylmaleimide-treated oat etioplast membranes were photoconverted in the presence of *N*-[^{3}H]phenylmaleimide and samples were removed after various time intervals for electrophoresis and fluorography. Control samples incubated for prolonged periods in darkness (tracks 1 and 2) or light (tracks 12 and 13) were also analysed. See text for details. The arrow indicates position of the reductase subunits.

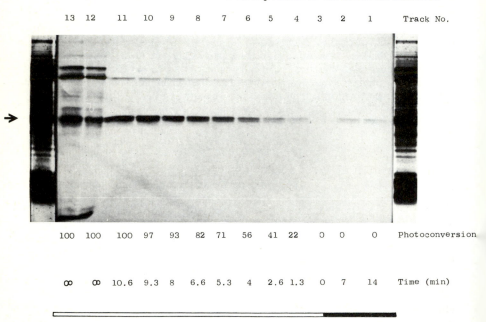

employed with barley membranes only a single peptide becomes labelled in contrast to the two labelled in both beans and oats. This would tend to support the authenticity of the results and indicates genuine species differences with respect to protochlorophyllide reductase. This identification gains further support from parallel work on co-purification of the activity of protochlorophyllide reductase with peptides related to the ones identified above (Apel *et al.*, 1980; Beer & Griffiths, 1981).

Recently, thanks largely to developments in instrumentation technology, intermediates in the photoconversion reaction have been detected. A non-fluorescent transient intermediate was detected by Dujardin and co-workers (Dujardin & Sironval, 1977; Dujardin & Correia, 1979; Frank, Dujardin & Sironval, 1980) in a study of the fluorescence spectra of etiolated leaves. Detection of the intermediate was facilitated by the fact that its lifetime was considerably increased, both in lyophilised tissue and in fresh tissue, at reduced temperatures. The intermediate was shown to absorb light maximally as a broad band centred at about 690 nm, and served as a strong energy sink, efficiently quenching the fluorescence of adjacent pigments (Dujardin & Sironval, 1977). Rapid absorption measurements on crude extracts from etiolated leaves have confirmed the existence of the long-wavelength-absorbing intermediate and have shown that the intermediate appeared immediately upon onset of actinic illumination with its decay ($t_{\frac{1}{2}} \sim 3.0\ \mu$s) coinciding with the appearance of chlide with a $t_{\frac{1}{2}}$ of 3.3 μs (Franck & Mathis, 1980). On an even shorter time scale, the emission spectra of several new transient intermediates in the photoconversion were very recently discovered in experiments where 10 ps laser excitation pulses were used, and emission spectra recorded (Dobek *et al.*, 1981). However, interpretation of these various transient spectroscopic intermediates in terms of a molecular mechanism for the photoconversion reaction has yet to be successfully made.

Effect of light on protochlorophyllide reductase activity

The effect of illumination of tissues on the amount of enzyme activity is rather unexpected. It might have been anticipated that the activity of the enzyme, in common with that of other chlorophyll biosynthetic enzymes, would increase on illumination of plants as the rate of chlorophyll synthesis increases. Surprisingly, however, illumination results in a dramatic decrease in the activity of the enzyme, this declining to 15% of the dark level within 30 min of illumination of 7-day old etiolated barley plants (Mapleston & Griffiths, 1977, 1980). In the same study it was shown that the light-induced effect could be largely reversed by returning the illuminated plants to darkness. It was also shown that the activity of the reductase could be correlated with the redox state of the NADPH/NADP$^+$ couple in the plastids

– a high ratio correlating with a high enzymic activity, and a lowered ratio, as found in plastids from illuminated plants, correlating with a reduced activity of the enzyme. On identification of the peptide of the enzyme it became apparent that illumination induced a reversible decrease in the actual amount of enzyme protein. It was shown, however, that illumination of dark-grown plants never brought about complete loss of the enzymic activity, and it was calculated that the residual enzyme in fully greened plants could still accommodate the observed rate of chlorophyll synthesis (Mapleston & Griffiths, 1980). The mechanism by which light controls the activity of the reductase presents interesting regulatory possibilities.

Apel and co-workers maintain that light via the phytochrome system regulates the amount of the enzyme at the mRNA level – with the amount of mRNA translatable into the reductase disappearing completely within 6 h illumination of etiolated barley plants (Apel, 1981; Santel & Apel, 1981). In contrast to our conclusions, these workers observe the complete disappearance of the reductase following illumination of dark-grown plants and speculate that an alternative mechanism of pchlide reduction is functional in green tissues. In our studies, however, we find a strong correlation, independent of illumination, between the amount and activity of the reductase and the pchlide (and probably NADPH) concentration in plastids, and envisage the presence of substrates, protecting the enzyme from proteolytic attack, as the principal mechanism for controlling the amount of the reductase in plants.

Finally, while we can account for most of the reactions accompanying pchlide photoconversion in leaves in terms of the properties of the enzyme protochlorophyllide reductase, it is inevitable that *in vivo* the enzyme must closely interact with many other macromolecules, e.g. those proteins producing pchlide and NADPH and esterifying chlide. *In vivo*, therefore, rather than existing in isolation, the enzyme is most probably part of a multifunctional complex, the complete structure and function of which will be fascinating to unravel.

References

Apel, K. (1981). The protochlorophyllide holochrome of barley (*Hordeum vulgare* L.). Phytochrome-induced decrease of translatable mRNA coding for the NADPH–protochlorophyllide oxidoreductase. *Eur. J. Biochem.*, **120**, 89–93.

Apel, K., Santel, H-J., Redlinger, T. E. & Falk, H. (1980). The protochlorophyllide holochrome of barley (*Hordeum vulgare* L.). Isolation and characterisation of the NADPH–protochlorophyllide oxidoreductase. *Eur. J. Biochem.*, **111**, 251–8.

Beer, N. S. & Griffiths, W. T. (1981). Purification of the enzyme NADPH: protochlorophyllide oxidoreductase. *Biochem. J.*, **195**, 83–92.

Boardman, N. K. (1962). Studies on a protochlorophyll–protein complex. Purification and molecular weight determination. *Biochim. biophys. Acta*, **62**, 63–79.

Bonner, B. A. (1969). A short-lived intermediate form in the *in vivo* conversion of protochlorophyllide 650 to chlorophyllide 684. *Pl. Physiol.*, **44**, 739–47.

Canaani, O. D. & Sauer, K. (1977). Analysis of the subunit structure of protochlorophyllide holochrome by sodium dodecyl sulphate polyacrylamide gel electrophoresis. *Pl. Physiol.*, **60**, 422–9.

Dobek, A., Dujardin, E., Franck, F., Sironval, C., Breton, J. & Roux, E. (1981). The first events of protochlorophyll(ide) photoreduction investigated in etiolated leaves by means of the fluorescence excited by short, 610 nm laser flashes at room temperature. *Photobiochem. Photobiophys.*, **2**, 35–40.

Dujardin, E. & Correia, M. (1979). Long wavelength absorbing pigment–protein complexes as fluorescence quenchers in etiolated leaves illuminated in liquid nitrogen. *Photobiochem. Photobiophys.*, **1**, 25–32.

Dujardin, E. & Sironval, C. (1977). Transitory pigment protein complexes similar to photosynthesis active centres during protochlorophyll(ide) photoreduction. *Pl. Sci. Lett.*, **10**, 347–55.

El Hamouri, B. & Sironval, C. (1980). NADP⁺/NADPH control of the protochlorophyllide, chlorophyllide-proteins in cucumber etioplasts. *Photobiochem. Photobiophys.*, **1**, 219–23.

Franck, F., Dujardin, E. & Sironval, C. (1980). Non-fluorescent short-lived intermediate in photoenzymatic protochlorophyllide reduction at room temperature. *Pl. Sci. Lett.*, **18**, 375–80.

Franck, F. & Mathis, P. (1980). A short-lived intermediate in the photo-enzymatic reduction of protochlorophyll(ide) into chlorophyll(ide) at a physiological temperature. *Photochem. Photobiol.*, **32**, 799–803.

Gassman, M. L. (1973). The conversion of photoactive protochlorophyllide to phototransformable protochlorophyllide in etiolated bean leaves treated with 5-aminolaevulinic acid. *Pl. Physiol.*, **52**, 295–300.

Gassman, M. L., Granick, S. & Mauzerall, D. (1968). A rapid spectral change in etiolated red kidney bean leaves following phototransformation of protochlorophyllide. *Biochem. biophys. Res. Commun.*, **32**, 295–309.

Griffiths, W. T. (1975). Characterisation of the terminal stages of chlorophyll(ide) synthesis in etioplast membrane preparations. *Biochem. J.*, **152**, 623–5.

Griffiths, W. T. (1978). Reconstitution of chlorophyllide formation by isolated etioplast membranes. *Biochem. J.*, **174**, 681–92.

Griffiths, W. T. (1981). Role of NADPH in chlorophyll synthesis by protochlorophyllide reductase. In *Photosynthesis V. Chloroplast Development*, ed. G. Akoyunoglou, pp. 65–71. Balaban International Science Services, Philadelphia.

Guignery, G., Luzzati, A. & Duranton, J. (1974). On the specific binding of protochlorophyllide and chlorophyll to different peptide chains. *Planta*, **115**, 227–43.

Henningsen, K. W. & Kahn, A. (1971). Photoactive subunits of protochlorophyllide holochrome. *Pl. Physiol.*, **47**, 685–90.

Henningsen, K. W. & Thorne, S. W. (1974). Esterification and spectral shifts of chlorophyll(ide) in wild-type and mutant seedlings developed in darkness. *Physiologia Pl.*, **30**, 82–9.

Horton, P. & Leech, R. M. (1972). The effect of ATP on photoconversion of protochlorophyllide in isolated etioplasts. *FEBS Lett.*, **26**, 277–80.

Kahn, A., Boardman, N. K. & Thorne, S. W. (1970). Energy transfer between protochlorophyllide molecules: evidence for multiple chromophores in the photoactive protochlorophyllide–protein complex *in vivo* and *in vitro. J. molec. Biol.*, **48**, 85–101.

Klockare, B. & Sundquist, C. (1977). Shifts in absorption and fluorescence maxima of chlorophyll(ide) in spectra of dark grown wheat leaves after irradiation. *Photosynthetica*, **11**, 189–99.

Krasnovsky, A. A. & Kosobutskaya, L. M. (1952). Spectral investigations of chlorophyll during its formation in plants and in colloidal solutions of material from etiolated leaves. *Dokl. Akad. Nauk. SSSR*, **85**, 177–80.

Mapleston, R. E. & Griffiths, W. T. (1977). Effects of illumination of whole barley plants on the protochlorophyllide activating system in the isolated plastids. *Biochem. Soc. Trans.*, **5**, 319–21.

Mapleston, R. E. & Griffiths, W. T. (1980). Light modulation of the activity of protochlorophyllide reductase. *Biochem. J.*, **189**, 125–33.

Oliver, R. P. & Griffiths, W. T. (1980). Identification of the polypeptides of NADPH–protochlorophyllide oxidoreductase. *Biochem. J.*, **191**, 277–80.

Redlinger, T. E. & Apel, K. (1980). The effect of light on four protochlorophyllide-binding polypeptides of barley. *Archs Biochem. Biophys.*, **200**, 253–60.

Santel, H. J. & Apel, K. (1981). The protochlorophyllide holochrome of barley (*Hordeum vulgare* L.). The effect of light on the NADPH:protochlorophyllide oxidoreductase. *Eur. J. Biochem.*, **120**, 95–103.

Schopfer, P. & Siegelman, H. W. (1968). Purification of protochlorophyllide holochrome. *Pl. Physiol.*, **43**, 990–6.

Shibata, K. (1957). Spectroscopic studies of chlorophyll formation in intact leaves. *J. Biochem. (Tokyo)*, **44**, 147–73.

Smith, J. H. C. (1952). Factors affecting the transformation of protochlorophyll to chlorophyll. *Carnegie Inst. Wash. Yearb.*, **51**, 151–3.

Thomas, P. E., Ryan, D. & Levin, W. (1976). An improved staining procedure for the detection of the peroxidase activity of cytochrome P-450 on sodium dodecyl sulphate polyacrylamide gels. *Analyt. Biochem.*, **75**, 168–76.

Walter, G. (1974). Spektrale Anderungen *in vivo* wahrend der chlorophyll bildung etiolierter weizenkeimpflanzen nach experimentall variiertum anteil der protochlorophyll(id)-holochrome P_{650} und P_{635} durch behandlung mit δ-aminolavulinsäure. *Photosynthetica*, **8**, 40–6.

PART IV

Photoregulation of chloroplast development

H. K. LICHTENTHALER AND D. MEIER

Regulation of chloroplast photomorphogenesis by light intensity and light quality

Plants possess the ability to adapt their growth and morphological develop-ment to the intensity of incident light. The high-light growth response is found at high quanta fluence rates (sun leaves, high-light (HL) plants) and the low-light growth response in shade leaves and low-light (LL) plants. This adaptive ability of plants and of chloroplasts is associated with distinct differences in the morphology, physiology and biochemistry of leaves and chloroplasts (Boardman *et al.*, 1975; Lichtenthaler & Buschmann, 1978; Lichtenthaler, 1979, 1981; Wild, 1979; Wild & Wolf, 1980; Buschmann & Lichtenthaler, 1982).

The photosynthetic apparatus of sun leaves and leaves from HL plants exhibits higher photosynthetic rates than shade leaves or leaves from LL plants (e.g. Boardman *et al.*, 1975; Wild, 1979; Lichtenthaler, 1981). Sun leaves and HL leaves possess HL chloroplasts (sun-type chloroplasts) which are very effective in photosynthetic quanta conversion and whose composition and arrangement of thylakoids are different from those of LL or shade-type chloroplasts.

HL chloroplasts are characterized by higher values of the chlorophyll a/b ratio, a higher proportion of β-carotene, higher prenylquinone levels (Lichtenthaler, 1979, 1981) and higher levels of cytochromes (Wild, 1979). That they possess a lower thylakoid content and lower grana stacks than LL chloroplasts can easily be shown by comparing electron micrographs of the two chloroplast types, and has been described by several authors (Ballantine & Forde, 1970; Skene, 1974; Boardman *et al.*, 1975; Meier & Lichtenthaler, 1981). These differences in ultrastructure between sun and shade chloroplasts, and between HL and LL chloroplasts, can be quantified by biometrical analysis, and this is described here together with new results on the chlorophyll–carotenoid–protein complexes.

Materials and methods
Radish (*Raphanus sativus* L. 'Saxa Treib'), wheat (*Triticum aestivum* L. 'Kolibri') and maize (*Zea mays* L. 'Protador') were grown under HL and LL conditions as described by Lichtenthaler *et al.*, (1981*b*).

The sun and shade leaves of the beech (*Fagus sylvatica* L.) were collected from a single standing tree at Karlsruhe. The barley seedlings (*Hordeum vulgare* L. 'Breuns Villa') were greened in red and blue light (Buschmann *et al.*, 1978).

Fixation of the plant material for electron microscopy was performed with glutaraldehyde and OsO_4 (Meier & Lichtenthaler, 1981). Thylakoid length, degree of membrane stacking and the ratio of appressed to non-appressed membranes (stroma thylakoids+end grana membranes) were determined by measuring the photosynthetic membranes with a mileage tracer.

The methods applied for the isolation of chlorophyll–proteins (sodium dodecyl sulphate (SDS) solubilization of chloroplasts, SDS–polyacrylamide gel electrophoresis (PAGE)) have been reported in detail (Lichtenthaler *et al.*, 1981*a*; Lichtenthaler *et al.*, 1982*b*). Chlorophyll was determined in diethylether (Ziegler & Egle, 1965). The carotenoids were separated by thin-layer chromatography and measured spectrophotometrically (Lichtenthaler, 1979).

Results and discussion

In the light the development of photosynthetically active chloroplasts during plant growth proceeds from undifferentiated proplastids. In dark-grown etiolated seedlings the accumulation of light-induced chlorophyll and thylakoids is blocked. The ultrastructure of etioplasts (Fig. 1) is characterized by prolamellar bodies, prothylakoids and varying amounts of osmiophilic plastoglobuli (e.g. Virgin, Kahn & Wettstein, 1963; Lichtenthaler, 1967; Weier & Brown, 1970). In recent years it was reported that in *Avena* seedlings, at least, the prolamellar body is made up of two saponins, and that these saponins can be taken as markers when separating prolamellar bodies from prothylakoids (Lütz & Klein, 1979). In a detailed study we could not detect the two presumed prolamellar saponins in intact etiolated *Avena* leaves, but found two other more hydrophilic saponins (Lichtenthaler, Bach & Wellburn, 1982*a*). This has been confirmed independently by Kesselmeier (1982), who showed that the two presumed prolamellar saponins are formed from more hydrophilic saponins during extraction of leaves with aqueous solvents. From this it is clear that the paracrystalline prolamellar body is not made up from saponins.

Upon illumination of etiolated leaves, thylakoid formation proceeds in parallel with the breakdown of the prolamellar body and an almost complete disappearance of the plastoglobuli (Sprey & Lichtenthaler, 1966; Engelbrecht & Weier, 1967). The possible two intermediate steps of tube transformation and vesicle dispersal before grana formation, as found under special conditions upon illumination of etiolated bean leaves (Virgin *et al.*, 1963), have not been

Fig. 1. Etioplasts from the primary leaves of 7-day-old dark grown plants with prolamellar body (plb), plastoglobuli (p) and prothylakoids (pt). (*a*) *Zea mais* L. 'Protador' (*b*) *Hordeum vulgare* L. 'Breuns Villa'.

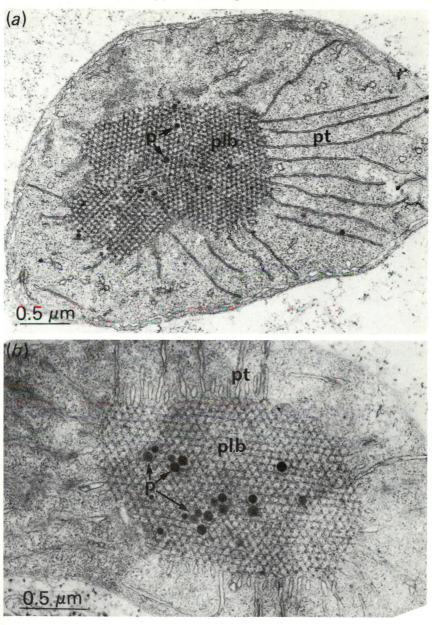

seen in other plants. At first stroma thylakoids are formed, then grana formation starts from initially short thylakoid overlaps (Wehrmeyer, 1964; Sprey & Lichtenthaler, 1966; Engelbrecht & Weier, 1967). At the same time as the occurrence of light-induced biogenesis of thylakoids, there is a change in the lipid metabolism of the etioplast with a concomitant accumulation of chlorophylls, carotenoids, prenylquinones, galactolipids and phospholipids, which is controlled by active phytochrome (for literature see references in Lichtenthaler & Buschmann, 1978).

High-light and low-light

Higher grana stacks with more thylakoids per granum are formed in shade leaves and LL plants than are formed in sun leaves and HL leaves (e.g. Ballantine & Forde, 1970; Skene, 1974; Boardman et al., 1975; Lichtenthaler, 1979, 1981; Meier & Lichtenthaler, 1981; Lichtenthaler et al., 1981b). The typical appearances of a HL chloroplast with a high starch content, and of a LL chloroplast with wide and high grana stacks are shown in Fig. 2. The difference between them suggests that the degree of grana formation is controlled by the light intensity.

The stacking degree of thylakoid membranes is higher in the shade leaf of the beech (Heumann et al., 1984), in the LL leaves of radish and wheat, and in the mesophyll chloroplasts of maize (Table 1), and has also been found to vary between different sun and shade plants (Guillot-Salomon et al., 1978). Another way of expressing these differences is in terms of the ratio of appressed to non-appressed chloroplast membranes. This ratio is significantly higher for the shade and LL leaves, which experience light shortage (Table 2).

A further ultrastructural parameter is the grana width. In early work on chloroplast ultrastructure it was shown that the grana stacks of LL plants are broader than those of HL plants (Abel, 1956). The data in Table 3 show that this is a general rule for the chloroplasts of sun and shade leaves and HL and LL plants. The number and range of thylakoids per granum is an additional parameter to describe the differences in fine structure between HL and LL chloroplasts (Table 4).

The greater thylakoid content (thylakoid frequency) of shade and LL chloroplasts, compared to sun and HL chloroplasts, can easily be quantified by determining the total length of thylakoids (in μm) per standard chloroplast section area (10 μm^2). This parameter indicates a significantly lower thylakoid content (which is consistent with a lower chlorophyll content) in the beech sun chloroplasts and the radish HL chloroplasts (Table 5). In wheat and maize leaves, however, we did not find differences in thylakoid frequency produced by growth at different light intensities. The differences between HL and LL chloroplasts described here are summarized in Fig. 3.

Fig. 2. Chloroplasts from the cotyledons of 8-day-old green radish seedlings (3 days' dark growth + 5 days' irradiation) grown in either high-light (HL; 90 W·m⁻²) or low-light (LL; 10 W·m⁻²) conditions; p = plastoglobuli.

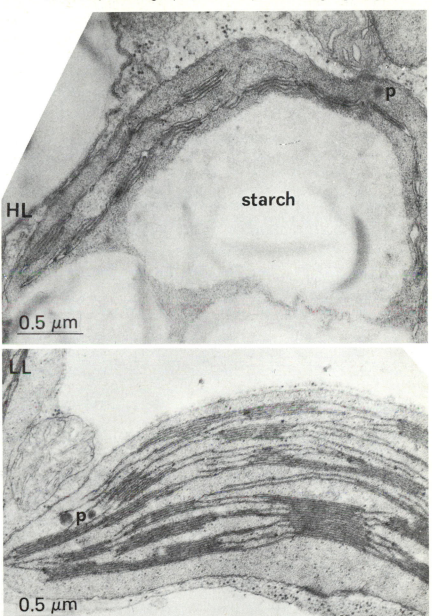

Blue light and red light

In plastid development, it is necessary to differentiate between (*a*) the blue light induction of chlorophyll synthesis and chloroplast development and (*b*) the blue light-induced modification of the photosynthetic apparatus. It has been shown for several non-green plant tissues that blue light is required for chlorophyll and chloroplast formation, and cannot be replaced by red light (Björn, Suzuki & Nilsson, 1963; Bergmann & Berger, 1966; Berger & Bergmann, 1967; Richter, 1969); this topic is reviewed by Björn (1980), and Sundqvist, Björn & Virgin (1980).

The first report on the modification of chloroplast ultrastructure by blue light and red light was given by Fasse-Franzisket (1956), who found a larger

Table 1. *Differences in the stacking of thylakoids in chloroplasts from sun and shade leaves* (Fagus), *HL and LL plants and from plants grown in blue* (*BL*) *and red light* (*RL*). *The values represent the percentage of stacked membranes*

Fagus sylvatica[a]	Sun leaf	Shade leaf
17 May	56 ± 6	72 ± 2
June–October	57 ± 6	82 ± 6
Raphanus sativus[b]	HL leaf	LL leaf
3 days dark growth		
+1 day light	52 ± 3	50 ± 4
+3 days light	55 ± 4	62 ± 3
+5 days light	55 ± 5	64 ± 3
Triticum aestivum[c]	HL leaf (4th)	LL leaf (3rd)
28-day-old plants		
medium leaf region	54 ± 5	73 ± 3
Zea mais (16-day-old)[c]	HL leaf (2nd)	LL leaf (2nd)
mesophyll chloroplasts		
upper leaf region	54 ± 3	72 ± 4
lower leaf region	55 ± 3	77 ± 3
Hordeum vulgare[d]	BL leaf (1st)	RL leaf (1st)
upper leaf region		
7 days dark growth		
+1 day light	39 ± 5	50 ± 4
+3 days light	47 ± 3	55 ± 3
Lower leaf region		
+1 day light	50 ± 3	53 ± 4
+3 days light	52 ± 4	59 ± 3

[a] Based on 6 (17 May) and 25 (June–October) median chloroplast sections ($P < 0.001$).
[b] Based on 40 chloroplast sections (3rd + 5th day, $P < 0.001$).
[c] Based on 10 chloroplast sections ($P < 0.001$).
[d] Based on 30 chloroplast sections (3rd day, $P < 0.001$).

number of grana stacks in *Agapanthus* plants in blue light, and fewer and broader grana in red light. A greater number of grana in blue light than in red light was also described for pea chloroplasts (Vlasova, Drozdova & Voskresenskaya, 1971). In the first hours of illumination of etiolated corn seedlings, greening and thylakoid formation proceeds faster in blue light than in red light; after 30 h red light the formation of larger grana in red-light chloroplasts is established (Shakhov & Balaur, 1969). It has been shown for barley chloroplasts that the faster greening in blue light is also associated with a higher prenylquinone content, higher rates of Hill activity, higher chlorophyll *a/b* values, fewer thylakoids and lower grana stacks (only few thylakoids per granum), whereas red-light chloroplasts possess high grana stacks and low chlorophyll *a/b* ratios (Lichtenthaler, 1977; Buschmann *et al.*, 1978). The typical ultrastructure of blue-light chloroplasts with low grana stacks and that

Table 2. *Differences in the ratio of appressed to non-appressed membranes of chloroplasts from sun and shade leaves* (Fagus), *HL and LL plants and from plants grown in blue* (BL) *and red light* (RL)

Fagus sylvatica[a]	Sun leaf	Shade leaf
17 May	1.3	2.6
June–October	1.3	4.7
Raphanus sativus[b]	HL leaf	LL leaf
3 days dark growth		
+1 day light	1.1	1.0
+3 days light*	1.2	1.6
+5 days light*	1.2	1.8
Triticum aestivum[c]	HL leaf (4th)	LL leaf (3rd)
28-day-old plants		
medium leaf region*	1.2	2.7
Zea mais (16-day-old)[c]	HL leaf (2nd)	LL leaf (2nd)
mesophyll chloroplasts		
upper leaf region*	1.2	2.7
lower leaf region*	1.2	3.3
Hordeum vulgare[d]	BL leaf (1st)	RL leaf (1st)
upper leaf region		
7 days dark growth		
+1 day light*	0.6	1.0
+3 days light*	0.9	1.2
Lower leaf region		
+1 day light	1.0	1.1
+3 days light*	1.1	1.4

[a] Based on 6 (17 May) and 25 (June–October) median chloroplast sections ($P < 0.001$).
[b] Based on 40, [c] based on 10 and [d] based on 30 median chloroplast sections (*$P < 0.001$).

Table 3. *Differences in the grana width (μm) of chloroplasts from sun and shade leaves* (Fagus), *HL and LL plants and from plants grown in blue* (BL) *and red light* (RL)

Fagus sylvatica[a] June–October	Sun leaf 0.23	Shade leaf 0.33
Raphanus sativus[b] 3 days dark grown	HL cotyledon	LL cotyledon
+3 days light	0.28	0.34
+5 days light	0.26	0.37
Triticum aestivum[c] 28-day-old plants	HL leaf (4th) 0.53	LL leaf (3rd) 0.63
Zea mais (16-day-old)[d] mesophyll chloroplasts	HL leaf (2nd) 0.49	LL leaf (2nd) 0.57
Hordeum vulgare[d] 7 days dark + 1 day light 7 days dark + 3 days light	BL leaf (1st) 0.21 0.32	RL leaf (1st) 0.33 0.40

[a] Mean values of 50 grana stacks ($P < 0.001$).
[b] Based on 800 (HL) and 1100 grana stacks (LL) of 40 median chloroplast sections ($P < 0.001$).
[c] Based on 200 grana stacks of 20 median chloroplast sections ($P < 0.001$).
[d] Based on 100 grana stacks of 10 median chloroplast sections ($P < 0.001$).

Table 4. *Differences in the average number (and range) of thylakoids per granum in chloroplasts from sun and shade leaves* (Fagus) *HL and LL plants* (Raphanus, Triticum, Zea mais) *and from plants grown in blue* (BL) *and red light* (RL)

Fagus sylvatica[a] May–October	Sun leaf 4 (2–30)	Shade leaf 9.5 (2–60)
Raphanus sativus[b] 3 days dark + 5 days light	HL cotyledon 2.7 (2–6)	LL cotyledon 4.5 (2–17)
Triticum aestivum[c] 28-day-old plants	HL leaf (4th) 8 (2–14)	LL leaf (3rd) 16 (2–45)
Zea mais (16-day-old)[c] mesophyll chloroplasts	HL leaf (3rd) 7.6 (2–40)	LL leaf (2nd) 16.8 (2–80)
Hordeum vulgare[a] 7 days dark + 3 days light	BL leaf (1st) 3.5 (2–9)	RL leaf (1st) 4.6 (2–13)

[a] Based on 30 median chloroplast sections with 500 grana.
[b] Based on 40 chloroplasts with 1300 grana.
[c] Based on 20 chloroplasts with 200 grana.

of red-light chloroplasts with wide and high grana stacks is shown in Fig. 4. The similarities between blue-light chloroplasts and HL chloroplasts, and between red-light chloroplasts and LL chloroplasts, and the importance of blue light for the development of sun-type chloroplasts are summarized by Lichtenthaler & Buschmann (1978) and Lichtenthaler, Buschmann & Rahmsdorf (1980 *b*). Higher photosynthetic rates of blue-light than of red-light chloroplasts, and a chemical composition similar to sun-chloroplasts, were also confirmed for blue-light chloroplasts of *Sinapis* (Wild, 1979; Wild & Holzapfel, 1980). An increased height of grana stacks was also seen in red-light chloroplasts of *Capsicum* plants (Horvath, Kalman & Titlianov, 1975). Special blue-light effects on photosynthetic carbon flow have already been summarized by Voskresenskaya (1972).

In our new biometrical analysis of the electron micrographs of blue-

Table 5. *Differences in thylakoid frequency (total length of thylakoids in μm per 10 μm² area of chloroplast section) between sun and shade leaves* (Fagus), *HL and LL plants and plants grown in blue* (BL) *and red light* (RL). *The length of thylakoids was determined per real stroma section; in the case of HL chloroplasts and chloroplasts from sun leaves the section area occupied by starch was deducted*

Fagus sylvatica[a]	Sun leaf	Shade leaf
17 May	153 ± 20	500 ± 21
June–October	148 ± 26	382 ± 56
Raphanus sativus[b]	HL leaf	LL leaf
3 days dark growth		
+3 days light*	107 ± 36	197 ± 28
+5 days light*	88 ± 17	220 ± 32
Triticum aestivum[c]	HL leaf (4th)	LL leaf (3rd)
28-day-old plants		
medium leaf region	201 ± 38	200 ± 18
Zea mais (16-day-old)[c]	HL leaf (2nd)	LL leaf (2nd)
mesophyll chloroplasts		
upper leaf region	311 ± 40	344 ± 61
lower leaf region	307 ± 61	319 ± 37
Hordeum vulgare[d]	BL leaf (1st)	RL leaf (1st)
upper leaf region		
7 days dark growth		
+1 day light*	78 ± 12	103 ± 10
+3 days light*	130 ± 20	180 ± 12

[a] Based on 6 (17 May) and 25 (June–October) median chloroplast sections ($P < 0.001$).
[b] Based on 14, [c] based on 10 and [d] based on 20 median chloroplast sections (*$P < 0.001$).

<u>etioplast</u>

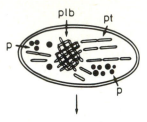

light induction
P_{fr} control

modified by:
high light-intensity
blue light

modified by:
low light-intensity
red light

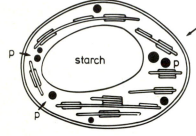

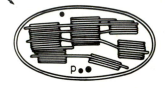

<u>sun-type chloroplast</u>
(high-light chloroplast)

<u>shade-type chloroplast</u>
(low-light chloroplast)

1. <u>low amounts of thylakoids</u> per chloroplast section

2. <u>narrow grana stacks</u>

3. <u>few thylakoids</u> per granum

4. <u>low stacking degree</u>

5. more <u>non-appressed</u> chloroplast membranes

6. higher proportion of <u>chlorophyll a-proteins</u> CPI, CPI *a*

7. large starch grains

8. many plastoglobuli

9. lower chlorophyll content

1. <u>high thylakoid frequency</u> per chloroplast section

2. <u>broad grana stacks</u>

3. <u>high grana stacks</u> with many thylakoids

4. <u>high stacking degree</u>

5. more <u>appressed</u> thylakoid regions

6. higher amounts of <u>light-harvesting chlorophyll-proteins</u> LHCP s

7. no starch

8. few plastoglobuli

9. more chlorophyll per chloroplast

Fig. 3.

light and red-light chloroplasts (determination of different ultrastructural parameters such as grana width, number of thylakoids per granum, total length of thylakoids per chloroplast section area, ratio of appressed to non-appressed membranes and stacking degree), we find the same typical differences between blue-light and red-light chloroplasts as between HL and LL chloroplasts (see Tables 1–5).

From the lower stacking degree of blue-light chloroplasts and their higher chlorophyll *a/b* ratios, one can predict lower amounts of light-harvesting chlorophyll proteins for blue-light chloroplasts than for red-light chloroplasts. This has in fact been shown directly for bean leaves when blue-light chloroplasts are compared to red-light chloroplasts (Akoyunoglou, Anni & Kalosakas, 1980). When plants are transferred from blue light to red light or vice versa, chloroplasts adapt to the new light conditions, as evidenced by changed pigment content and pigment ratios (Lichtenthaler *et al.*, 1980*b*).

Chlorophyll–carotenoid–proteins

Within the thylakoids, chlorophylls and carotenoids are bound to different pigment-proteins. These can be separated by SDS–PAGE of SDS-solubilized chloroplasts as shown in Fig. 5. CPI and CPI*a* represent the chlorophyll–proteins of photosystem I, which contain β-carotene as the main carotenoid (Lichtenthaler, Prenzel & Kuhn, 1982*e*). We also found high amounts of β-carotene in the chlorophyll *a* protein CP*a*, the presumed reaction center of photosystem II. In contrast, the different light-harvesting proteins, LHCP$_1$, LHCP$_2$, LHCP$_3$ and the minor component LHCP$_y$, contain chlorophyll *b* (*a/b* ratio 1.1–1.3) and the xanthophylls lutein and neoxanthin. The preferential association of the individual pigments with specific pigment-proteins is summarized in Fig. 6.

The differences in pigment ratios between HL and LL leaves and sun and shade leaves (Table 6) reflect differences in the proportions of chlorophyll *a*

Fig. 3. Scheme for the formation of sun-type and shade-type chloroplasts showing the ontogenetic adaptation of the photosynthetic apparatus to high or low light intensity. p = plastoglobuli, plb = prolamellar body; pt = prothylakoids. The formation of high-light (sun-type) chloroplasts can be simulated with blue light (Buschmann *et al.*, 1978; Lichtenthaler *et al.*, 1980*b*) and by application of cytokinins (Buschmann & Lichtenthaler, 1977, 1982; Lichtenthaler & Buschmann, 1978). Low-light chloroplasts are formed in red light and under the influence of photosystem II herbicides, e.g. bentazon (Lichtenthaler, 1979; Lichtenthaler *et al.*, 1980*a*).

Preliminary investigations indicate that plants, when transferred from low-light to high-light growth condition or vice versa, will gradually adapt to the new light conditions within 2 to 5 days depending on the individual parameter.

Fig. 4. Chloroplasts from the primary leaves of 10-day-old barley seedlings (7 days' dark growth + 3 days' irradiation) greened in either blue light (λ_{max} 450 nm; 1.5 J·m^{-2}·s^{-1}) or red light (λ_{max} 660 nm; 1.0 J·m^{-2}·s^{-1}) of low intensity and equal quanta fluence rates (5.5 μmol·m^{-2}·s^{-1}); p = plastoglobuli.

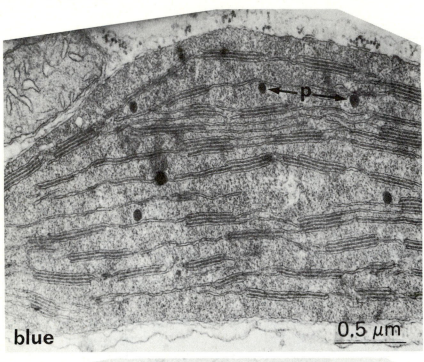

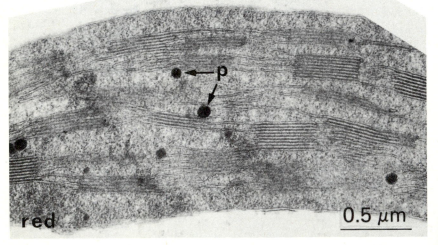

Table 6. *Pigment ratios in HL and LL cotyledons of 7-day-old radish seedlings (3 days dark growth+4 days light) and in sun and shade leaves of the beech (July)*

	Pigment ratios		
	a/b	a/c	x/c
Raphanus sativus			
High light	3.7	14	3.4
Low light	2.8	20	4.3
Fagus sylvatica			
Sun leaf	3.5	9.0	1.6
Shade leaf	2.8	13.1	2.2

a/b = ratio of chlorophyll *a* to chlorophyll *b*; a/c = ratio of chlorophyll *a* to β-carotene; x/c = ratio of xanthophylls to β-carotene.

Fig. 5. Densitometer scan of chlorophyll-proteins from SDS-solubilized chloroplasts of 7-day-old low-light radish seedlings (3 days' dark growth + 4 days' irradiation) separated by PAGE. CPI and CPIa = chlorophyll a/β-carotene-proteins of photosystem I; LHCP$_1$, LHCP$_2$, LHCP$_3$ and LHCP$_y$ = light-harvesting chlorophyll a/b–xanthophyll–proteins; CPa = chlorophyll a/β-carotene–protein of photosystem II; FC = free chlorophylls.

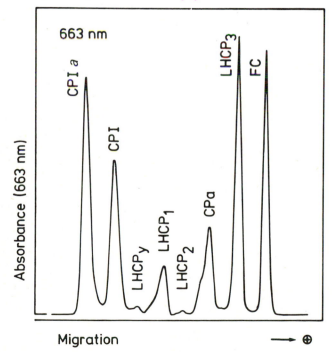

to β-carotene and of LHCPs. The lower values for chlorophyll a/b ratio, and the higher values for chlorophyll a/β-carotene (a/c) and xanthophylls/β-carotene (x/c) ratios in shade leaves and LL-plants, predict a higher level of LHCPs. This has been shown directly for radish seedlings. The LL chloroplasts contain a lower proportion of chlorophyll in the photosystem I proteins CPI + CPIa, and higher amounts of chlorophyll in the LHCPs (Table 7). The

Table 7. *Differences in percentage distribution of total chlorophyll $a+b$ in the chlorophyll a–proteins of photosystem I (CPI+CPIa) and in the light-harvesting chlorophyll a/b–proteins (LHCPs) between HL and LL radish chloroplasts from 7-day-old seedlings (3 days dark growth+4 days light).*

	% chlorophyll $a+b$	
	HL	LL
Growth on peat[a]		
CPI + CPIa	28.7	24.2
LHCPs	44.0	50.7
Growth on water[b]		
CPI + CPIa	42	31
LHCPs	34	39

[a] Data from Lichtenthaler *et al.*, 1982a and 1982b.
[b] Data from Lichtenthaler *et al.*, 1982c.

Fig. 6. Scheme for the organization of the seven chlorophyll–carotenoid–proteins in the thylakoids in relation to the photosynthetic electron transport chain. The preferential localization of chlorophyll b, lutein and neoxanthin in the light-harvesting chlorophyll–proteins, and of β-carotene (with chlorophyll a) in CPI, CPIa and CPa is indicated (after Lichtenthaler *et al.*, 1982e). PQ = plastoquinone; Cyt f = cytochrome f; PC = plastocyanin; X = xanthophyll; Fd = ferrodoxin; other abbreviations as in Fig. 5.

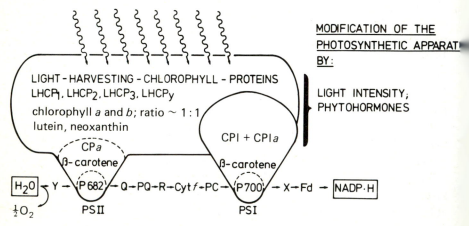

levels of CP*a* are little changed. Plants grown on peat are better developed and contain more total chlorophyll than those grown on water; the differences in LHCP content between HL and LL growth conditions are evident in both cases. Higher percentages of LHCPs were also detected in the shade plants *Malaxis*, *Alocasia* and *Helmholtzia* when compared with the sun plants spinach, pea and barley (Anderson, 1980; Yun-ling *et al.*, 1980).

These data show that the increased grana height and higher stacking degree of thylakoid membranes in LL and shade chloroplasts is associated with an increased proportion of chlorophyll in the LHCPs. This points to a participation of the LHCPs in grana formation and membrane stacking.

It is known that stacking of thylakoids is controlled by monovalent and divalent cations (e.g. McDonnel & Staehelin, 1980). There is further support for the participation of LHCPs in thylakoid stacking from results using freeze-fracture techniques, and from work with isolated and trypsinated LHCPs (Armond & Arntzen, 1977; Steinback *et al.*, 1978; Carter & Staehelin, 1980; Ryrie, Anderson & Goodchild, 1980). The stacked membrane regions contain predominantly photosystem II units (Andersson & Anderson, 1980; Anderson, 1981). Studies on the distribution of photosystem II activity between grana and stroma thylakoids showed that about 80% of photosystem II are localized in the stacked membrane region, and that the photosystem II units of stroma lamellae contain a smaller antenna with only a partial complement of LHCP (Armond & Arntzen, 1977; Arntzen, 1978). These results and the participation of LHCP in membrane stacking are summarized in Fig. 7.

Effects of photosystem II herbicides

Inhibitors of photosynthetic electron transport at photosystem II, such as bentazon and diuron, not only block photosynthetic electron transport by binding to the Q_B-protein, but also induce the formation of shade-type (LL-type) chloroplasts. This induction is reflected in many parameters, e.g. in lower values for the chlorophyll *a*/*b* ratio (Lichtenthaler, 1979; Lichtenthaler *et al.*, 1980*a*), a lower percentage of the chlorophyll *a* proteins CPI + CPI*a*, and a higher proportion of the LHCPs (Lichtenthaler *et al.*, 1982*c*). Further characteristics of chloroplasts from plants treated with photosystem II-type inhibitors are an increased stacking degree of thylakoids, wider grana stacks, a higher number of thylakoids per granum and a higher thylakoid frequency per unit chloroplast section (Meier & Lichtenthaler, 1981; Lichtenthaler *et al.*, 1982*c*).

This shade-type modification of the photosynthetic apparatus by photosystem II herbicides such as bentazon proceeds at medium, and particularly at high, light intensities; there is, however, little effect under low light growth conditions (Meier & Lichtenthaler, 1981). The block of the HL-induced formation of HL-chloroplasts, which at light saturation are more efficient in photosynthetic quantum conversion than LL chloroplasts, contributes to the effectiveness of the diuron-type photosystem II herbicides (Lichtenthaler *et al.*, 1982*d*).

This work was sponsored by a grant from the Deutsche Forschungsgemein-schaft. We wish to thank Miss A. Ehrfeld, Mrs W. Meier, Mrs S. Pfirrmann and Mrs U. Prenzel for excellent assistance.

Summary

Plants possess the capacity for an ontogenetic adaptation of their photosynthetic apparatus to the incident light intensity. Development of chloroplasts in sun leaves and at HL growth conditions (20 klx, 90 W·m⁻²) results in the formation of HL (sun-type) chloroplasts which differ in ultrastructure, composition and photosynthetic

Fig. 7. Model for the differential localization of photosystem I (PSI) units and of photosystem II (PSII) units with different content of light-harvesting proteins in the grana and stroma thylakoids showing the participation of light-harvesting chlorophyll *a/b*–proteins (LHCPs) in membrane stacking. (Based on Staehelin & Arntzen, 1979, revised according to the results of McDonnel & Staehelin, 1980; Anderson, 1981; and Lichtenthaler *et al.*, 1982*b*, 1982*c*.)

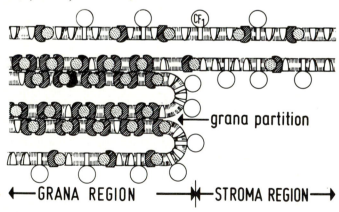

grana partition

←—GRANA REGION ——→⊣⊢—STROMA REGION—→

- PSI PHOTOSYSTEM I UNITS
- PSII CORE
- PSII CORE WITH PARTIAL COMPLEMENT OF LHCP (EXPOSED MEMBRANES)
- PSII CORE WITH FULL COMPLEMENT OF LHCP (STACKED MEMBRANES)
- (CF₁) COUPLING FACTOR (CF₁ + CF₀)
- ←CF₀

capacity from LL (shade-type) chloroplasts of shade leaves and of seedlings grown at LL intensity (1 klx; $10 \ W \cdot m^{-2}$). The formation of HL chloroplasts can be simulated by blue light and that of LL chloroplasts by red light.

High-light chloroplasts

HL chloroplasts are characterized by a low thylakoid frequency per chloroplast section (μm thylakoids per 10 μm^2 section area), by narrow and low grana stacks (only few thylakoids per granum), a low stacking degree of membranes, a higher proportion of non-appressed membranes (stroma thylakoids + end grana membranes), and a high starch content.

Their higher photosynthetic capacity is associated with a higher proportion of chlorophyll a and β-carotene (higher values for a/b and lower values for a/c and x/c ratios – see Table 6), which is caused by increased levels of the chlorophyll a/β-carotene–proteins CPI and CPIa of photosystem I and a reduced level of LHCPs.

Low-light chloroplasts

LL chloroplasts contain no starch, but possess a high thylakoid frequency and high and broad grana stacks. Their higher stacking degree is associated with an increased level of the LHCPs (LHCP$_1$, LHCP$_2$, LHCP$_3$, LHCP$_y$). They exhibit a larger pigment antenna (larger photosynthetic unit size), which is also seen from higher levels of chlorophyll b and xanthophylls.

References

Abel, B. (1956). Über die Beeinflussung der Chloroplastenstruktur durch Licht bei *Antirrhinum majus* (haploid). *Naturwissenschaften*, **43**, 136–7.

Akoyunoglou, G., Anni, H. & Kalosakas, K. (1980). The effect of light quality and the mode of illumination on chloroplast development in etiolated bean leaves. In *The Blue Light Syndrome*, ed. H. Senger, pp. 473–84. Springer Verlag, Berlin.

Anderson, J. M. (1980). Chlorophyll–protein complexes of higher plant thylakoids: distribution, stoichiometry and organization in the photosynthetic unit. *FEBS Lett.*, **117**, 327–31.

Anderson, J. M. (1981). Consequences of spatial separation of photosystems I and II in thylakoid membranes of higher plant chloroplasts. *FEBS Lett.*, **124**, 1–10.

Andersson, B. & Anderson, J. M. (1980). Lateral heterogeneity in the distribution of chlorophyll–protein complexes of the thylakoid membranes of spinach chloroplasts. *Biochim. biophys. Acta*, **593**, 427–40.

Armond, P. A. & Arntzen, C. J. (1977). Localization and characterization of photosystem II in grana and stroma lamellae. *Pl. Physiol.*, **59**, 398–404.

Arntzen, C. J. (1978). Dynamic structural features of chloroplast lamellae. *Curr. Top. Bioenerget.*, **8**, 111–60.

Ballantine, J. E. M. & Forde, B. J. (1970). The effect of light intensity and temperature on plant growth and chloroplast ultrastructure in soybean. *Am. J. Bot.*, **57**, 1150–9.

Berger, C. & Bergmann, L. (1967). Farblicht und Plastidendifferenzierung im Speichergewebe von *Solanum tuberosum* L. *Z. Pfl. Physiol.*, **56**, 439–45.

Bergmann, L. & Berger, C. (1966). Farblicht und Plastidendifferenzierung in Zellkulturen von *Nicotiana tabacum* var. 'Samsun'. *Planta*, **69**, 58–69.

Björn, L. O. (1980). Blue light effects on plastid development in higher plants. In *The Blue Light Syndrome*, ed. H. Senger, pp. 455–64. Springer Verlag, Berlin.

Björn, L. O., Suzuki, Y. & Nilsson, J. (1963). Influence of wavelength on the light response of excised wheat roots. *Physiologia Pl.*, **16**, 132–41.

Boardman, N. K., Björkman, O., Anderson, J. M., Goodchild, D. J. & Thorne, S. W. (1975). Photosynthetic adaptation of higher plants to light intensity: relationship between chloroplast structure, composition of the photosystems and photosynthetic rates. In *Proc. 3rd Int. Congr. Photosynthesis*, vol. 3, ed. M. Avron, pp. 1809–27. Elsevier, Amsterdam.

Buschmann, C. & Lichtenthaler, H. K. (1977). Hill-activity and P700 concentration of chloroplasts isolated from radish seedlings treated with β-indoleacetic acid, kinetin or gibberellic acid. *Z. Naturfors.*, **32c**, 798–802.

Buschmann, C. & Lichtenthaler, H. K. (1982). The effect of cytokinins on growth and pigment accumulation of radish seedlings (*Raphanus sativus* L.) grown in the dark and at different light quanta fluence rates. *Photochem. Photobiol.*, **35**, 217–21.

Buschmann, C., Meier, D., Kleudgen, H. K. & Lichtenthaler, H. K. (1978). Regulation of chloroplast development by red and blue light. *Photochem. Photobiol.*, **27**, 195–8.

Carter, D. P. & Staehelin, L. A. (1980). Proteolysis of chloroplast thylakoid membranes. II. Evidence for the involvement of the light-harvesting chlorophyll *a*/*b*–protein complex in thylakoid stacking and for effects of magnesium ions on photosystem II-light-harvesting complex aggregates in the absence of membrane stacking. *Arch. Biochem. Biophys.*, **200**, 374–86.

Engelbrecht, A. H. P. & Weier, T. E. (1967). Chloroplast development in the germinating safflower (*Carthamus tinctorius*) cotyledon. *Am. J. Bot.*, **54**, 844–56.

Fasse-Franzisket, U. (1956). Die Teilung der Proplastiden und Chloroplasten bei *Agapanthus umbellatus* 'l'Hérit'. *Protoplasma*, **45**, 194–227.

Guillot-Salomon, T., Tuquet, C., Lubac, M. de, Hallais, M.-F. & Signol, M. (1978). Analyse comparative de l'ultrastructure et de la composition lipidique des chloroplastes de plantes d'ombre et de soleil. *Cytobiologie*, **17**, 442–52.

Heumann, H. G., Lichtenthaler, H. K., Meier, D. & Peveling, E. (1984). Ultrastructure of chloroplasts from sun and shade leaves of *Fagus sylvatica*. *Protoplasma*, in press.

Horvath, I., Kalman, F. & Titlyanov, E. A. (1975). The influence of monochromatic blue and red light on the electron microscope structure of chloroplasts. *Acta bot. Acad. Sci. hung.*, **21**, 273–8.

Kesselmeier, J. (1982). Steroidal saponins in etiolated, greening and green leaves and in isolated etioplasts and chloroplasts of *Avena sativa*. *Protoplasma*, **112**, 127–32.

Lichtenthaler, H. K. (1967). Beziehungen zwischen Zusammensetzung und Struktur der Plastiden in grünen und etiolierten Keimlingen von *Hordeum vulgare* L. *Z. Pfl. Physiol.*, **56**, 273–81.

Lichtenthaler, H. K. (1977). Regulation of prenylquinone synthesis in higher plants. In *Lipids and Lipid Polymers in Higher Plants*, ed. M. Tevini & H. K. Lichtenthaler, pp. 231–58. Springer Verlag, Berlin.

Lichtenthaler, H. K. (1979). Effect of biocides on the development of the photosynthetic apparatus of radish seedlings grown under strong and weak light conditions. *Z. Naturfors.*, **34C**, 936–40.

Lichtenthaler, H. K. (1981). Adaptation of leaves and chloroplasts to high quanta fluence rates. In *Photosynthesis VI*, ed. G. Akoyunoglou, pp. 278–88. Balaban International Sciences Services, Philadelphia.

Lichtenthaler, H. K., Bach, T. J. & Wellburn, A. R. (1982a). Cytoplasmic and plastidic isoprenoid compounds of oat seedlings and their distinct labelling from ^{14}C-mevalonate. In *Biochemistry and Metabolism of Plant Lipids*, ed. J. F. G. M. Wintermans & P. J. C. Kuiper, pp. 489–500. Elsevier Biomedical Press B.V., Amsterdam.

Lichtenthaler, J. K., Burkard, G., Grumbach, H. K. & Meier, D. (1980a). Physiological effects of photosystem II-herbicides on the development of the photosynthetic apparatus. *Photosyn. Res.*, **1**, 29–43.

Lichtenthaler, H. K., Burkard, G., Kuhn, G. & Prenzel, U. (1981a). Light-induced accumulation and stability of chlorophylls and chlorophyll–proteins during chloroplast development. *Z. Naturfors.*, **36C**, 421–30.

Lichtenthaler, H. K. & Buschmann, C. (1978). Control of chloroplast development by red light, blue light and phytohormones. In *Chloroplast Development: Developments in Plant Biology*, vol. 2, ed. G. Akoyunoglou & J. H. Argyroudi-Akoyunoglou, pp. 801–16. Elsevier, Amsterdam.

Lichtenthaler, H. K., Buschmann, C., Döll, M., Fietz, H.-J., Bach, T., Kozel, U., Meier, D. & Rahmsdorf, U. (1981b). Photosynthetic activity, chloroplast ultrastructure, and leaf characteristics of high-light and low-light plants and of sun and shade leaves. *Photosyn. Res.*, **2**, 115–41.

Lichtenthaler, H. K., Buschmann, C. & Rahmsdorf, U. (1980b). The importance of blue light for the development of sun-type chloroplasts. In *The Blue Light Syndrome*, ed. H. Senger, pp. 485–94. Springer Verlag, Berlin.

Lichtenthaler, H. K., Kuhn, G., Prenzel, U., Buschmann, C. & Meier, D. (1982b). Adaptation of chloroplast ultrastructure and of chlorophyll–protein levels to high-light and low-light growth condition. *Z. Naturfors.*, **37C**, 464–75.

Lichtenthaler, H. K., Kuhn, G., Prenzel, U. & Meier, D. (1982c). Chlorophyll–protein levels and stacking degree of thylakoids in radish chloroplasts from high-light, low-light and bentazon-treated plants. *Physiologia Pl.*, **56**, 183–8.

Lichtenthaler, H. K., Meier, D., Retzlaff, G. & Hamm, R. (1982*d*). Distribution and effects of bentazon in crop plants and weeds. *Z. Naturfors.*, **37C**, 889–97.

Lichtenthaler, H. K., Prenzel, U. & Kuhn, G. (1982*e*). Carotenoid composition of chlorophyll–carotenoid–proteins from radish chloroplasts. *Z. Naturfors.*, **36C**, 10–12.

Lütz, C. & Klein, S. (1979). Biochemical and cytological observations on chloroplast development. VI. Chlorophylls and saponins in prolamellar bodies and prothylakoids separated from etioplasts of etiolated *Avena sativa* L. leaves. *Z. Pfl. Physiol.*, **95**, 227–37.

McDonnel, A. & Staehelin, L. A. (1980). Adhesion between liposomes mediated by the chlorophyll *a/b* light-harvesting complex isolated from chloroplast membranes. *J. Cell Biol.*, **84**, 40–56.

Meier, D. & Lichtenthaler, H. K. (1981). Ultrastructural development of chloroplasts in radish seedlings grown at high and low light conditions and in the presence of the herbicide bentazon. *Protoplasma*, **107**, 195–207.

Richter, G. (1969). Chloroplastendifferenzierung in isolierten Wurzeln. *Planta*, **86**, 229–300.

Ryrie, I. J., Anderson, J. M. & Goodchild, D. J. (1980). The role of the light-harvesting chlorophyll *a/b*–protein complex in chloroplast membrane stacking. *Eur. J. Biochem.*, **107**, 345–54.

Shakhov, A. A. & Balaur, N. S. (1969). Action spectrum and structure of chloroplasts. *Dokl. Akad. Nauk SSSR*, **186**, 1206–9.

Skene, D. S. (1974). Chloroplast structure in mature apple leaves grown under different levels of illumination and their response to changed illumination. *Proc. R. Soc. Ser. B*, **186**, 75–8.

Sprey, B. & Lichtenthaler, H. K. (1966). Zur Frage der Beziehungen zwischen Plastoglobuli und Thylakoidgenese in Gerstenkeimlingen. *Z. Naturfors.*, **21B**, 697–9.

Staehelin, L. A. & Arntzen, C. J. (1979). Effects of ions and gravity forces on the supramolecular organization and excitation energy distribution in chloroplast membranes. In *Chlorophyll Organization and Energy Transfer in Photosynthesis*, ed. G. Wolstenholme & D. W. Fitzsimons, Ciba Foundation Symposium 61, pp. 147–69. Excerpta Medica, Amsterdam.

Steinback, K. E., Burke, J. J., Mullet, J. E. & Arntzen, C. J. (1978). The role of the light-harvesting complex in cation-mediated grana formation. In *Chloroplast Development*, ed. G. Akoyunoglou & J. H. Argyroudi-Akoyunoglou, pp. 389–400. Elsevier, Amsterdam.

Sundqvist, C., Björn, L. O. & Virgin, H. I. (1980). Factors in chloroplast differentiation. In *Chloroplasts*, ed. J. Reinert, pp. 201–4. Springer Verlag, Heidelberg.

Virgin, H. I., Kahn, A. & von Wettstein, D. (1963). The physiology of chlorophyll formation in relation to structural changes in chloroplasts. *Photochem. Photobiol.*, **2**, 83–91.

Vlasova, M. P., Drozdova, I. S. & Voskresenskaya, N. P. (1971). Changes in fine structure of the chloroplasts in pea plants greening under blue and red light. *Fiziol. Rast.*, **18**, 5–11.

Voskresenskaya, N. P. (1972). Blue light and carbon metabolism. *A. Rev. Pl. Physiol.*, **23**, 219–34.

Weier, T. E. & Brown, D. L. (1970). Formation of the prolamellar body in 8-day, dark-grown seedlings. *Am. J. Bot.*, **57**, 267–75.

Wehrmeyer, W. (1964). Über Membranbildungsprozesse im Chloroplasten. 2. Mitteilung: Zur Entstehung der Grana durch Membranüberschiebung. *Planta*, **63**, 13–30.

Wild, A. (1979). Physiology of photosynthesis in higher plants. The adaptation of photosynthesis to light intensity and light quality in higher plants. *Ber. dt. bot. Ges.*, **92**, 341–64.

Wild, A. & Holzapfel, A. (1980). The effect of blue and red light on the content of chlorophyll, cytochrome *f*, soluble reducing sugars, soluble proteins and the nitrate reductase activity during growth of the primary leaves of *Sinapis alba*. In *The Blue Light Syndrome*, ed. H. Senger, pp. 444–51. Springer Verlag, Berlin.

Wild, A. & Wolf, G. (1980). The effect of different light intensities on the frequency and size of stomata, the size of cells, the number, size and chlorophyll content of chloroplasts in the mesophyll and the guard cells during the ontogeny of primary leaves of *Sinapis alba*. *Z. Pfl. Physiol.*, **97**, 325–42.

Yun-ling, D., Zhong-xi, C., Chun-hui, X., Da-zhang, M. & Fu-hong, Z. (1980). Studies on the chlorophyll–protein complexes from several sun and shade plants. In *5th int. Congress of Photosynthesis*, ed. G. Akoyunoglou, Abstractbook, p. 636. Athens. Balban International Science Services, Philadelphia.

Ziegler, R. & Egle, K. (1965). Zur quantitativen Analyse der Chloroplastenpigmente. I. Kritische überprüfung der spektralphotometrischen chlorophyllbestimmung. *Beitr. Biol. Pfl.*, **41**, 11–37.

M. EDELMAN, A. K. MATTOO AND
J. B. MARDER

Three hats of the rapidly metabolized 32 kD protein of thylakoids

Molecular control

In 1974 Eaglesham and Ellis discovered a group of quantitatively minor membrane components synthesized in the light by intact, isolated pea chloroplasts. Most prominently labelled among these was a 32000 Dalton (32 kD) polypeptide called peak D. The characteristics of peak D in isolated chloroplasts (Eaglesham & Ellis, 1974; Siddell & Ellis, 1975) and of a rapidly metabolized, 32 kD protein synthesized *in vivo* in *Spirodela* (Edelman & Reisfeld, 1978; Weinbaum *et al.*, 1979) have since been extensively studied.

Synthesis in vivo

In *Spirodela oligorhiza* the 32 kD protein is the main membrane component synthesized by the chloroplast protein-synthesizing system. Evidence for this is its synthesis *in organello* (Reisfeld, 1979), and the insensitivity of its synthesis to cycloheximide but its sensitivity to D-*threo*-chloramphenicol *in vivo* (Fig. 1).

The 32 kD protein is produced in quantities approaching those of the vastly abundant large subunit of ribulose-1,5-bisphosphate carboxylase (Edelman & Reisfeld, 1980); however, it does not accumulate, turning over at a rate some 50 times more rapid than that of the large subunit of carboxylase (LS) or that of the apoprotein of the chlorophyll *a/b* light-harvesting complex (LHCP) (Edelman & Reisfeld, 1980; Hoffman-Falk, 1980). This is demonstrated in Fig. 2 where the results of a pulse-chase experiment with [³H]leucine-labelled fronds are shown.

Comparing template activity and protein synthesis

The rapidly metabolized 32 kD protein of *Spirodela* is synthesized from a discrete, poly(A)⁻, chloroplast messenger RNA of 500 kD (Rosner *et al.*, 1977) as a 33.5 kD precursor protein molecule which is subsequently processed to 32 kD. This translation has been studied *in vitro* in heterologous protein-synthesizing systems (Reisfeld, Jakob & Edelman, 1978). From these

283

studies it is clear that this mRNA is among the most abundant transcription products of the cell.

In order to understand the regulation of synthesis of the 32 kD protein, the question of the levels of its macromolecular control has been approached through experiments both *in vivo* and *in vitro*. *Spirodela* colonies can grow in the light as autotrophs, or indefinitely in the dark as heterotrophs when

Fig. 1. Steady-state (stain) and pulse-labelled (autoradiogram) proteins from chloroplast membranes of *Spirodela*. *Spirodela* fronds were labelled for 3 h with [^{35}S]methionine. Chloramphenicol (200 μg ml^{-1}) was present where indicated from 1 h before labelling, until the end of labelling. Chloroplast extraction, polyacrylamide gel electrophoresis (PAGE), staining with Coomassie blue, and autoradiography were as described in Mattoo *et al.* (1982). Molecular weights (in kD) are given on the left. The arrow indicates the labelled 32 kD band. Lanes 1, control; lanes 2, chloramphenicol-treated; lanes 3, cycloheximide-treated.

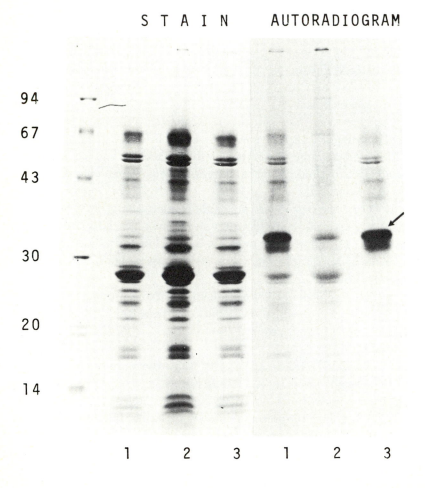

phototrophic media are supplemented with sucrose (Edelman & Reisfeld, 1980). Under these steady-state conditions there is a correlation between the rate of 32 kD protein formation and the level of its *in vitro* mRNA activity. Thus, 32 kD protein and transcriptional activity for the 33.5 kD precursor are hardly evident during steady-state, heterotrophic dark growth (Fig. 3, dark, + sucrose); on the other hand considerable 32 kD protein synthesis and

Fig. 2. Chase of [³H]leucine-labelled proteins from whole cell homogenates of *Spirodela*. Plants were labelled with 20 μCi ml⁻¹ [³H]leucine for 3 h, rinsed with mineral medium containing 100-fold unlabelled leucine, and chased in the same medium for different periods of time as indicated. Whole-cell homogenate proteins were prepared and electrophoresed on 20 cm long SDS–PAG. Equal amounts of radioactivity were applied to the slots, the gels processed for fluorography, and optical density scans made at 500 nm. The LS peak represents an optical density of approximately 1, and marks the position of the large subunit of ribulose bisphosphate carboxylase. LHCP marks the apoprotein of the light-harvesting chlorophyll *a/b*–protein complex. (From Hoffman-Falk, 1980.)

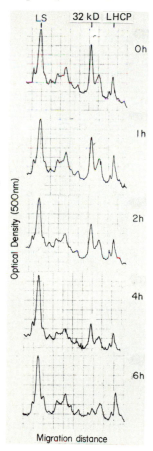

mRNA activity is evidenced during steady-state phototrophic growth (Fig. 3, +light). A similar correlation was found during early greening stages (Reisfeld *et al.*, 1978), and between juvenile versus mature fronds, as regards synthesis *in vivo* of 500 kD plastid mRNA and 32 kD protein (Weinbaum *et al.*, 1979). However, when steady-state, dark-grown *Spirodela* were starved

Fig. 3. Coupling and uncoupling of template activity and protein synthesis for the 32 kD polypeptide in *Spirodela*. Steady-state light-grown *Spirodela* were labelled for 3 h with [³⁵S]methionine and the membranes isolated. Concomitantly, RNA was extracted from these cultures and translated in wheat-germ extracts. Steady-state dark-grown *Spirodela* in the presence of sucrose, and dark-grown *Spirodela* starved for sucrose for 17 days (by changing the medium to a mineral growth without sucrose), were labelled with [³⁵S]methionine (100 μCi ml⁻¹) for 5 h, and whole cell membranes were isolated. RNA was extracted from these cultures and translated in wheat-germ extracts. The various samples were then electrophoresed and fluorographed. Lanes 1 – membrane proteins labelled *in vivo*. Lanes 2 – products translated *in vitro*. Molecular weights in kD are marked as 32 (mature form) and 33.5 (precursor form). (From Reisfeld, 1979.)

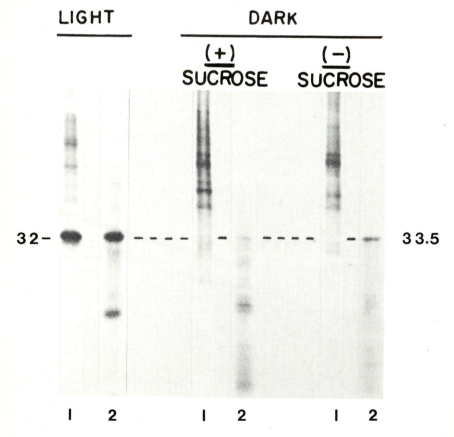

of a carbon source, an uncoupling of transcription from translation seemed to occur resulting in an accumulation of mRNA activity without concomitant synthesis of product (Fig. 3, dark, −sucrose). Thus, under this stress condition, regulation of 32 kD synthesis appears to be at some post-transcriptional level.

Processing of a precursor in vivo

A 33.5 kD translation product was also found *in vivo* in experiments where *Spirodela* fronds were labelled with radioactive amino acids for very short periods of time. Indeed, labelling periods of a few minutes proved sufficient clearly to identify newly and rapidly synthesized chloroplast proteins in this aquatic plant. Figure 4 shows the results of an experiment

Fig. 4. Synthesis, processing and localization *in vivo* of the precursor form of the 32 kD protein in *Spirodela*. Light-grown cultures were labelled *in vivo* with 500 μCi ml^{-1} of [^{35}S]methionine for 6 min (lane 1), 10 min (lanes 2, 6, 7), 12 min (lane 3), and 55 min (lane 4), or with 400 μCi ml^{-1} of [^{3}H]lysine for 10 min (lane 5). The sample shown in lane 7 was preincubated with 5 μg ml^{-1} of tunicamycin for 5 min prior to labelling. Whole cell (lanes 1–4) or chloroplast (lanes 5–7) membranes were analyzed. Lane 8 shows the [^{35}S]methionine-labelled products of *Spirodela* chloroplast poly(A)$^{-}$-RNA translated *in vitro* in a wheat-germ system. Samples were electrophoresed on SDS–PAG and fluorographed. LHCP marks the apoprotein of the light-harvesting chlorophyll *a/b* complex. The numbers on the left are the apparent molecular weights. (Adapted from Reisfeld *et al.*, 1982.)

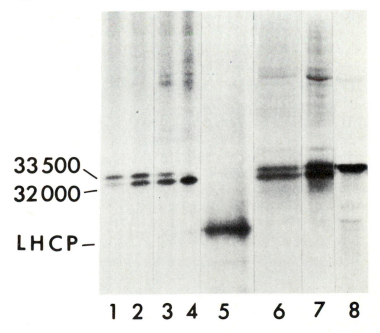

in which phototrophically grown fronds were incubated with [^{35}S]methionine for periods ranging from 6 to 55 min. Polyacrylamide gel electrophoresis (PAGE) of the membrane proteins showed two main bands, at 33.5 kD and 32 kD. With increasing time of labelling, the radioactivity in the latter built up, apparently at the expense of the former (lanes 1–4). Neither of these proteins incorporated [^{3}H]lysine (lane 5), a characteristic associated with the 32 kD protein of *Spirodela* (Edelman & Reisfeld, 1978; Reisfeld, Mattoo & Edelman, 1982) and other organisms (Zurawski *et al.*, 1982). Both protein bands were localized to the chloroplast membrane fraction (lane 6). In this fraction their appearance as separate bands, as well as their mobilities, was not affected by prior treatment of fronds *in vivo* with tunicamycin, an inhibitor of protein glycosylation (Hidetaka & Elbein, 1981) (lane 7). In lane 8, the slower-moving band synthesized *in vivo* is seen to have co-electrophoresed with the 33.5 kD polypeptide translated *in vitro* on a template of chloroplast mRNA.

In *bona fide* pulse-chase experiments (Reisfeld *et al.*, 1982), the 33.5 kD protein of *Spirodela* is processed *in vivo*, in the light, with a half-life of about 3 min. This processing occurs even in the presence of chloramphenicol or cycloheximide. Thus protein maturation is uncoupled from synthesis, i.e. the precursor is processed into the mature 32 kD form only after completion of the polypeptide chain and insertion into the membrane. Processing has recently been shown to consist of removal of an oligopeptide from the carboxy-terminus of the precursor molecule (Marder, Goloubinoff & Edelman, 1984).

Widespread presence in photosynthetic organisms

Thylakoid proteins, rapidly synthesized in the light and having a molecular weight of 32 kD have also been described in maize (Grebanier, Steinback & Bogorad, 1979), spinach (Bottomley, Spencer & Whitfield, 1974), tobacco (Lescure, 1978), and other organisms (for a comprehensive listing and references see Edelman, Marder & Mattoo, 1983). We compared several structural properties in rapidly labelled, 32 kD polypeptides from chloroplast membranes of various angiosperms, such as *Spirodela*, pea, tobacco, maize, lettuce, *Solanum* and *Bromus*, and the alga *Chlamydomonas* (Hoffman-Falk *et al.*, 1982). Partial proteolytic mapping of the 32 kD polypeptide showed similar patterns for all species tested (Fig. 5). In some cases, a 33.5 kD precursor polypeptide was identified which also displayed interspecific structural homologies. Lastly, trypsinization of the surface-exposed 32 kD protein *in situ* in the thylakoids, yielded a common pattern of products for the organisms studied (cf. Fig. 11). These findings point toward a broad distribution and phylogenetic similarity of the 32 kD thylakoid protein of the chloroplast at levels of precursor maturation, membrane orientation, and primary structure.

Summary

In summary, 32 kD mRNA and protein are major products in eukaryotic chloroplasts. In *Spirodela*, the flow of information culminating in production of the mature 32 kD polypeptide can apparently be regulated at transcriptional or translational levels, depending on the physiological state of the plant. Two unusual features characteristic of this thylakoid protein have been shown: 1. Although not vectorally transported across the chloroplast envelope, the initial translation product, while on the thylakoid membrane, undergoes carboxy-terminal processing to its mature form. 2. In steady-state phototrophic conditions the mature 32 kD protein is rapidly degraded.

Physiological role

The function of the 32 kD thylakoid protein initially eluded discovery. Although light-induced and rapidly metabolized, it was not rate-limiting for

Fig. 5. Partial proteolytic digestion of the 32 kD protein from *Spirodela* (S), pea (P), and *Chlamydomonas* (C). *Spirodela* and *Chlamydomonas* were labelled *in vivo* with [³⁵S]methionine and then chloroplast membranes were isolated. Pea chloroplasts were labelled *in organello* for 30 min. Membrane samples were electrophoresed on successive SDS–PAG in the presence of proteolytic enzymes as indicated. Each pair of lanes was digested at the same enzyme concentration. The numbers represent the sizes in kD of two of the proteolytic fragments. (Adapted from Hoffman-Falk *et al.*, 1982.)

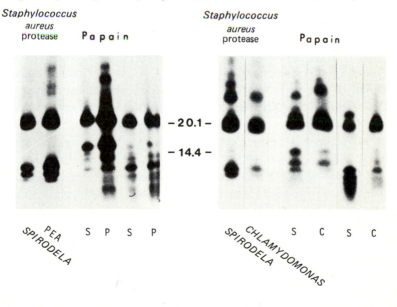

CO_2 fixation; nor was it involved in thylakoid biogenesis, being poorly synthesized in expanding juvenile tissue but vigorously synthesized in mature plants (Weinbaum *et al.*, 1979). The unravelling of the mystery surrounding its physiological role came only recently, following a re-evaluation, in *Spirodela*, of its possible role in photosynthesis (Mattoo *et al.*, 1981) and the appearance of mutants in maize lacking the protein (Leto & Miles, 1980; Leto, Keresztes & Arntzen, 1982).

Renger's proteinaceous-shield model

The use of mild trypsinization to probe structure–function relationships of membrane-surface proteins (Regitz & Ohad, 1975) led G. Renger in 1976 to infer the existence of a 'proteinaceous shield' covering a primary electron acceptor of photosystem (PS) II and acting as a regulator of electron flow between PSII and PSI in the chloroplast. In Renger's model (Renger, 1976), trypsin is assumed effectively to digest the proteinaceous shield in isolated thylakoids. This digestion leads to a functional inhibition of electron transport between a primary electron acceptor of PSII (Q) and the plastoquinone pool. At the same time, digestion of the shield bares Q to external redox agents. The model assumes that light-induced conformational changes occur *in vivo* in the proteinaceous shield permitting allosteric regulation of electronic interaction between Q and the plastoquinone pool. Renger further hypothesized that the well-known effect of the herbicide diuron (DCMU: 3-(3,4-dichlorophenyl)-1,1-dimethylurea) in inhibiting electron transport in photosynthesis (see Moreland, 1980) was a result of reversible binding of the herbicide to a site on, or close to, the shield, leading to allosteric inhibition of electron transfer from the primary acceptor to its connector molecule (plastoquinone B). After Renger (1976) had put forward the principal idea of the proteinaceous shield, developments came from the laboratories of A. Trebst (Trebst & Draber, 1979) and C. J. Arntzen (Pfister, Radosevich & Arntzen, 1979), who developed a more detailed view of overlapping binding sites on the shield for various herbicide inhibitors of PSII electron transport. Further evidence supporting the idea of a protein shield covering B then came from trypsin-based studies reported by several groups (Boger & Kunert, 1979; Croze, Kelly & Horton, 1979; Steinback, Burke & Arntzen, 1979; Trebst, 1979; Van Rensen & Kramer, 1979). The current hypothesis suggests that the 32 kD protein is the Q/B plastoquinone oxidoreductase (Arntzen & Duesing, 1983).

Relationship of the 32 kD protein to photosystem II activities

In Table 1 we list several factors which relate the rapidly metabolized 32 kD protein of *Spirodela* to Renger's proteinaceous shield.

Thylakoids depleted *in vivo* of a large percentage of their 32 kD protein (Weinbaum *et al.*, 1979) were used to delineate the reducing (acceptor) side of PSII as the region in the electron transport chain affected by the protein. Table 2 lists rates for several partial election transport reactions of control and 32 kD-depleted thylakoids. Only reactions passing through the site of DCMU inhibition were affected in the depleted thylakoids (Mattoo *et al.*,

Table 1. *Factors identifying the 32 kD protein of* Spirodela *with Renger's (1976) proteinaceous shield*

1. Exposed at the outer surface of the thylakoid membrane.
2. Removal of a portion of about 1 kD from the molecule makes Q accessible to external redox reagents.
3. 32 kD proteolysis is light-modulated.
4. In 32 kD-depleted thylakoids, electron transport is inhibited at the reducing side of PSII.
5. DCMU (diuron) alters trypsinization of the 32 kD protein.

Evidence for the above is given in Mattoo *et al.*, 1981.

Table 2. *Electron transport in control and 32 kD-depleted thylakoids of* Spirodela

Reaction measured	Initial rate of electron transport μequiv. (mg chl)$^{-1}$ h^{-1}		
	Control thylakoids	32 kD-depleted thylakoids	Depleted/ control
$H_2O \rightarrow$ silicomolybdate[a]	339	474	1.4
$H_2O \rightarrow$ phenylenediamine[b]	218	90	0.4
$H_2O \rightarrow$ ferricyanide[a]	705	385	0.5
$H_2O \rightarrow$ methylviologen[a]	204	92	0.5
Diaminodurene + ascorbate $\rightarrow$ methylviologen[a]	854	819	1.0
DCPIP[c] + ascorbate $\rightarrow$ methylviologen[a]	660	864	1.3

[a] Data from Mattoo *et al.*, 1981.
[b] Rates measured as ferricyanide reduction, followed as the light-induced decrease in absorbance at 420 nm. Measurements were carried out in a Cary 14 spectrophotometer. Saturating illumination was provided from a tungsten lamp through a Corning CS2-58 filter. The photomultiplier was protected from actinic light by Wratten 34 and Corning CS4-96 filters. Reaction mixtures contained 30 mmol l^{-1} Tricine–NaOH pH 7.8 : 40 mmol l^{-1} KCl : 2 mmol l^{-1} NH$_4$Cl : 1 mmol l^{-1} K$_3$Fe(CN)$_6$: 0.2 mmol l^{-1} phenylenediamine and thylakoids equivalent to 16 μg of chlorophyll in a total volume of 2.5 ml.
[c] Dichlorophenol-indophenol.

1981). Figure 6 shows fluorescence induction curves for control and depleted *Spirodela* fronds *in vivo*. The reduction in variable fluorescence in depleted fronds correlates the loss of the 32 kD protein with changes at the reducing side of PSII.

A photomorphogenic association involving the 32 kD protein appears to exist. PSII activity and particle formation are known to be late events in chloroplast development (Grumbach, 1981), and this correlates with the late development of a PSII-located, DCMU-sensitive site in developing chloroplasts (Smillie *et al.*, 1977). Figure 7 indicates a sharply changing ratio between LHCP and 32 kD synthesis during frond maturation. In the youngest frond (lane 4) a major activity is the laying down of the photosynthetic membranes. Here, synthesis of LHCP, the structural protein of the thylakoid, dominates. In mature fronds (lane 1) with fully developed chloroplasts (and electron transport activities), 32 kD synthesis predominates.

Fig. 6. Fluorescence induction curves for intact *Spirodela* fronds. Samples were incubated for 50 h in the dark or light (3000 lx), in the presence or absence of 20 μg ml^{-1} chloramphenicol (CAP), as indicated. Light-incubated samples were dark-adapted for at least 30 min before measurements were made. Time scales: 200 ms/division (solid lines) and 20 ms/division (dashed lines). Repeat measurements were made after a 2 min relaxation period. Induction was carried out according to the method of Malkin (1981), using a sample cell suitable for small colonies of fronds.

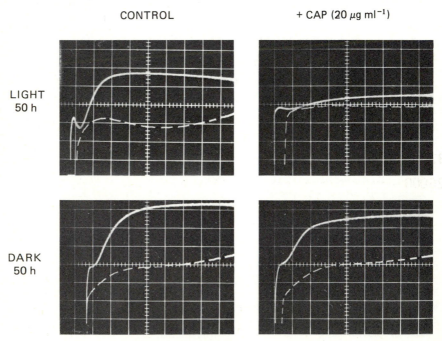

Delineating structural domains

Trypsin is a large, water-soluble enzyme which acts on surface-exposed portions of membrane proteins (Regitz & Ohad, 1975). From our proteolysis studies of isolated thylakoids (Mattoo *et al.*, 1981) an initial delineation of the structural domains of the 32 kD protein can be made; a 20 kD section of the molecule acts as a tightly bound anchor in the membrane while the remaining 12 kD section extends beyond the outer thylakoid surface into the stroma. Upon cleavage at one of the sites within the stromal domain, the primary acceptor of PSII (Q) becomes accessible to external redox reagents. Another nearby site becomes susceptible to proteolysis upon incubation of the thylakoids with DCMU. The 32 kD protein apparently undergoes a conformational change in the presence of DCMU such that as one trypsin-ization site is newly revealed, other sites become more resistant to enzymatic attack.

Fig. 7. Fractionation of newly synthesized, membrane-bound proteins of *Spirodela* during frond maturation. Steady-state, light-grown plants were radiolabelled for 1 h with [^{35}S]methionine at 2500 lx, washed, and a colony (lane 5) microdissected into individual fronds of different ages as in the schematic drawing (lanes 1–4). The sample in lane 6 is from a colony labelled in the presence of 200 μg ml^{-1} chloramphenicol (CAP). The sample in lane 7 is from the roots of a colony. Each lane received equal amounts of radioactivity. The numbers 32000 and 26000 mark the positions of the 32 kD protein and the apoprotein of LHCP, respectively.

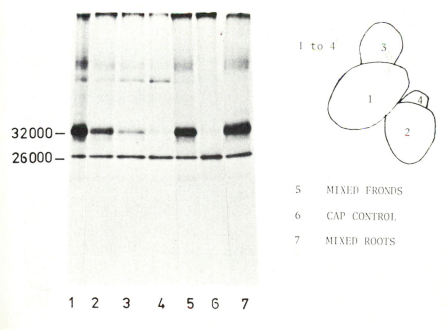

1 to 4

3

1

4

2

5   MIXED FRONDS

6   CAP CONTROL

7   MIXED ROOTS

32000 —

26000 —

1 2 3 4 5 6 7

The extents of the anchor and stromal domains have now been mapped, and the two portions of the molecule have been shown to differ markedly in their amino-acid compositions (Marder, Goloubinoff & Edelman, 1984). The anchor domain includes the amino-terminus and the stromal domain the carboxy-terminus of the protein. The site of the cleavage which results in the exposure of Q has tentatively been identified near the carboxy-terminus.

Fig. 8. Synthesis of the 32 kD protein with increase in light intensity. Light (2000 lx) grown *Spirodela* plants were pulse-labelled for 30 min with 350 μCi ml^{-1} of [^{35}S]methionine at 400, 3000 or 6000 lx of white light. After labelling, mature fronds were separated away from colonies and membrane fractions isolated. Samples with approximately equal amounts of starting material were subjected to electrophoresis and fluorographed. Symbols as in Fig. 4. (From Mattoo *et al.*, 1984.)

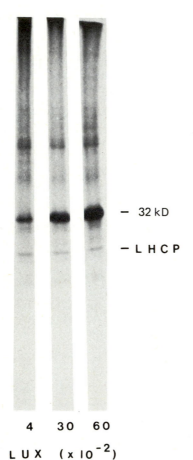

— 32 kD

— L H C P

4   30   60

L U X   (x 10^{-2})

The metabolism of the 32 kD protein is photoregulated

A feature of the 32 kD protein is its rapid metabolism. In *Spirodela* this feature can be shown to be finely attuned to incident light reaching the frond. Not only is the rate of 32 kD protein synthesis greater at increased light intensities (Fig. 8) but so too is its degradation (Fig. 9). Conversely, in the dark, synthesis (cf. Fig. 3) and degradation (Fig. 9) are minimal. We find it significant that in the presence of PSII inhibitors, such as DCMU, metabolism of the shield protein is minimal, even at high light intensities (Fig. 10). This coordination between the rate of synthesis and breakdown of the 32 kD protein may be one of the mechanisms by which the chloroplast adapts to changing light conditions. Light adaptations are known to be associated with specific alterations at several levels in the chloroplast, resulting in plants with

Fig. 9. Degradation of the 32 kD protein as a function of light intensity. Light-grown *Spirodela* plants were pulsed for 3 h with 50 μCi ml^{-1} of [^{35}S]methionine at 3000 lx, then washed and incubated with fresh sucrose-free mineral medium containing 10 mmol l^{-1} unlabelled methionine in the dark or at the light intensities indicated. Samples were removed after 2 h (lanes 1), 5 h (lanes 2), 10 h (lanes 3) and 20 h (lanes 4) of the chase. Cell membranes were isolated and subjected to electrophoresis and fluorography. Samples with equal amounts of radioactivity were loaded on the gels. LHPC refers to the apoprotein of the light-harvesting chlorophyll *a/b* complex. (From Hoffman-Falk, 1980.)

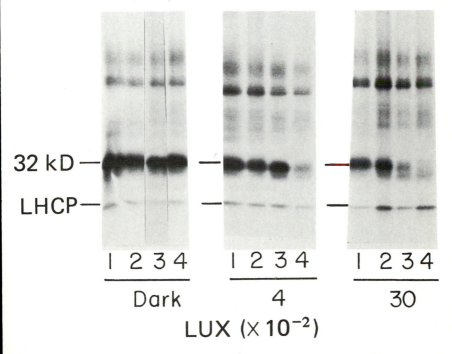

shade-type or sun-type plastids (cf. Malkin & Fork, 1981; Lichtenthaler & Meier, this volume). Herbicides that inhibit electron flow through PSII also induce shade-type chloroplasts in plants grown in high light (Lichtenthaler, 1979; Lichtenthaler & Meier, this volume), while high light intensity can also be detrimental to certain plants, causing photoinhibition (Critchley, 1981). Under normal growth conditions a mechanism may exist to protect the chloroplast against excessive illumination.

Our working model sees the *metabolism* of the 32 kD protein as being tightly photoregulated and coupled to electron transport. At its site of action the 32 kD protein may, in addition, serve a scavenging function preventing activated electrons from occasionally reducing the wrong component

Fig. 10. The effect of DCMU (diuron) on degradation of the 32 kD protein. Light-grown *Spirodela* plants were pulse-labelled with 50 μCi ml^{-1} of [^{35}S]methionine for 4 h at 3000 lx of white light. The plants were washed and resuspended in 3 ml of sucrose-free mineral medium containing 10 mmol l^{-1} unlabelled methionine in the absence (control) or presence of 10 μmol l^{-1} DCMU (diuron). Samples were removed at time 0 (lane 0), and after 1 h, 3 h, 7 h and 13 h of the chase. Cell membranes were isolated and analysed on an equal radioactivity basis by PAGE. Symbols as in Fig. 4. (From Mattoo *et al.*, 1984.)

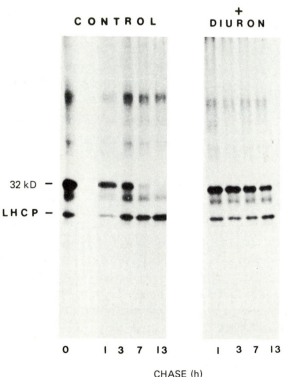

(Ellis, 1981). As a consequence of its functions, the 32 kD protein becomes denatured and is consumed. Hence its need for constant replenishment at a rate which is attuned to its rate of degradation.

Agricultural role
The 32 kD polypeptide as a herbicide binding protein
Direct evidence for the binding of herbicides to the 32 kD shield protein has come from studies (Pfister *et al.*, 1981) using azido-atrazine (2-azido-4-ethylamino-6-isopropylamino-*s*-triazine) as a covalent, photo-affinity label to identify the herbicide receptor protein. [^{14}C]Azido-atrazine covalently attaches to thylakoid membranes of triazine-sensitive *Amaranthus hybridus* but not to the membranes of a triazine-resistant mutant biotype. Autoradiographic analysis showed the binding to be to the 32 kD protein in the sensitive species. Previously, competitive non-covalent binding studies conducted with radioactive herbicides suggested that *s*-triazines, triazones, pyridazinones, biscarbamates, phenylureas, *N*-phenylcarbamates, uracils, acrylamides, cyclic ureas and thiadiazoles all interfered with the same photosynthetic electron transport site at the reducing side of PSII (Pfister *et al.*, 1979; Trebst, 1979; Van Rensen & Kramer, 1979; Moreland, 1980). These studies also demonstrated a single binding site per electron transport chain. To account for the results of the competitive binding studies the various families of herbicides were suggested to have both specific and overlapping binding sites on the shield protein (Pfister *et al.*, 1979; Trebst & Draber, 1979). It is thought that by binding to the PSII complex, herbicide inhibitors, such as diuron or atrazine, induce a change in the redox potential of the quinone cofactor of B, thus making the transfer of electrons from the acceptor, Q, thermodynamically unfavorable (Pfister *et al.*, 1981).

Expression of the gene for the 32 kD protein in triazine-resistant plants
In recent years, along with massive application of the commercially successful *s*-triazine herbicides, resistance has appeared in several weed species in the most intensively sprayed areas (cf. Moreland, 1980). Mathematical considerations suggest resistance to be a suddenly appearing monogenic trait (Gressel & Segal, 1978). This is in line with evidence involving chloroplast membrane alterations in triazine-resistant species which also suggest genetically controlled modification of the herbicide target site (Arntzen, Ditto & Brewer, 1979). Genetic studies show the resistance to be a cytoplasmic, maternally inherited trait in *Brassica campestris*, correlated to the maternal inheritance of PSII components which confer atrazine sensitivity (Darr, Souza-Machado & Arntzen, 1981).

Results of comparative studies have shown that thylakoid membranes of triazine-resistant biotypes, which have recently appeared in susceptible species, have a greatly reduced affinity for atrazine (Pfister *et al.*, 1979; Arntzen *et al.*, 1979) and azido-atrazine (Pfister *et al.*, 1981). This prompted us to investigate whether atrazine-resistant plants synthesize the 32 kD protein (Mattoo *et al.*, 1982). Figure 11 shows the PAGE patterns of [35S]methionine-labelled proteins of thylakoids from resistant (r) and suscep-tible (s) biotypes of *Brassica campestris*, *Bromus tectorum* and *Chenopodium album* treated with trypsin for different periods of time. In all cases digestion of the 32 kD protein coincides with the appearance of fragment A, which on prolonged incubation was further digested to fragment B. The results of these comparisons indicate that (1) a close similarity in structure and membrane topography exists between the newly synthesized 32 kD polypeptides of the weed species investigated here and those more fully characterized in *Spirodela*;

Fig. 11. Distribution of rapidly labelled proteins after trypsin digestion of thylakoids from triazine-susceptible (s) and resistant (r) biotypes. [35S[Methionine-labelled thylakoid membranes, equivalent to 175 μg chlorophyll ml^{-1}, were digested with trypsin at 50 μg ml^{-1} in 10 mmol l^{-1} Tris-HCl, pH 7.6, at 22 °C for 5 min (lanes 2), 15 min (lanes 3) or 30 min (lanes 4); undigested samples are in lanes 1. Trypsin digestion was stopped by a 10-fold dilution of samples in a solution of lima-bean trypsin inhibitor (2.5 mg ml^{-1}) and phenylmethylsulfonylfluoride (1 mmol l^{-1}). Samples were electrophoresed on SDS–PAG and autoradiographed. The positions of the 32 kD protein and its digestion products (A and B) are indicated. (Adapted from Mattoo *et al.*, 1982.)

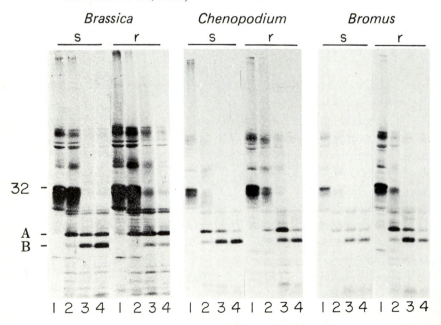

and (2) the same 32 kD polypeptide is synthesized by both the triazine r and s biotypes.

From our data it is clear that atrazine resistance does not result from a loss of expression of the gene for the 32 kD protein, as synthesis of the 32 kD polypeptide occurs at similar rates in both triazine r and s biotypes; 32 kD proteins in both biotypes have a similar mobility on denaturing polyacrylamide gels, and possess common cleavage sites for three different proteases (Steinback *et al.*, 1981; Mattoo *et al.*, 1982). Thus it appears that the primary structures of triazine r and s 32 kD proteins are not grossly different. However, point differences, e.g. a change in a single amino acid, would not be revealed by these experiments unless it occurred fortuitously at a cleavage site. Recently it has been reported that the polypeptide sequence decoded from the gene for the 32 kD protein in *Amaranthus* r and s biotypes indeed differ in a single amino acid residue (Hirschberg & McIntosh, 1983).

The collaborations of Avi Reisfeld and Hedda Hoffman-Falk in the experiments summarized in Fig. 3, and Figs. 2 and 9, respectively, are gratefully acknowledged. This work was supported in part by a US–Israel BARD grant.

References

Arntzen, C. J., Ditto, C. L. & Brewer, P. E. (1979). Chloroplast membrane alterations in triazine-resistant *Amaranthus retroflexus* biotypes. *Proc. natn. Acad. Sci. USA*, **76**, 278–82.

Arntzen, C. J. & Duesing, J. H. (1983). Chloroplast-encoded herbicide resistance. In *Advances in Gene Technology: Molecular Genetics of Plants and Animals*, ed. F. Ahmand *et al.*, pp. 273–91. Academic Press, New York.

Boger, P. & Kunert, K. J. (1979). Differential effects of herbicides upon trypsin-treated chloroplasts. *Z. Naturfors.*, **34C**, 1015–20.

Bottomley, W., Spencer, D. & Whitfeld, P. R. (1974). Protein synthesis in isolated chloroplasts. Comparison of light-driven and ATP-driven synthesis. *Archs. Biochem. Biophys.*, **164**, 106–17.

Critchley, C. (1981). Studies on the mechanism of protoinhibition in higher plants. Effects of high light intensity on chloroplast activities in cucumber adapted to low light. *Pl. Physiol.*, **67**, 1161–5.

Croze, E., Kelly, M. & Horton, P. (1979). Loss of sensitivity to diuron after trypsin digestion of chloroplast photosystem II particles. *FEBS Lett.*, **103**, 22–6.

Darr, S., Souza-Machado, V. & Arntzen, C. J. (1981). Uniparental inheritance of a chloroplast photosystem II polypeptide controlling herbicide binding. *Biochim. biophys. Acta*, **634**, 219–28.

Eaglesham, A. R. J. & Ellis, R. J. (1974). Protein synthesis in chloroplasts. II. Light-driven synthesis of membrane proteins by isolated pea chloroplasts. *Biochim. biophys. Acta*, **335**, 396–407.

Edelman, M., Marder, J. B. & Mattoo, A. K. (1983). A compendium of characteristics for the rapidly-metabolized 32 kD protein of the chloroplast membrane. In *Structure and Function of Plant Genomes*, ed. O. Ciferri & L. Dure, pp. 187–92. Plenum Press, New York.

Edelman, M. & Reisfeld, A. (1978). Characterization, translation and control of the 32 000 Dalton membrane protein in *Spirodela*. In *Chloroplast Development*, ed. G. Akoyunoglou & J. H. Argyroudi-Akoyunoglou, pp. 641–52. Elsevier/North-Holland, Amsterdam.

Edelman, M. & Reisfeld, A. (1980). Synthesis, processing and functional probing of P-32000, the major membrane protein translated within the chloroplast. In *Genome Organization and Expression in Plants*, ed. C. J. Leaver, pp. 353–62. Plenum Press, New York.

Ellis, R. J. (1981). Chloroplast proteins: synthesis, transport and assembly. *A. Rev. Pl. Physiol.*, **32**, 111–37.

Grebanier, A. E., Steinback, K. E. & Bogorad, L. (1979). Comparison of the molecular weight of proteins synthesized by isolated chloroplasts with those which appear during greening in *Zea mays*. *Pl. Physiol.*, **63**, 436–9.

Gressel, J. & Segal, L. A. (1978). The paucity of genetic adaptive resistance of plants to herbicides: possible biological reasons and implications. *J. theoret. Biol.*, **75**, 349–71.

Grumbach, K. H. (1981). Formation of photosynthetic pigments and quinones and development of photosynthetic activity. *Physiologia Pl.*, **51**, 53–62.

Hidetaka, H. & Elbein, A. D. (1981). Tunicamycin inhibits protein glycosylation in suspension cultured soybean cells. *Pl. Physiol.*, **67**, 882–6.

Hirschberg, J. & McIntosh, L. (1983). Molecular basis of herbicide resistance in *Amaranthus hybridus*. *Science*, **222**, 1346–9.

Hoffman-Falk, H. (1980). Ubiquity, metabolism and function of the 32 kD chloroplast membrane protein. M.Sc. Thesis, The Weizmann Institute of Science, Rehovot, Israel.

Hoffman-Falk, H., Mattoo, A. K., Marder, J. B., Edelman, M. & Ellis, R. J. (1982). General occurrence and structural similarity of the rapidly synthesized 32000-Dalton protein of the chloroplast membrane. *J. biol. Chem.*, **257**, 4583–7.

Lescure, A.-M. (1978). Chloroplast differentiation in cultured tobacco cells: *in vitro* protein synthesis efficiency of plastids at various stages of their evolution. *Cell Different.*, **7**, 139–52.

Leto, K. J., Keresztes, A. & Arntzen, C. J. (1982). Nuclear involvement in the appearance of a chloroplast-encoded 32000 Dalton thylakoid membrane polypeptide integral to the photosystem II complex. *Pl. Physiol.*, **69**, 1450–8.

Leto, K. J. & Miles, D. (1980). Characterization of three photosystem II mutants in *Zea mays* lacking a 32000 Dalton lamellar polypeptide. *Pl. Physiol.*, **66**, 18–24.

Lichtenthaler, H. K. (1979). Effect of biocides on the development of the photosynthetic apparatus of radish seedlings grown under strong and weak light conditions. *Z. Naturfors.*, **34C**, 936–40.

Malkin, S. (1981). Control of photosynthetic electron transfer from the reaction center to electron carriers of photosystem II studied by fluorescence induction. *Israel J. Chem.*, **21**, 306–15.

Malkin, S. & Fork, D. C. (1981). Photosynthetic units of sun and shade plants. *Pl. Physiol.*, **67**, 580–3.

Marder, J. B., Goloubinoff, P. & Edelman, M. (1984). Molecular architecture of the rapidly metabolized 32-kilodalton protein of photosystem II. *J. biol. Chem.*, **259**, 3900–8.

Mattoo, A. K., Hoffman-Falk, H., Marder, J. B. & Edelman, M. (1984). Regulation of protein metabolism: coupling of photosynthetic electron transport to *in vivo* degradation of the rapidly metabolized 32-kilodalton protein of the chloroplast membranes. *Proc. natn. Acad. Sci. USA*, **81**, 1380–4.

Mattoo, A. K., Marder, J. B., Gressel, J. & Edelman, M. (1982). Presence of the rapidly-labelled 32000-Dalton chloroplast membrane protein in triazine-resistant biotypes. *FEBS Lett.*, **140**, 36–40.

Mattoo, A. K., Pick, U., Hoffman-Falk, H. & Edelman, M. (1981). The rapidly metabolized 32000 Dalton polypeptide is the proteinaceous shield regulating photosystem II electron transport and mediating diuron herbicide sensitivity in chloroplasts. *Proc. natn. Acad. Sci. USA*, **78**, 1572–6.

Moreland, D. E. (1980). Mechanisms of action of herbicides. *A. Rev. Pl. Physiol.*, **31**, 597–638.

Pfister, K., Radosevich, S. R. & Arntzen, C. J. (1979). Modification of herbicide binding to photosystem II in two biotypes of *Senecio vulgaris* L. *Pl. Physiol.*, **64**, 995–9.

Pfister, K., Steinback, K. E., Gardner, G. & Arntzen, C. J. (1981). Photoaffinity labeling of a herbicide receptor protein in chloroplast membranes. *Proc. natn. Acad. Sci. USA*, **78**, 981–5.

Regitz, G. & Ohad, I. (1975). Changes in the protein organization in developing thylakoids of *Chlamydomonas reinhardtii* Y-1 as shown by sensitivity to trypsin. In *Proc. 3rd int. Congr. Photosynthesis*, vol. 3, ed. M. Avron, pp. 1615–25. Elsevier, Amsterdam.

Reisfeld, A. (1979). Characterization, translation and control of the chloroplast proteins P-32000 and LS-carboxylase from *Spirodela*. Ph.D. Thesis, The Weizmann Institute of Science, Rehovot, Israel.

Reisfeld, A., Gressel, J., Jakob, K. M. & Edelman, M. (1978). Characterization of the 32000 Dalton membrane protein. I. Early synthesis during photoinduced plastid development of *Spirodela*. *Photochem. Photobiol.*, **27**, 161–5.

Reisfeld, A., Jakob, K. M. & Edelman, M. (1978). Characterization of the 32000 Dalton chloroplast membrane protein. II. The molecular weight of chloroplast messenger RNAs translating the precursor to P-32000 and full-size RuBP carboxylase large subunit. In *Chloroplast Development*, ed. G. Akoyunoglou & J. H. Argyroudi-Akoyunoglou, pp. 669–74. Elsevier/ North-Holland, Amsterdam.

Reisfeld, A., Mattoo, A. K. & Edelman, M. (1982). Processing of a chloroplast-translated membrane protein *in vivo*. Analysis of the rapidly synthesized 32000-Dalton shield protein and its precursor in *Spirodela oligorhiza*. *Eur. J. Biochem.*, **124**, 125–9.

Renger, G. (1976). Studies on the structural and functional organization of system II of photosynthesis. The use of trypsin as a structurally selective inhibitor at the outer surface of the thylakoid membrane. *Biochim. biophys. Acta*, **440**, 287–300.

Rosner, A., Reisfeld, A., Jakob, K. M., Gressel, J. & Edelman, M. (1977). Shifts in the RNA and protein metabolism of *Spirodela* (duckweed). In *Acides Nucléiques et Synthèse des Protéines chez les Vegetaux*, eds. L. Bogorad & J. H. Weil, pp. 305–11. C.N.R.S., Paris.

Siddell, S. G. & Ellis, R. J. (1975). Protein synthesis in chloroplasts. VI. Characteristics and products of protein synthesis *in vitro* in etioplasts and developing chloroplasts from pea leaves. *Biochem. J.*, **146**, 675–85.

Smillie, R. M., Nielsen, N. C., Henningsen, K. W. & von Wettstein, D. (1977). Development of photochemical activity in chloroplast membranes. I. Studies with mutants of barley grown under a single environment. *Aust. J. Pl. Physiol.*, **4**, 415–49.

Steinback, K. E., Burke, J. J. & Arntzen, C. J. (1979). Evidence for the role of surface-exposed segments of the light-harvesting complex in cation-mediated control of chloroplast structure and function. *Archs. Biochem. Biophys.*, **195**, 546–57.

Steinback, K. E., McIntosh, L., Bogorad, L. & Arntzen, C. J. (1981). Identification of the triazine receptor protein as a chloroplast gene product. *Proc. natn. Acad. Sci. USA*, **78**, 7463–7.

Trebst, A. (1979). Inhibition of photosynthetic electron flow by phenol and diphenylether herbicides in control and trypsin-treated chloroplasts. *Z. Naturfors.*, **34C**, 986–91.

Trebst, A. & Draber, W. (1979). Structure-activity correlations of recent herbicides in photosynthetic reactions. *Adv. Pesticide Sci.*, **2**, 223–34.

Van Rensen, J. J. S. & Kramer, H. J. M. (1979). Short-circuit electron transport insensitive to diuron-type herbicides induced by treatment of isolated chloroplasts with trypsin. *Pl. Sci. Lett.*, **17**, 21–7.

Weinbaum, S. A., Gressel, J., Reisfeld, A. & Edelman, M. (1979). Characterization of the 32 000 Dalton chloroplast membrane protein. III. Probing its biological function in *Spirodela. Pl. Physiol.*, **64**, 828–32.

Zurawski, G., Bohnert, H. J., Whitfeld, P. R. & Bottomley, W. (1982). Nucleotide sequence of the gene for the M_r 32 000 thylakoid protein from *Spinacia oleracea* and *Nicotiana debneyi* predicts a totally conserved primary translation product of M_r 38 950. *Proc. natn. Acad. Sci. USA*, **79**, 7699–703.

T. F. GALLAGHER, S. M. SMITH, G. I. JENKINS AND R. J. ELLIS

Photoregulation of transcription of the nuclear and chloroplast genes for ribulose bisphosphate carboxylase

Chloroplast development requires the co-ordinated activities of both the chloroplast and nuclear genetic systems (Ellis, 1981). Some chloroplast proteins are encoded in the nucleus, whereas others are encoded in chloroplast DNA. Ribulose bisphosphate carboxylase–oxygenase (RuBP carboxylase, EC 4.1.1.39) is an example of a protein whose synthesis requires the participation of both the cytoplasmic and chloroplast protein-synthesizing systems. RuBP carboxylase is a multimeric protein consisting of eight large subunits (LSU), each of molecular weight about 52000, and eight small subunits (SSU), each of molecular weight about 14000 (Jensen & Bahr, 1977). The LSU is encoded in chloroplast DNA (Coen *et al.*, 1977; Bedbrook *et al.*, 1979), and is synthesized within the chloroplast (Blair & Ellis, 1973), while the SSU is encoded in the nucleus (Kawashima & Wildman, 1972; Kung, 1976) and is synthesized on cytoplasmic ribosomes as a precursor of molecular weight about 20000 (Dobberstein, Blobel & Chua, 1977; Highfield & Ellis, 1978; Chua & Schmidt, 1978). The SSU is post-translationally transported into the chloroplast, where it is cleaved to its mature size (Chua & Schmidt, 1978; Highfield & Ellis, 1978; Smith & Ellis, 1979) by a soluble processing enzyme which has been purified (Ellis & Robinson, 1984).

The LSU and SSU genes are present in markedly different amounts in plant cells. The LSU gene is present as a single copy per chloroplast DNA molecule (Bedbrook *et al.*, 1979), but there are several thousand copies of this DNA molecule in each leaf cell (Bedbrook & Kolodner, 1979). The small subunit is encoded in a small multigene family, and is present as a few (6–12) copies per nucleus (Cashmore, 1979; Berry-Lowe *et al.*, 1982; Broglie *et al.*, 1983; Dunsmuir, Smith & Bedbrook, 1983). It is not understood how the expression of genes present at such different dosages is co-ordinated to produce similar amounts of the two subunit polypeptides. In higher plants the synthesis of the two subunits is not tightly co-ordinated, but can be uncoupled by either inhibitors or heat-treatment (Feierabend & Wildner, 1978; Barraclough & Ellis, 1979).

One mechanism by which the synthesis of the two subunits is regulated involves the action of light. Exposure to light induces changes in the abundance of the majority of chloroplast proteins. In pea (*Pisum sativum*), light acting through the photoreceptor phytochrome stimulates the accumulation of RuBP carboxylase, measured either in terms of enzymic activity (Graham, Grieve & Smillie, 1968) or as the abundance of the two subunits (Jenkins, Hartley & Bennett, 1983). The amount of translatable messenger RNA (mRNA) for the small subunit is greater in light-grown pea seedlings than in plants grown in darkness (Bedbrook, Smith & Ellis, 1980; Sasaki *et al.*, 1981). Light also causes an increase in SSU mRNA activity in *Lemna gibba* (Tobin & Suttie, 1980), and an increase in both LSU and SSU translatable mRNA contents in *Cucumis sativus* (Walden & Leaver, 1981).

In this article we describe the results of experiments with *Pisum sativum* which indicate that the stimulatory effect of light on the accumulation of RuBP carboxylase is achieved, to a major extent, through an increase in transcription. These experiments involve the use of hybridization probes to measure both the content of specific RNA transcripts in seedlings exposed to various light treatments, and the rate of synthesis of SSU transcripts by isolated nuclei. In addition, we report both slow and rapid effects of light on the rate of transcription of SSU genes in different developmental situations.

Measurement of LSU and SSU transcripts in *Pisum* leaves

The amounts of particular gene transcripts in different cellular compartments can be determined by quantitative hybridization analysis using specific cloned DNA probes. RNA isolated from different organelles is resolved by agarose gel electrophoresis, and then transferred to diazobenzyloxymethyl (DBM) paper or nitrocellulose according to the procedure of Alwine, Kemp & Stark (1977). The paper is incubated under hybridization conditions with a specific, ^{32}P-labelled cloned DNA probe, washed to remove non-hybridized DNA, and then analysed by autoradiography. Figure 1 shows such a determination of the abundance of both LSU and SSU transcripts in RNA samples from *Pisum sativum*. Total leaf, nuclear or chloroplast RNA, isolated from plants grown for 9 days under a 12 h photoperiod, was transferred to DBM paper and incubated with hybridization probe sequences complementary to either the LSU or the SSU gene transcripts. The SSU probe was an equimolar mixture of plasmids pSSU60 and pSSU160 (Bedbrook *et al.*, 1980), a combination of cloned DNAs which represents most of the structural sequence of the mature *Pisum* SSU polypeptide. The LSU probe was a 580 base-pair sequence cleaved from a cloned BamH1 fragment of maize chloroplast DNA using the restriction endonuclease Pst1; this Pst1 fragment is contained entirely within the structural region of the LSU gene (Coen *et al.*, 1977). The LSU genes of pea and maize are sufficiently conserved to allow cross-hybridization to occur.

Sequences which hybridize to the SSU copy DNA (cDNA) probes can be detected in both total leaf and nuclear RNA (Fig. 1 A). The size class of the hybridizing species is similar in both cases and is about 900–1000 nucleotides (Fig. 1 A, a). Additional bands of higher molecular weight are seen in the nuclear RNA fraction (Fig. 1 A, b and c), and may represent precursor forms.

Fig. 1. Identification of RuBP carboxylase transcripts in RNA isolated from pea leaves. RNA was denatured by glyoxylation and fractionated by electrophoresis on agarose gels. The fractionated RNA was transferred to DBM paper and hybridized with nick-translated [^{32}P]DNA probes specific for the SSU (A) and the LSU (B) gene transcripts. The figure shows an autoradiograph of the DBM paper. T, total leaf RNA; N, nuclear RNA; C, chloroplast RNA. The symbol a indicates the mRNA for the SSU, and b and c indicate possible precursors of this mRNA present in nuclear RNA. The mobilities of pea ribosomal RNA are shown for comparison in the ethidium bromide-stained gel, photographed under ultraviolet (UV) light. O = origin. (Reprinted with permission from Smith & Ellis, 1981.)

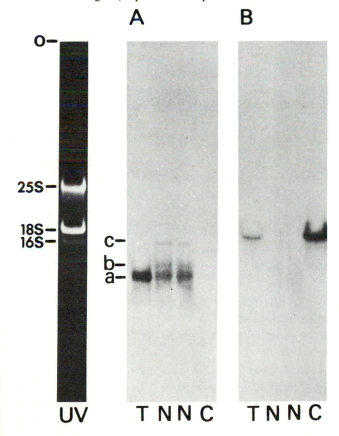

LSU sequences can be detected in chloroplast RNA, and to a lesser extent in total leaf RNA, but not in nuclear RNA (Fig. 1 B). These observations are consistent with the known locations of the LSU and SSU genes (Coen *et al.*, 1977; Kawashima & Wildman, 1972).

Light-stimulation of the accumulation of LSU and SSU gene transcripts in *Pisum* leaves

Bedbrook *et al.* (1980) demonstrated that a much greater amount of SSU mRNA, measured by translation in a heterologous cell-free system, was present in light-grown pea leaves than in dark-grown apical buds. As shown in Fig. 2, light causes a similar large increase in the abundance of transcripts for both LSU and SSU, as determined by hybridization analysis. These transcripts are barely detectable in RNA extracted from plants grown in darkness. This observation shows that the light-induced increase in translatable mRNA activity reported earlier was not the result of a conversion of mRNA into a form which could be translated in the heterologous protein-synthesizing system used, but rather resulted from an increase in the actual content of these transcripts in the RNA sample.

The concentration of SSU transcripts in RNA extracted from dark-grown apical buds has been analysed quantitatively. The degree of hybridization to the SSU probe on DBM paper is approximately proportional to the amount of RNA applied to the gel prior to transfer (Fig. 3, tracks c–f). The poly(A)$^+$-RNA fraction was used in these experiments to increase the sensitivity of the method. The large difference in SSU transcript content between light-grown and dark-grown tissue is again apparent (Fig. 3, tracks a, b), but when large amounts of RNA from dark-grown apical buds are applied to the gel a very small amount of SSU mRNA can be detected (Fig. 3, track g). The difference in SSU mRNA content between dark-grown and light-grown tissue is at least 30-fold. This difference is consistent with recent measurements using dot-blot hybridization which show that the concentration of SSU mRNA in RNA extracted from dark-grown *Pisum* apical buds is 1–2% of that in samples from buds illuminated with continuous white light for 48 h (Jenkins *et al.*, 1983).

It is possible that the transcription of SSU genes occurs at equal rates in both dark-grown and light-grown tissue, and that the effect of light is to cause the post-transcriptional selection and transport of these transcripts from nucleus to cytoplasm. However, analysis of the nuclear fraction shows that the SSU transcript is present in nuclear RNA isolated from light-grown tissue but not in nuclear RNA from dark-grown tissue (Fig. 2C).

The observations described above indicate that the light-stimulated accumulation of RuBP carboxylase in *Pisum sativum* is achieved to a major extent through an increase in the abundance of both LSU and SSU mRNAs. The question thus arises as to how this increase in transcript content is produced. There are two possibilities. Transcripts could be synthesized at equal rates in light and darkness, and the effect of light could be to inhibit an increased rate of degradation of transcripts that occurs in darkness. Alternatively, light could cause a stimulation of transcription. Hybridization analysis measures only the steady-state concentration of a specific transcript in an RNA sample,

Fig. 2. Transcripts of LSU and SSU genes in RNA isolated from light-grown pea leaves or dark-grown apical buds. Electrophoresis, transfer and hybridization were carried out as described in Fig. 1 using DNA probes for either the SSU (A and C) or the LSU (B). In A and B, total RNA (20 μg) from light-grown leaves (L) or dark-grown apical buds (D) was applied to the gel; in C, nuclear RNA (20 μg) from light-grown (L) or dark-grown (D) plants was used. The control track (A+) consisted of total RNA (20 μg) from dark-grown apical buds mixed with poly(A)+-RNA (0.1 μg) from light-grown leaves. Ribosomal RNA markers are shown on an ultraviolet (UV) photograph as in Fig. 1. O = origin. (Reprinted with permission from Smith & Ellis, 1981.)

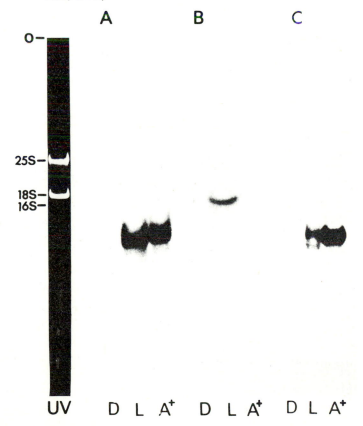

Fig. 3. Measurement of the relative amounts of SSU mRNA in light-grown pea leaves and dark-grown apical buds. Poly(A+)-RNA was isolated, fractionated and hybridized as in Fig. 1. Track a, light-grown RNA (10 μg) mixed with dark-grown RNA (10 μg); track b, dark-grown RNA (10 μg); tracks c, d, e, and f contain 10, 5, 2 and 1 μg of light-grown RNA, respectively; track g, 30 μg dark-grown RNA. O = origin.

O

a   b   c   d   e   f   g

and not its rate of synthesis or degradation, and therefore is not capable of distinguishing between these two possibilities; another experimental approach must be used.

Transcription of small subunit genes in isolated nuclei

One way to distinguish between the above possibilities in the case of the small subunit is to measure the rate of transcription of these genes in isolated nuclei. It has been observed in other species that transcription in isolated nuclei represents elongation of already-engaged polymerases rather than the combined result of elongation and initiation (Tsai *et al.*, 1978; Dermon *et al.*, 1981).

Nuclei were isolated from light-grown and dark-grown pea seedlings by a modification of the method of Luthe & Quatrano (1980*a*). Nuclei prepared by this method appear devoid of contamination by other cellular organelles when sections are examined by electron microscopy (Fig. 4). The nuclei lack membranes, probably because of the use of detergent in the isolation procedure. These isolated nuclei incorporate [³H]UMP into RNA when incubated with nucleoside triphosphates (Table 1). Transcription is reduced by 90% when the unlabelled nucleoside triphosphates are omitted, and by 75% when actinomycin D is added. The addition of α-amanitin at 10 μg ml^{-1}, a concentration which completely inhibits transcription by isolated wheat-germ RNA polymerase II (Jendrisak & Guilfoyle, 1978), inhibits RNA synthesis by 36%. The rate of transcription by *Pisum* leaf nuclei is comparable with those rates reported for nuclei isolated from plant tissues by other workers (Wilson & Bennett, 1976; Slater, Venis & Grierson, 1978; Luthe & Quatrano, 1980*b*).

The RNA synthesized by isolated pea leaf nuclei is heterodisperse in size

Table 1. *Characteristics of transcription by isolated* Pisum *leaf nuclei*[a]

Incubation mixture	[³H]UMP incorporated into acid-insoluble fraction		
	c.p.m. per 5 μl	pmol (100 μg DNA)$^{-1}$	%
Complete	19 689	52.1	100
Minus unlabelled nucleoside triphosphates	1 784	4.7	9
Minus nuclei	170	—	0.9
Plus 10 μg ml^{-1} actinomycin D	4 887	12.9	25
Plus 10 μg ml^{-1} α-amanitin	12 545	33.2	64

[a] Nuclei were incubated with [³H]UTP for 10 min; reprinted with permission from Gallagher & Ellis, 1982.

(Fig. 5). A decrease in the mean size of the transcripts occurs when nuclei are incubated for a further 60 min in the presence of unlabelled UTP and actinomycin D, but this decrease is not accompanied by the loss of acid-precipitable radioactivity. Small subunit transcripts can be detected among the RNA synthesized by isolated leaf nuclei, as shown by hybridization to SSU probes immobilized on nitrocellulose filters. There is a linear relationship between the amount of labelled RNA applied and the amount of hybridization to the SSU probe (Fig. 6).

The SSU gene is transcribed at a much greater rate in nuclei isolated from

Fig. 4. Electron micrograph of a section through a pellet of *Pisum* leaf nuclei purified on a Percoll gradient. (Reprinted with permission from Gallagher & Ellis, 1982.)

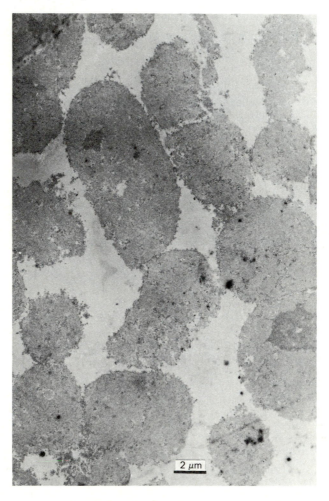

light-grown pea leaves than in nuclei isolated from dark-grown apical buds (Table 2). The rate of transcription of the SSU gene in nuclei from dark-grown tissue is very low. To obtain accurate counts, filters are counted for one hour. The low level of transcription of the SSU gene in nuclei isolated from dark-grown apical buds has been observed in more than ten separate experiments.

These results suggest that the difference in steady-state concentration of SSU gene transcripts in RNA isolated from light-grown and dark-grown pea

Fig. 5. Size of RNA synthesized by isolated pea leaf nuclei. Nuclei isolated from dark-grown apical buds were incubated with [^{32}P]UTP for 20 min (track A) or for 20 min followed by a further 60 min in the presence of 10 mmol l^{-1} unlabelled UTP and 10 μg ml^{-1} actinomycin D (track B).

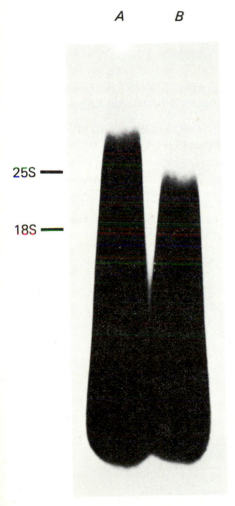

A B

25S —

18S —

seedlings is produced by differences in the rate of transcription of the SSU genes. It is possible, however, that the SSU genes are transcribed at equal rates in dark-grown and light-grown plants, but the transcripts are then rapidly degraded in darkness, so that the low degree of hybridization observed represents the residual transcript. The stability of the small subunit transcript in nuclei isolated from dark-grown apical buds was therefore tested. Nuclei were incubated for 10 min with [³²P]UTP and aliquots were incubated

Table 2. *Transcription of SSU genes in nuclei isolated from light-grown pea leaves and dark-grown apical buds*[a]

Source of [³H]RNA	Input [³H]RNA c.p.m. × 10⁻⁶	c.p.m. bound to filter[b]	% hybridization × 10³
Light-grown nuclei	4.8	239	4.98
Light-grown nuclei incubated with α-amanitin	2.39	19	0.79
Dark-grown nuclei	4.36	12	0.28

[a] Nuclei were incubated with [³H]UTP for 10 min and the labelled RNA hybridized to filters bearing 5 μg of SSU cDNA.
[b] c.p.m. bound to the SSU cDNA filters minus c.p.m. bound to control filters bearing 5 μg of plasmid DNA lacking SSU inserts; counted for 1 h.

Fig. 6. Quantitative hybridization to SSU cDNA of RNA synthesized by nuclei isolated from light-grown pea leaves. Increasing amounts of [³²P]RNA were hybridized to filters bearing 5 μg of the SSU probe. Counts binding to filters bearing 5 μg of plasmid DNA lacking SSU inserts were subtracted. (Reprinted with permission from Gallagher & Ellis, 1982.)

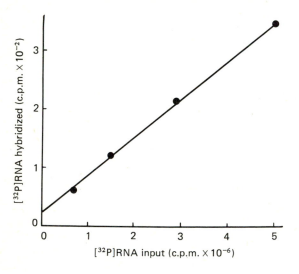

for a further 60 and 120 min in the presence of an excess of unlabelled UTP and actinomycin D. Table 3 shows that the small amount of hybridization of the SSU probes to RNA synthesized by nuclei from dark-grown plants persists during the chase period. In a parallel incubation, carried out in the presence of α-amanitin, no hybridization to the SSU probe was seen, confirming that the observed transcripts are produced by the activity of RNA polymerase II.

These observations indicate that an increase in the rate of synthesis of SSU transcripts rather than a decrease in their rate of degradation is responsible for the much greater abundance of SSU mRNA in light-grown pea seedlings. Transcriptional control of SSU gene expression is therefore of major importance in accounting for the light-induced increase in RuBP carboxylase accumulation. Similar experiments should now be undertaken with isolated plastids to investigate whether light has a corresponding effect on the transcription of the LSU genes.

Slow and rapid effects of light on SSU gene transcription

The experiments described above were concerned only with the difference in the rate of SSU gene transcription between light-grown and dark-grown plants. But how rapidly does transcription start to increase once plants are illuminated? If rapid changes can be observed then there is some possibility of extending the experimental approach using isolated nuclei to investigate *in vitro* the mechanisms by which light stimulates transcription. We have therefore started to analyse the kinetics of the increase in the rate of SSU gene transcription when *Pisum* plants are transferred from darkness to light.

When dark-grown etiolated pea seedlings are transferred to continuous white light, the rate of transcription of the SSU genes, measured in nuclei isolated from the apical buds, increases only slowly during the first 24 h

Table 3. *Stability of SSU mRNA transcripts in nuclei isolated from dark-grown pea leaves and apical buds*

Source of [^{32}P]RNA	Input [^{32}P]RNA c.p.m. $\times 10^{-6}$	c.p.m. bound to filter	c.p.m. bound as % input $\times 10^4$
Pulse[a]	5.42	5.6	1.03
Chase[b] 60 min	6.35	7.1	1.12
Chase[b] 120 min	3.61	4.0	1.11

[a] Nuclei from dark-grown apical buds were incubated with [^{32}P]UTP for 10 min.
[b] Aliquots of (*a*) were incubated with 10 mmol l^{-1} unlabelled UTP and 10 μg ml^{-1} actinomycin D for a further 60 min and 120 min.

of illumination; a more rapid increase is then observed up to about 42 h (Fig. 7). At 42–48 h the rate of SSU gene transcription is comparable to that in plants grown under the normal 12 h photoperiod. This time-course of the rate of SSU gene transcription closely parallels the slow increase in the concentration of SSU mRNAs in total RNA extracts measured by dot hybridization (Thompson *et al.*, 1983; Jenkins *et al.*, 1984).

To investigate rapid light-induced changes in transcription we have used plants which are already capable of high rates of SSU transcript synthesis, and not the etiolated plants which respond only slowly to illumination. When dark-grown plants which have been illuminated for 36 h with white light are returned to darkness, the rate of SSU gene transcription declines markedly within 20 min and approaches the rate observed in etiolated plants (Fig. 8). This low rate of transcription is maintained for at least 5 h in darkness. It should be emphasized that the initial stages of nuclear isolation from the seedlings are carried out in darkness. When these plants are again transferred to white light, the rate of transcription increases within 20 min, and is close to that measured after 36 h of continuous illumination (Fig. 8). These observations clearly demonstrate that light–dark transitions can induce rapid changes in the rate of transcription of the SSU genes measured in

Fig. 7. Changes in the rate of transcription of SSU genes during illumination of etiolated pea plants. Pea plants were grown in darkness for 6 days and then transferred to continuous light ($200 \ \mu\text{mol m}^{-2} \text{ s}^{-1}$; 400–700 nm). Apices were harvested and nuclei isolated at various times over a 48 h period. The rate of transcription of the SSU genes was determined as described in Table 2.

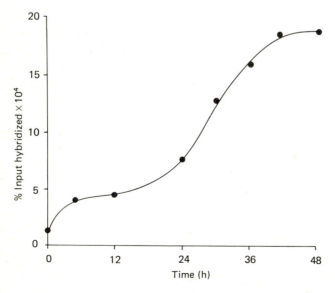

isolated nuclei. This contrasts with the slow change in the rate of transcription observed when etiolated plants are first illuminated, and suggests that light has dual effects on transcription. Light causes the development of the ability to transcribe rapidly under the appropriate conditions of illumination but, in addition, it rapidly modulates the rate of transcription in competent plants. The molecular basis of these effects may well be different and is not yet understood in either case.

The mechanism of photoregulation

We can now draw a number of conclusions about the levels at which light controls the expression of genes encoding the LSU and SSU polypeptides in *Pisum sativum*. It is clear that both LSU and SSU transcripts are more abundant in light-grown pea leaves than in dark-grown apical buds; indeed the concentration of SSU mRNA in extracts of etiolated seedlings approaches the limit of detection by hybridization analysis. The light-stimulated accumulation of SSU mRNA in the cytosol is not controlled primarily at the level of post-transcriptional selection and transport from the nucleus since SSU transcripts are more abundant in nuclei from light-grown leaves (Fig. 2C).

Fig. 8. Effects of changes in the light regime on the rate of transcription of the SSU genes in de-etiolated pea plants. Pea plants were grown in darkness (D) for 6 days, transferred to continuous white light (L) (see Fig. 7) for 36 h, and then returned to darkness for 20 min, 1 h or 5 h. Plants kept in darkness for 5 h were then transferred to white light for 20 min, 1 h or 4 h. Apical buds were harvested at various times during the treatments and nuclei were isolated. Plants were harvested and the nuclei isolated in darkness. The rate of transcription of SSU genes was determined as described in Table 2.

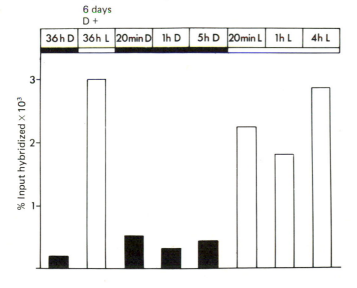

We conclude that the light-induced increase in the steady-state concentration of SSU transcripts in both the nuclei and cytosol of pea seedlings is mediated mainly at the level of transcription (Tables 2 and 3). This conclusion does not necessarily indicate that SSU transcripts are stable in the cytoplasm; in both pea (G. I. Jenkins, unpublished data) and *Lemna gibba* (Tobin & Suttie, 1980) there is evidence for breakdown of SSU mRNA when light-grown plants are transferred to darkness.

The rate of accumulation of the SSU polypeptide during greening of pea seedlings is closely correlated with the abundance of SSU mRNA in the apical buds (Jenkins *et al.*, 1984). Hence there is little reason at present to suppose that light controls accumulation of the SSU at either the translational or post-translational levels. The situation with regard to the LSU is much less clear, however, and detailed investigations of transcriptional and other possible levels of control are required. These studies should include an evaluation of the possible role of the LSU binding protein (Ellis, 1981) in regulating the accumulation of RuBP carboxylase.

The effect of light on the expression of genes encoding the LSU and SSU polypeptides must be mediated by specific photoreceptor pigments. There is now a large body of evidence which indicates that the ubiquitous, regulatory plant photoreceptor phytochrome detects the light that controls the expression of a number of plant genes, including those encoding the light-harvesting chlorophyll *a/b* binding protein and the LSU and SSU polypeptides (Apel, 1979; Tobin, 1981; and Bennett *et al.* and Tobin *et al.*, this volume). Although light acting through phytochrome has a large stimulatory effect on the accumulation of RuBP carboxylase in pea, it is not an essential requirement for carboxylase accumulation in a number of cereal species; for example, dark-grown barley plants contain 50% as much RuBP carboxylase as light-grown plants (Kleinkopf, Huffaker & Matheson, 1970).

The mechanism of phytochrome action in any developmental plant system is not yet known in molecular detail, and thus the reactions which link photoreception by phytochrome with events concerned with gene expression must remain the subject of speculation. Our finding that the extent to which light stimulates SSU mRNA accumulation *in vivo* is reflected in the rate of SSU gene transcription in isolated nuclei, coupled with the fact that rapid effects of light–dark transitions can be observed, opens up the possibility of an analysis *in vitro* of the mechanism by which light regulates transcription.

A number of models for the control of transcription of specific genes have been developed by workers with other systems. These include methylation of non-expressed genes (Razin & Riggs, 1980) and alterations in chromatin configuration (Weintraub & Goudine, 1976). A survey of a range of plant species which express the SSU genes to different extents should be undertaken to determine if such regulatory mechanisms are likely to operate. In addition, the development *in vitro* of transcriptional systems from plant tissues of the type available from animal cells (Manley *et al.*, 1980) may help to identify factors from light-grown tissue which specifically stimulate transcription of light-regulated genes.

References

Alwine, J. C., Kemp, P. J. & Stark, G. R. (1977). Method for the detection of specific RNAs in agarose gels by transfer to diazobenzyloxymethyl paper and hybridization with DNA probes. *Proc. natn. Acad. Sci. USA*, **74**, 5350–4.

Apel, K. (1979). Phytochrome-induced appearance of mRNA activity for the apoprotein of the light-harvesting chlorophyll *a/b* protein of barley (*Hordeum vulgare*). *Eur. J. Biochem.*, **97**, 183–8.

Barraclough, B. R. & Ellis, R. J. (1979). The biosynthesis of ribulose bisphosphate carboxylase. Uncoupling of the synthesis of large and small subunits in isolated soybean leaf cells. *Eur. J. Biochem.*, **94**, 165–77.

Bedbrook, J. R., Coen, D. M., Beaton, A. R., Bogorad, L. & Rich, A. (1979). Location of the single gene for the large subunit of ribulose bisphosphate carboxylase on the maize chloroplast chromosome. *J. Biol. Chem.*, **254**, 905–10.

Bedbrook, J. R. & Kolodner, R. (1979). The structure of chloroplast DNA. *A. Rev. Pl. Physiol.*, **30**, 593–620.

Bedbrook, J. R., Smith, S. M. & Ellis, R. J. (1980). Molecular cloning and sequencing of DNA encoding the precursor to the small subunit of chloroplast ribulose-1,5-bisphosphate carboxylase. *Nature*, **287**, 692–7.

Berry-Lowe, S. L., McKnight, T. D., Shah, D. M. & Meagher, R. B. (1982). The nucleotide sequence, expression and evolution of one member of a multigene family encoding the small subunit of ribulose-1,5-bisphosphate carboxylase in soybean. *J. molec. appl. Genet.*, **1**, 483–98.

Blair, G. E. & Ellis, R. J. (1973). Protein synthesis in chloroplasts. I. Light-driven synthesis of the large subunit of Fraction I protein by isolated pea chloroplasts. *Biochim. biophys. Acta*, **319**, 223–34.

Broglie, R., Coruzzi, G., Lamppa, G., Keith, B. & Chua, N.-H. (1983). Structural analysis of nuclear genes coding for the precursor to the small subunit of wheat ribulose-1,5-bisphosphate carboxylase. *Biotechnology*, **1**, 55–61.

Cashmore, A. R. (1979). Reiteration frequency of the gene coding for the small subunit of ribulose-1,5-bisphosphate carboxylase. *Cell*, **17**, 383–8.

Chua, N.-H. & Schmidt, G. W. (1978). Post-translational transport into intact chloroplasts of a precursor to the small subunit of ribulose-1,5-bisphosphate carboxylase. *Proc. natn. Acad. Sci. USA*, **75**, 6110–14.

Coen, D. M., Bedbrook, J. R., Bogorad, L. & Rich, A. (1977). Maize chloroplast DNA encoding the large subunit of ribulose bisphosphate carboxylase. *Proc. natn. Acad. Sci. USA*, **74**, 5487–91.

Derman, E., Krauter, K., Walling, L., Weinberger, C., Roy, M. & Darnell,

J. E. (1981). Transcriptional control in the production of liver specific mRNA. *Cell*, **23**, 731–9.

Dobberstein, B., Blobel, G. & Chua, N.-H. (1977). *In vitro* synthesis and processing of a putative precursor for the small subunit of ribulose-1,5-bisphosphate carboxylase of *Chlamydomonas reinhardtii*. *Proc. natn. Acad. Sci. USA*, **74**, 1082–5.

Dunsmuir, P., Smith, S. M. & Bedbrook, J. R. (1983). A number of different nuclear genes for the small subunit of RuBPCase are transcribed in *Petunia*. *Nucleic Acids. Res.*, **11**, 4177–83.

Ellis, R. J. (1981). Chloroplast proteins: synthesis, transport and assembly. *A. Rev. Pl. Physiol.*, **32**, 111–37.

Ellis, R. J. & Robinson, C. (1984). Post-translational transport and processing of cytoplasmically-synthesized precursors of organellar proteins. In *The Enzymology of the Post-Translational Modification of Proteins*, vol. II, ed. R. B. Freedman & H. C. Hawkins. Academic Press, London and New York. (In press.)

Feierabend, J. & Wildner, G. (1978). Formation of the small subunit of RuBP carboxylase in 70S ribosome-deficient rye leaves. *Archs. Biochem. Biophys.*, **186**, 283–91.

Gallagher, T. F. & Ellis, R. J. (1982). Light-stimulated transcription of genes for two chloroplast polypeptides in isolated pea leaf nuclei. *EMBO J.*, **1**, 1493–8.

Graham, D., Grieve, A. M. & Smillie, R. M. (1968). Phytochrome as the primary photoregulator of the synthesis of Calvin cycle enzymes in etiolated pea seedlings. *Nature*, **218**, 89–90.

Highfield, P. E. & Ellis, R. J. (1978). Synthesis and transport of the small subunit of chloroplast ribulose bisphosphate carboxylase. *Nature*, **271**, 420–4.

Jendrisak, J. J. & Guilfoyle, T. J. (1978). Eukaryotic RNA polymerases: comparative subunit structures, immunological properties, and α-amanitin sensitivities of the class II enzymes from higher plants. *Biochemistry*, **17**, 1322–7.

Jenkins, G. I., Gallagher, T. F., Hartley, M. R., Bennett, J. & Ellis, R. J. (1984). Photoregulation of gene expression during chloroplast biogenesis. In *Advances in Photosynthesis Research*, ed. C. Sybesma, vol. IV, pp. 863–72. Junk, The Hague.

Jenkins, G. I., Hartley, M. R. & Bennett, J. (1983). Photoregulation of chloroplast development: transcriptional, translational and post-translational controls? *Phil. Trans. R. Soc. Lond. B*, **303**, 419–31.

Jensen, R. G. & Bahr, J. T. (1977). Ribulose-1,5-bisphosphate carboxylase oxygenase. *Ann. R. Pl. Physiol.*, **28**, 379–400.

Kawashima, N. D. & Wildman, S. G. (1972). Studies on Fraction I protein. IV. Mode of inheritance of primary structures in relation to whether chloroplast or nuclear DNA contains the code for a chloroplast protein. *Biochim. biophys. Acta*, **262**, 42–9.

Kleinkopf, G. E., Huffaker, R. C. & Matheson, A. (1970). Light induced *de novo* synthesis of ribulose bisphosphate carboxylase in greening leaves of barley. *Pl. Physiol.*, **46**, 416–18.

Kung, S. D. (1976). Tobacco Fraction I protein: a unique genetic marker. *Science*, **191**, 429–34.

Luthe, D. S. & Quatrano, R. S. (1980*a*). Transcription in isolated wheat nuclei. I. Isolation of nuclei and elimination of endogenous ribonuclease activity. *Pl. Physiol.*, **65**, 305–8.

Luthe, D. S. & Quatrano, R. S. (1980*b*). Transcription in isolated wheat nuclei. II. Characterization of RNA synthesized *in vitro*. *Pl. Physiol.*, **65**, 309–13.

Manley, J. L., Fire, A., Cano, A., Sharp, P. A. & Gefter, M. C. (1980). DNA-dependent transcription of adenovirus genes in a soluble cell extract. *Proc. natn. Acad. Sci. USA*, **77**, 3855–9.

Razin, A. & Riggs, A. D. (1980). DNA methylation and gene expression. *Science*, **210**, 604–10.

Sasaki, Y., Ishiye, M., Sakihama, T. & Kamikubo, T. (1981). Light induced increase in mRNA activity coding for the small subunit of ribulose-1,5-bisphosphate carboxylase. *J. biol. Chem.*, **256**, 2315–20.

Slater, R. J., Venis, M. A. & Grierson, D. (1978). Characteristics of ribonucleic acid synthesized by nuclei isolated from *Zea mays*. *Planta*, **144**, 89–93.

Smith, S. M. & Ellis, R. J. (1979). Processing of small subunit precursor of ribulose bisphosphate carboxylase and its assembly into whole enzyme are stromal events. *Nature*, **278**, 662–4.

Smith, S. M. & Ellis, R. J. (1981). Light-stimulated accumulation of transcripts of nuclear and chloroplast genes for ribulose bisphosphate carboxylase. *J. molec. appl. Genet.*, **1**, 127–37.

Thompson, W. F., Everett, M., Polans, N. O., Jorgensen, R. A. & Palmer, J. D. (1983). Phytochrome control of RNA levels in developing pea and mung bean leaves. *Planta*, **158**, 487–500.

Tobin, E. M. (1981). Phytochrome-mediated regulation of messenger RNAs for the small subunit of ribulose-1,5-bisphosphate carboxylase and the light-harvesting chlorophyll *a/b* protein in *Lemna gibba*. *Pl. molec. Biol.*, **1**, 35–51.

Tobin, E. M. & Suttie, J. L. (1980). Light effects on the synthesis of ribulose-1,5-bisphosphate carboxylase in *Lemna gibba*. *Pl. Physiol.*, **65**, 641–7.

Tsai, S. Y., Roop, D. R., Tsai, M. J., Stein, J. P., Mears, A. R. & O'Malley, B. N. (1978). Effect of estrogen on gene expression in the chick oviduct. Regulation of the ovomucoid gene. *Biochemistry*, **17**, 5773–80.

Walden, R. & Leaver, C. J. (1981). Synthesis of chloroplast proteins during germination and early development of cucumber. *Pl. Physiol.*, **67**, 1090–6.

Weintraub, H. & Goudine, M. (1976). Chromosomal subunits in active genes have an altered conformation. *Science*, **193**, 848–56.

Wilson, P. W. & Bennett, J. (1976). Transcription in nuclei isolated from a higher plant. *Biochem. Soc. Trans.*, **4**, 812–13.

E. M. TOBIN, J. SILVERTHORNE, W. J. STIEKEMA AND C. F. WIMPEE

Phytochrome regulation of the synthesis of two nuclear-coded chloroplast proteins

Phytochrome has been known for many years to affect various aspects of chloroplast development in higher plants. Early reports (Marcus, 1960; Margulies, 1965; Feierabend & Pirson, 1966; Filner & Klein, 1968; Graham, Grieve & Smillie, 1968; Klein, 1969) demonstrated that red light could cause increases in the level of activity of a number of chloroplast enzymes. For example, Marcus (1960) found that red light could cause an increase in the enzymic activity of NADPH–glyceraldehyde-3-phosphate dehydrogenase in red kidney bean plants.

One of the most intensively studied chloroplast enzymes has been ribulose-1,5-bisphosphate carboxylase/oxygenase (RuBP carboxylase). Graham *et al.* (1968) demonstrated that the red-light induced increase in the activity of this enzyme in pea seedlings was partially reversed by far-red illumination. Filner & Klein (1968) found a similar result in bean, and this response was shown to have an unusually long 'escape time' from phytochrome control. The increase in carboxylase activity could be halted by far-red illumination given 24 h after the initial red illumination. Feierabend (1969) showed that in etiolated rye seedlings red light acted synergistically with a cytokinin to increase carboxylase activity. Later studies (e.g. Kleinkopf, Huffaker & Matheson, 1970; Tobin, 1978; Kobayashi, Asami & Akazawa, 1980) used antibodies to demonstrate that the light-induced increases in activity represented increases in the amount of carboxylase protein.

More recently, studies with isolated mRNAs translated *in vitro* have shown that the amount of the mRNA encoding the small subunit (SSU) of RuBP carboxylase (Tobin, 1978, 1981*c*) as well as the light-harvesting chlorophyll *a/b* protein (chl *a/b*-protein) and some other nuclear-coded chloroplast-proteins (Tobin, 1981*c*; Apel, 1979, 1981) change rapidly in response to phytochrome action. Our laboratory has found that in *Lemna gibba* the amounts of translatable mRNA encoding the SSU and the chl *a/b*-protein decline to a low level in the dark, while a single minute of red illumination results in a substantial increase in both mRNAs within 2 h. This effect of red light is reversible by far-red illumination.

321

In the case of etiolated barley leaves, phytochrome action induces an increase in the translatable mRNA encoding the chl a/b-protein and a decrease in that for the NADPH–protochlorophyllide reductase (Apel, 1979, 1981). The SSU mRNA is present in substantial amounts in etiolated barley leaves. In addition, there are reports of differences in mRNA levels for specific proteins between etiolated and green or greening plants of other species (Bedbrook *et al.*, 1978; Highfield & Ellis, 1978; Reisfeld *et al.*, 1978; Bedbrook, Smith & Ellis, 1980; Hague & Sims, 1980; Cuming & Bennett, 1981; Smith & Ellis, 1981; Tobin, 1981*a*) and phytochrome has been shown to play a regulatory role in these cases as well (Jenkins, Hartley & Bennett, 1983). A recent report by Link (1982) has suggested the involvement of phytochrome in the regulation of the chloroplast-coded 32000 Dalton polypeptide in mustard. In this case repeated brief red illuminations over several days resulted in an increase in the content of the mRNA sequence encoding this polypeptide. Another report has suggested the involvement of a blue-absorbing photoreceptor in the regulation of the mRNA for a similar 32000 Dalton protein in *Spirodela oligorhiza* (Gressel, 1978).

The work so far has established that there are species differences in the effects of phytochrome on the synthesis of specific mRNAs. These differences are reflected both in the proteins which are affected and in the timing of the responses. However, an understanding of the mechanism of the specific response of one species may be expected to elucidate the mechanisms operating in other species. In order to be able to understand the phytochrome response in *Lemna gibba* in molecular detail, we have cloned hybridization probes for the SSU of RuBP carboxylase and for the chl a/b-protein, two specific polypeptides whose levels of translatable mRNAs have been shown to be rapidly regulated by phytochrome in *Lemna* (Tobin, 1981*c*).

In *Lemna* the chl a/b-protein contains at least two separate polypeptides (Tobin, 1981*b*). However, in contrast to the situation in peas (Cuming & Bennett, 1981; Schmidt *et al.*, 1981), the antiserum raised against these two polypeptides immunoprecipitates only a single band (on a one-dimensional gel) from translation products of poly(A)-RNA. We have not established the connection between the immunoprecipitated precursor(s) and the final form of the apoproteins in *Lemna*, and for convenience we will refer to the chl a/b-protein as if it were a single polypeptide. The cloned hybridization probes ultimately should help to clarify the situation.

We have also studied the synthesis of the proteins *in vivo* by labeling intact fronds with [^{35}S]methionine in order to complete the picture of the regulation of the expression of these genes by light. Although the synthesis of the SSU *in vivo* seems to reflect the levels of its mRNA in either the light or the dark (Tobin & Suttie, 1980), the situation for the chl a/b-protein is more complex.

When the plants are grown in the dark on sucrose with intermittent red light (2 min every 8 h), they contain high levels of translatable chl *a/b*-protein mRNA. However, *in vivo* labeling experiments with [^{35}S]methionine show that the protein is apparently being synthesized at only one-twentieth the rate of that in the light-grown plants (Slovin & Tobin, 1982); that is, these plants incorporate only one-twentieth of the label into the chl *a/b*-protein in 1 h as compared to a 1 h labeling of light-grown plants. Similar results have been reported for a mutant of barley lacking chlorophyll *b* (Apel & Kloppstech, 1978, 1980) and peas grown with intermittent illumination (Cuming & Bennett, 1981). Since in all three cases the mRNA can be found to be associated with polyribosomes, and the protein product or precursor cannot be found in the expected amounts in either the soluble or membrane fraction, it has been suggested that, in the dark in the absence of chlorophyll synthesis, the newly synthesized chl *a/b*-protein is rapidly degraded. We have measured the turnover of the newly synthesized chl *a/b*-protein in *Lemna* grown in the dark with brief intermittent red light, and found that its half-life is approximately 8–10 h. This evidence supports the idea that the protein is degraded more rapidly in the absence of chlorophyll synthesis, but this is not a sufficient explanation of the observed 20-fold difference in labeling in a 1 h period. Therefore we have proposed an additional light-regulated step which is necessary for the effective translation of this protein (Slovin & Tobin, 1982).

In this article we report the results of experiments designed to test the hypothesis that the changes in translatable mRNA levels induced by phytochrome in *Lemna* are produced by changes in transcription.

Results and discussion

Construction and selection of hybridization probes

A chloroplast (cp) DNA library was constructed from poly(A)-RNA isolated from light-grown *Lemna* plants. Complementary DNA was prepared by reverse transcription and converted to double-stranded DNA using fragment A of polymerase I. The hairpin loop was removed with S1 nuclease and the DNA sized on a 5–30% sucrose gradient. Double-stranded cp-DNA having a length greater than 200 base-pairs was tailed with dCTP and annealed to Pst 1-linearized pBR322 DNA tailed with dGTP. The annealed DNA was used to transform *Escherichia coli* strain HB101, and Tet[r], Amp[s] colonies were selected.

A set of genomic clones was constructed from *Lemna* nuclear DNA in the bacteriophage lambda derivative Charon 4. The plant DNA was extracted from isolated nuclei and partially digested with EcoR1. Fractions of approximately 10–20 kilobases from a sucrose gradient were pooled, ligated to the lambda Charon 4 arms, and packaged into recombinant phage.

The recombinant clones were screened by two different strategies in order to identify those encoding the SSU of RuBP carboxylase and the chl *a/b*-protein. Probes enriched for the sequence desired were prepared by first enriching the mRNA for the appropriate sequence. Poly(A)-RNA was fractionated on a 1% agarose/methylmercury hydroxide gel, and the RNA eluted from slices of the gel. An aliquot of each fraction was translated *in vitro*, and one fraction was found to contain predominantly RNA encoding the SSU of RuBP carboxylase and another was enriched for RNA encoding the chl *a/b*-protein. These two fractions were used to make ^{32}P-labelled copy-DNA (cDNA) for use as hybridization probes of the cloned DNA. In addition we took advantage of the observation that the levels of the translatable RNAs decline in the dark, and prepared cDNA against poly(A)-RNA isolated from dark-treated plants. We used this ^{32}P-labelled cDNA in a differential screening of the cDNA clones by comparing the hybridization of this probe with the hybridization of a probe made from RNA from light-grown plants to replicate sets of colonies of transformants. An example of such screening with triplicate half-filters probed with the labeled cDNA probes (Hoeijmakers *et al.*, 1980) is shown in Fig. 1. Several colonies can be seen which reacted extensively with the probes made both from the RNA enriched for the SSU message and from the RNA isolated from light-grown plants, but not so extensively with the probe made from RNA isolated from dark-treated plants. In addition two colonies can be seen to hybridize preferentially with the probe made from RNA from the dark-treated plants. The screenings have identified one group of colonies which contain sequences which are more abundant in the light and another group containing sequences more abundant in the dark. We have not identified any of the dark-abundant sequences, but Apel (1981) has shown that the mRNA for the NADPH–chlorophyllide oxidoreductase is more abundant in etiolated barley leaves than in plants exposed to light.

Fig. 1. Differential screening of cDNA clones. Triplicate half filters of recombinant colonies probed with ^{32}P-labeled cDNA made from: poly(A)-RNA isolated from light-grown plants (L); poly(A)-RNA isolated from plants placed into the dark for 5 days (D); poly(A)-RNA enriched for the SSU mRNA by electrophoresis (SSU).

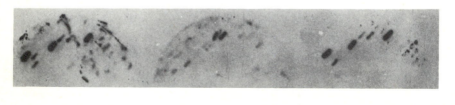

L   D   S S U

Several transformants containing sequences hybridizing to the SSU-encoding sequence have been identified among the cDNA clones. The identity of these has been confirmed by hybridization-selection and translation (Ricciardi, Miller & Roberts, 1979), followed by immunoprecipitation with antibodies to the *Lemna* SSU (Tobin & Suttie, 1980). Figure 2 shows the results of such an experiment for the clone designated pLgSSU1. Restriction enzyme analysis of two of the SSU clones of lengths of approximately 870 base-pairs gave identical maps. It is probable that these are full-length or nearly full-length clones which should contain the information for the sequence of the precursor polypeptide of the SSU. The clone used in the experiments to be described here was pLgSSU1.

The genomic clones for both the SSU and the chl *a/b*-protein were selected

Fig. 2. Identification of pLgSSU1. Plasmid DNA isolated from the clone was used to select complementary RNA by hybridization with total poly(A)-RNA. RNA was translated in a rabbit reticulocyte lysate system. ENDOGENOUS = translation products with no added RNA; TOTAL = translation products of total poly(A)-RNA; SELECTED = translation products of RNA selected by hybridization with the cloned DNA. IMMUNOPPT = translation products of the selected RNA immunoprecipitated with antiserum to the SSU.

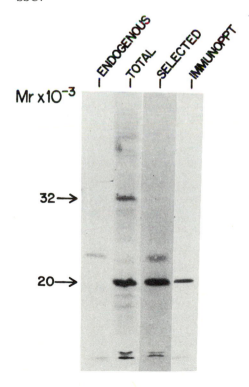

by the method of Woo (1979), using the probes enriched for the sequence of interest. Figure 3 shows an example of such a screening with the probe enriched for the chl *a/b*-protein sequence. From such screenings, one set of putative SSU clones and another set of putative chl *a/b*-protein clones were selected from the genomic library. Over 95% of the clones selected in this way have proven to be correctly identified. The selected clones were rescreened

Fig. 3. Screening of genomic clones. In this example the clones were hybridized with ^{32}P-labeled cDNA made from poly(A)-RNA enriched for the chl *a/b*–protein mRNA by electrophoresis. M, markers. The arrow and circle indicate a strongly hybridizing clone which contains DNA encoding the chl *a/b*–protein.

a/b ENRICHED
cDNA PROBE

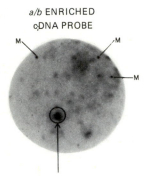

Fig. 4. Characterization of genomic clones for the chl *a/b*–protein by EcoR1 digestion. The clones isolated could be divided into three classes by restriction mapping. An example of each class is shown.

EcoR1 digest of λ*a/b* 19,
λ*a/b* 26, and λ*a/b* 30

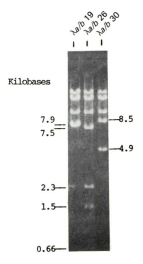

with a *Lemna* SSU cDNA clone or with a *Pisum sativum* chl *a/b*-protein cDNA clone (Broglie *et al.*, 1981) to confirm their identity. The *Lemna* genomic clones for the two proteins were analyzed with restriction enzymes and each set was found to fall into a limited number of categories. The SSU clones thus far identified all represent the same cloned gene(s). The chl *a/b*-clones fall into three classes when digested with EcoR1 (Fig. 4). In each case, the largest fragment of inserted DNA contains the coding sequence (Fig. 5). The DNA was purified from one clone selected from each of these three categories, and used in the hybrid selection and translation assay. Each DNA hybridized to RNA which, when translated, produced immunoprecipitable precursor to the chl *a/b*-polypeptides. A Hind III fragment of 1900 base-pairs containing the coding sequence from one of the chl *a/b*-protein genomic clones, designated LgAB19, was subcloned into pBR322. This clone was then used as a hybridization probe in some of the experiments to be described below.

Fig. 5. Localization of the coding sequence in the EcoR1 fragments of the genomic clones for the chl *a/b*–protein. EcoR1 digests of the DNA of the representative clones were transferred to nitrocellulose after electrophoresis and hybridized to ^{32}P-labeled cDNA made from poly(A)-RNA isolated from light-grown plants. The numbers refer to the sizes of the fragments, in kilobases.

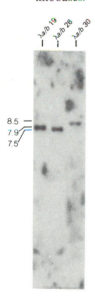

'Northern' blot analysis of changes in the sequence content in response to phytochrome action

The isolated probes have been used to investigate the possible ways by which phytochrome can affect the levels of translatable mRNAs for these two proteins. Earlier reports had indicated that red light might affect the translation machinery (e.g. Travis, Key & Ross, 1974) or the mobilization of mRNA onto polysomes (Giles, Grierson & Smith, 1977). Therefore it was important to establish whether the changes observed in translatable mRNAs were the same as the changes in the actual content of these RNA sequences.

The RNA sequences present in plants given various conditions of illumination were detected by hybridization with the cloned probes. Cloned DNA was labeled with ^{32}P by nick-translation. The labeled DNA was then hybridized with the RNA after electrophoresis on formaldehyde–agarose gels and transfer onto nitrocellulose (Thomas, 1980). The sequences for the chl *a/b*-protein and the SSU could be detected in RNA from light-grown plants, and appear to be of length about 1100 and 900 nucleotides, respectively. Light-grown plants which were put into the dark for 7–8 days contained very low levels of the sequences in comparison to the light-grown control plants (Fig. 6).

Fig. 6. Abundance of the RNA encoding the chl *a/b*–protein in light-grown and dark-treated plants. Poly(A)-RNA was isolated from plants grown in continuous light (L) or plants put into the dark for 7 days (D). After electrophoresis, the RNA was hybridized with [^{32}P]DNA from the genomic clone λ*a/b*19. 14S marks the position of the mRNA for the chl *a/b*–protein.

HYBRIDIZATION OF CLONED
a/b SEQUENCE TO 'LIGHT'
vs 'DARK' POLY(A)-RNA

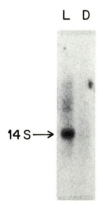

Plants which are grown heterotrophically in the dark with intermittent red illumination (2 min every 8 h) also contain high levels of both sequences (Tobin, 1981c). As with light-grown plants, transfer of such plants to complete darkness results in a decrease of both sequences to a low level. This remaining level in complete darkness is higher in these plants than in the light-grown plants after an equivalent treatment, because of the presence of a cytokinin (kinetin) in the medium (Tobin & Turkaly, 1982). When these dark-treated plants are illuminated with a single minute of red light, the amounts of both sequences increase substantially within 2 h (Stiekema *et al.*, 1983). This effect of red light is reversible by far-red illumination given immediately after the red, and far-red illumination alone does not cause an increase. The increase can be detected within 30 min of red illumination. Thus, the sequence content of the RNAs encoding both the SSU of RuBP carboxylase and the chl *a/b*-protein is under phytochrome control in *Lemna gibba*. These results suggest that phytochrome action can rapidly affect the transcription of the genes for the two proteins, although the possibility that the stability of the transcripts is primarily affected is not excluded by these experiments. Others have now also demonstrated phytochrome-regulated changes in the sequence content of the chl *a/b*-protein in barley (Gollmer & Apel, 1983) and of the SSU in pea (Sasaki *et al.*, 1983). In pea the increase was observed two days after illumination, but in barley the increase appears to be much more rapid.

Use of isolated nuclei to study the effects of phytochrome action on transcription

Nuclei can be isolated from plant cells and used for transcriptional studies *in vitro*. Such experiments have the advantage of enabling one to look at newly synthesized RNA transcripts rather than at the total accumulation of sequences. Thus, isolated nuclei can give a more sensitive indication of changes in the transcription of particular genes than can assays of sequence content. The possibility of differences in (nuclear) degradation of particular transcripts under different conditions of illumination is difficult to eliminate rigorously, even in these experiments. However, transcription *in vitro* from isolated nuclei is widely used as an assay of gene transcription, and to date there is no indication of varying degradation rates for a specific transcript *in vitro*.

Nuclei were isolated from *Lemna* on Percoll gradients by the procedures of Luthe & Quatrano (1980a, 1980b). Newly synthesized RNA was assayed by incubation of the nuclei with [^{32}P]UTP as well as with the other nucleoside triphosphates, creatine phosphate and creatine phosphokinase, and the appropriate concentrations of $MnCl_2$ and $(NH_4)_2SO_4$ for optimal incorporation in *Lemna* nuclei. The incorporation was found to be linear for about 2 h. Examination of the labeled RNA transcripts by electrophoresis and autoradiography reveals substantial amounts of newly synthesized products of size up to about 28S. The acid-precipitable radioactivity reached a plateau after about 2 h and did not decline over the next hour; however, auto-radiography demonstrated that the size of the products declines with time. Therefore, some nuclease activity is present, and incubations of 30 min were

Fig. 7. Hybridization of transcripts from isolated nuclei to the SSU sequence. Nuclei were isolated from light-grown plants (L), from plants which had been put into the dark for 5 days (D), and from plants put into the dark for 5 days, given 2 min red illumination, and returned to the dark for 2 h (R). DNA containing the SSU sequence was excised from pLgSSU1 with Pst 1 and transferred to nitrocellulose after electrophoresis. The ethidium bromide-stained DNA is shown in the first lane (S). The other lanes were hybridized to aliquots of the nuclear transcripts of each sample which contained equal amounts of acid-precipitable counts incorporated into RNA. Lanes L, D, and R show the fluorographs obtained. pBR indicates the DNA of the cloning vector.

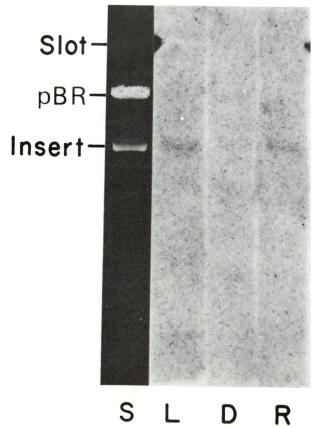

used for further experiments. We do not yet know whether any initiation of transcripts occurs, or whether we are observing only completion of already initiated transcripts. The incorporation of UTP into RNA is totally abolished by actinomycin D ($5\,\mu g\ ml^{-1}$), but is only partly dependent on the added nucleoside triphosphates. The incorporation is differentially sensitive to α-amanitin: low levels ($0.1\,\mu g\ ml^{-1}$) abolish nearly 60% of the incorporation, indicating that most of the incorporation is a result of polymerase II activity.

In order to examine the effects of phytochrome on the transcription observed in isolated nuclei, light-grown plants were placed into the dark for five days, given brief illuminations with red and/or far-red light, and then returned to the dark. Kinetin (3×10^{-5} mol l^{-1}) was added to all cultures on the fourth day of the dark treatment because it may be necessary for the action of phytochrome (Tobin, unpublished observations). The plants were harvested 2 h after the illumination, and the nuclei from these dark-treated plants were isolated in the dark under dim-green illumination from a safe-light. The nuclei were frozen in liquid nitrogen until assayed for transcription. Aliquots of the incubation mixtures, containing equal amounts of radioactivity after purification of the RNA, were used to compare the effects of the various illumination conditions on the synthesis of specific transcripts. The newly synthesized RNA was isolated from the aliquots and hybridized to the cloned sequence of interest. In the example shown (Fig. 7) the pLgSSU1 DNA was cut with Pst 1, which releases the inserted DNA from the plasmid sequences. Four aliquots of this DNA were separated on an agarose gel. After electrophoresis, one sample (S) was stained with ethidium bromide. The three remaining samples were transferred to nitrocellulose, and each was hybridized with one of the samples of RNA transcribed *in vitro*. The amount of radioactivity bound to the 'insert' DNA indicates the relative amount of newly transcribed RNA complementary to the sequence encoding the SSU of RuBP carboxylase. The experiment demonstrates that nuclei isolated from dark-treated plants produce much lower levels of the SSU sequence than nuclei from light-grown plants. Similar results have been reported for nuclei isolated from light-grown and dark-grown plants of *Pisum sativum* (Gallagher & Ellis, 1982; Gallagher *et al.*, this volume). In addition, if the dark-treated *Lemna* plants are treated with a single minute of red light, within 2 h their nuclei produce the SSU sequence in amounts equivalent to the nuclei isolated from light-grown plants. Similar results have been obtained with the chl *a/b*-protein sequence (Tobin *et al.*, 1983). In both cases, far-red light can reverse the effect of red light. The low level of transcripts found in the nuclei of dark-treated plants are not subject to rapid degradation. Thus our results demonstrate that phytochrome action can rapidly change transcriptional events in the intact plant.

Conclusions

The use of cloned probes for specific transcripts has now made possible the investigation of phytochrome action on a new level. We have reported here the first such use of specific probes to demonstrate that phytochrome action can rapidly change the sequence content of transcripts for at least two nuclear-encoded chloroplast protein genes in *Lemna*. The hypothesis that the changes in sequence content are a result of changes in transcription of the genes is supported by the studies of transcription *in vitro* with isolated nuclei. We have shown that phytochrome action can change the production of specific RNA transcripts in the nucleus as early as 30 min after red illumination. Because the results of these changes can be observed *in vitro*, we hope that such an in-vitro system can be utilized to elucidate the mechanism of action of phytochrome in the plant. Ultimately we hope to understand by what chain of events the photoconversion of the phytochrome chromophore is linked to the changes in the expression of these genes in the nucleus.

This work was supported by funds from US Public Health Service Grant GM23167 and by the Science and Education Administration of the US Department of Agriculture under Grant No. 5-901-0410-8-0177-0.

References

Apel, K. (1979). Phytochrome-induced appearance of mRNA activity for the apoprotein of the light-harvesting chlorophyll *a/b* protein of barley (*Hordeum vulgare*). *Eur. J. Biochem.*, **97**, 183–8.

Apel, K. (1981). The protochlorophyllide holochrome of barley (*Hordeum vulgare* L.). Phytochrome-induced decrease of translatable mRNA coding for the NADPH:protochlorophyllide oxidoreductase. *Eur. J. Biochem.*, **120**, 89–93.

Apel, K. & Kloppstech, K. (1978). The plastid membranes of barley (*Hordeum vulgare*). Light-induced appearance of mRNA coding for the apoprotein of the light-harvesting chlorophyll *a/b*-protein. *Eur. J. Biochem.*, **85**, 581–8.

Apel, K. & Kloppstech, K. (1980). The effect of light on the biosynthesis of the light-harvesting chlorophyll *a/b* protein. Evidence for the requirement of chlorophyll *a* for the stabilization of the apoprotein. *Planta*, **150**, 426–30.

Bedbrook, J. R., Link, G., Coen, D. M., Bogorad, L. & Rich, A. (1978). Maize plastid gene expressed during photoregulated development. *Proc. natn. Acad. Sci. USA*, **75**, 3060–4.

Bedbrook, J. R., Smith, S. M. & Ellis, R. J. (1980). Molecular cloning and sequencing of cDNA encoding the precursor to the small subunit of chloroplast ribulose-1,5-bisphosphate carboxylase. *Nature*, **287**, 692–7.

Broglie, R., Bellemare, G., Bartlett, S. G., Chua, N.-H. & Cashmore, A. R. (1981). Cloned DNA sequences complementary to mRNAs encoding precursors to the small subunit of ribulose-1,5-bisphosphate carboxylase and a chlorophyll *a/b* binding polypeptide. *Proc. natn. Acad. Sci. USA*, **78**, 7304–8.

Cuming, A. C. & Bennett, J. (1981). Biosynthesis of the light-harvesting chlorophyll *a/b* protein. Control of messenger RNA activity by light. *Eur. J. Biochem.*, **118**, 71–80.

Feierabend, J. (1969). Der Einfluss von Cytokinnen auf die Bildung von Photosyntheseenzymen in Roggenkeimlingen. *Planta*, **84**, 11–29.

Feierabend, J. & Pirson, A. (1966). Die Wirkung des Lichts auf die Bildung von Photosyntheseenzymen in Roggenkeimlingen. *Z. Pfl. Physiol.*, **55**, 235–45.

Filner, B. & Klein, A. O. (1968). Changes in enzymatic activities in etiolated bean seedling leaves after a brief illumination. *Pl. Physiol.*, **43**, 1587–96.

Gallagher, T. F. & Ellis, R. J. (1982). Light-stimulated transcription of genes for two chloroplast polypeptides in isolated pea leaf nuclei. *EMBO J.*, **1**, 1493–8.

Giles, A. B., Grierson, D. & Smith, H. (1977). *In vitro* translation of messenger-RNA from developing bean leaves. Evidence for the existence of stored messenger RNA and its light-induced mobilization into poly-ribosomes. *Planta*, **136**, 31–6.

Gollmer, I. & Apel, K. (1983). The phytochrome-controlled accumulation of mRNA sequences encoding the light-harvesting chlorophyll *a/b*–protein of barley (*Hordeum vulgare* L.). *Eur. J. Biochem.*, **133**, 309–13.

Graham, D., Grieve, A. M. & Smillie, R. M. (1968). Phytochrome as the primary photoregulator of the synthesis of Calvin cycle enzymes in etiolated pea seedlings. *Nature*, **218**, 89–90.

Gressel, J. (1978). Light requirements for the enhanced synthesis of a plastid mRNA during *Spirodela* greening. *Photochem. Photobiol.*, **27**, 167–9.

Hague, D. R. & Sims, T. L. (1980). Evidence for light-stimulated synthesis of phosphoenolpyruvate carboxylase in leaves of maize. *Pl. Physiol.*, **66**, 505–9.

Highfield, P. E. & Ellis, R. J. (1978). Synthesis and transport of the small subunit of chloroplast ribulose bisphosphate carboxylase. *Nature*, **271**, 420–4.

Hoeijmakers, J. H. J., Borst, P., van der Burg, J., Weissman, C. & Cross, G. A. M. (1980). The isolation of plasmids containing DNA complementary to messenger RNA for variant surface glycoproteins of *Trypanosoma brucei*. *Gene*, **8**, 391–417.

Jenkins, G. I., Hartley, M. R. & Bennett, J. (1983). Photoregulation of chloroplast development: transcriptional, translational and post-transl-ational controls? *Phil. Trans. R. Soc. Lond. B*, **303**, 419–31.

Klein, A. O. (1969). Persistent photoreversibility of leaf development. *Pl. Physiol.*, **44**, 897–902.

Kleinkopf, G. E., Huffaker, R. C. & Matheson, A. (1970). Light-induced *de novo* synthesis of ribulose 1,5-diphosphate carboxylase in greening leaves of barley. *Pl. Physiol.*, **46**, 416–18.

Kobayashi, H., Asami, S. & Akazawa, T. (1980). Development of enzymes involved in photosynthetic carbon assimilation in greening seedlings of maize (*Zea mays*). *Pl. Physiol.*, **65**, 198–203.

Link, G. (1982). Phytochrome control of plastid mRNA in mustard (*Sinapis alba* L.) *Planta*, **154**, 81–6.

Luthe, D. S. & Quatrano, R. S. (1980*a*). Transcription in isolated wheat nuclei. I. Isolation of nuclei and elimination of endogenous ribonuclease activity. *Pl. Physiol.*, **65**, 305–8.

Luthe, D. S. & Quatrano, R. S. (1980*b*). Transcription in isolated wheat nuclei. II. Characterization of RNA synthesized *in vitro*. *Pl. Physiol.*, **65**, 309–13.

Marcus, A. (1960). Photocontrol of formation of red kidney bean leaf TPN-linked triosephosphate dehydrogenase. *Pl. Physiol.*, **35**, 126–8.

Margulies, M. M. (1965). Relationship between red-light mediated glyceraldehyde-3-phosphate dehydrogenase formation and light-dependent development of photosynthesis. *Pl. Physiol.*, **40**, 57–61.

Reisfeld, A., Gressel, J., Jakob, K. M. & Edelman, M. (1978). Characterization of the 32 000 Dalton membrane protein. I. Early synthesis during photoinduced plastid development of *Spirodela*. *Photochem. Photobiol.*, **27**, 161–5.

Ricciardi, R. P., Miller, J. S. & Roberts, B. E. (1979). Purification and mapping of specific mRNAs by hybridization-selection and cell-free translation. *Proc. natn. Acad. Sci. USA*, **76**, 4927–31.

Sasaki, Y., Sakihama, T., Kamikubo, T. & Shinozaki, K. (1983). Phytochrome-mediated regulation of two mRNAs, encoded by nuclei and chloroplasts, of ribulose-1,5-bisphosphate carboxylase/oxygenase. *Eur. J. Biochem.*, **133**, 617–20.

Schmidt, G., Bartlett, S. G., Grossman, A. R., Cashmore, A. R. & Chua, N.-H. (1981). Biosynthetic pathways of two polypeptide subunits of the light-harvesting chlorophyll *a/b*–protein complex. *J. Cell Biol.*, **91**, 468–78.

Slovin, J. P. & Tobin, E. M. (1982). Synthesis and turnover of the light-harvesting chlorophyll *a/b*–protein in *Lemna gibba* grown with intermittent red light: possible translational control. *Planta*, **154**, 465–72.

Smith, S. M. & Ellis, R. J. (1981). Light-stimulated accumulation of transcripts of nuclear and chloroplast genes for ribulose bisphosphate carboxylase. *J. molec. appl. Genet.*, **1**, 127–37.

Stiekema, W. J., Wimpee, C. F., Silverthorne, J. & Tobin, E. M. (1983). Phytochrome control of the expression of two nuclear genes encoding chloroplast proteins in *Lemna gibba* L. G-3. *Pl. Physiol.*, **72**, 717–24.

Thomas, P. S. (1980). Hybridization of denatured RNA and small DNA fragments transferred to nitrocellulose. *Proc. natn. Acad. Sci. USA*, **77**, 5201–5.

Tobin, E. M. (1978). Light regulation of specific mRNA species in *Lemna gibba* L. G-3. *Proc. natn. Acad. Sci. USA*, **75**, 4749–53.

Tobin, E. M. (1981*a*). White light effects on the mRNA for the light-harvesting chlorophyll *a/b*–protein in *Lemna gibba*. L. G-3. *Pl. Physiol.*, **67**, 1078–83.

Tobin, E. M. (1981*b*). Light regulation of the synthesis of two major nuclear-coded chloroplast polypeptides in *Lemna gibba*. *Photosynthesis V*,

Chloroplast Development, ed. G. Akoyunoglou, pp. 949–59. Balaban International Science Services, Philadelphia.

Tobin, E. M. (1981c). Phytochrome-mediated regulation of messenger RNAs for the small subunit of ribulose 1,5-bisphosphate carboxylase and the light-harvesting chlorophyll *a/b*–protein in *Lemna gibba*. *Pl. molec. Biol.*, **1**, 35–51.

Tobin, E. M. & Suttie, J. L. (1980). Light effects on the synthesis of ribulose 1,5-bisphosphate carboxylase in *Lemna gibba* L. G-3. *Pl. Physiol.*, **65**, 641–7.

Tobin, E. M. & Turkaly, E. (1982). Kinetin affects rates of degradation of mRNAs encoding two major chloroplast proteins in *Lemna gibba* L. G-3. *J. Pl. Growth Reg.*, **1**, 3–13.

Tobin, E. M., Wimpee, C. F., Stiekema, W. J., Neumann, G. N., Silverthorne, J. & Thornber, J. P. (1984). Phytochrome regulation of the expression of two nuclear-coded chloroplast proteins. In *Biosynthesis of the Photosynthetic Apparatus: Molecular Biology, Development, and Regulation*, UCLA Symposium on Molecular and Cellular Biology, New Series, vol. 13, ed. R. Hallick, L. A. Staehelin & J. P. Thornber. New York: Alan R. Liss, Inc., New York. (In press.)

Travis, R. L., Key, J. L. & Ross, C. W. (1974). Activation of 80S maize ribosomes by red light treatment of dark-grown seedlings. *Pl. Physiol.*, **53**, 28–31.

Woo, S. L. C. (1979). A sensitive and rapid method for recombinant phage screening. *Meth. Enzymol.*, **68**, 389–95.

Index

346   *Index*

  galactolipid synthesis 205, 211
  light-induced biogenesis 263, 264
  location 2, 3
  membrane complexes 148–56
  membrane polypeptide labelling 148, 149
  mesophyll polypeptide in *Zea mais* 54
  nuclear contribution to biogenesis 61
  photoregulation of biosynthesis 167–87
  polypeptide composition in *Zea mais* 53–7
  primary 175
  structure in virescent mutants 33, 34
  sulphur *Nicotiana* sp mutants 36–8
transcriptions
  chloroplast DNA 142, 143
  chloroplast RNA 109–13
    components 109–13
    problems 117
    promotor sequences 112–15
    requirements 112
    systems, need for 117
    terminator sequences 113, 116
  control, models 316
  features, isolated *Pisum* sp nuclei 309–13
  photoregulation mechanism 315–17
  phytochrome and isolated nuclei 329–31
  products, chloroplast 70
transcription-processing model 93
translation
  chloroplast DNA 142, 143
  chloroplast RNA, cell-free systems 141, 142
  products
    chloroplast 70
    32 kD protein 283
    poly(A)-mRNA 174, 283, 284
triazine resistance, 32 kD protein gene expression 297–9

Triticum aestivum (wheat), high and low light leaf thylakoids 264–70

UDP-gal:diacylglycerolgalactosyl transferase 206, 207

Vicia faba (broad bean)
  chloroplast DNA properties 68, 70
  photosystem I mutant 33
  transfer RNA features 98–103
violoxanthin 171, 235

wheat germ translation system 142

xanthophyll 31, 35

Zea mais (maize)
  anatomical features and C_4 51
  bundle sheath photosystem II polypeptide 54–6
  bundle sheath polypeptide 54, 55
    analysis 51, 52
  carbon dioxide fixation processes 51
  chlorophyll content and temperature 28
  chlorophyll synthesis mutants 24, 25
  and DNA transcriptional unit 95
  hcf mutant effects 34–6
  high and low light leaf thylakoids 264–70
  mesophyll polypeptide features 54, 55
  oil-yellow, yellow-green 36, 37, 39
  olive necrotic mutants 37
  poly(A)-RNA translation 57–60
  promotor sequences 114, 115
  protoplast analysis 52
  ribosomal RNA sizes 89, 90
  RNA polymerase 111
  thylakoid polypeptides 53–7
  terminator structure 116
  tRNA features 98–103
  tRNA gene organization 72